**Operation and Control of Renewable Energy Systems**

# Operation and Control of Renewable Energy Systems

*Mukhtar Ahmad*
Retired Professor from Aligarh Muslim University (AMU), India

*Registered Office(s)*
John Wiley & Sons, Inc., 111 River Street, Hoboken, NJ 07030, USA
John Wiley & Sons Ltd, The Atrium, Southern Gate, Chichester, West Sussex, PO19 8SQ, UK

*Editorial Office*
The Atrium, Southern Gate, Chichester, West Sussex, PO19 8SQ, UK

For details of our global editorial offices, customer services, and more information about Wiley products visit us at www.wiley.com.

Wiley also publishes its books in a variety of electronic formats and by print-on-demand. Some content that appears in standard print versions of this book may not be available in other formats.

*Library of Congress Cataloging-in-Publication Data applied for*

Hardback 9781119281689

Cover Design: Wiley
Cover Images: © William C. Y. Chu/Getty Images; © Henrik Sorensen/Getty Images

Set in 10/12pt WarnockPro by SPi Global, Chennai, India
Printed and bound in Malaysia by Vivar Printing Sdn Bhd

10  9  8  7  6  5  4  3  2  1

# Contents

# Preface

The demand for energy supply, especially electricity, is an ever-increasing trend as the population grows and in order to meet the social and economic development of the world. This means we require more consumption of fossil fuels to produce electricity. Fossil fuels comprise mainly coal, oil and gas. These sources were formed millions of years ago from the decomposed bodies of dead plants and animals. These sources are likely to exhaust after few years as demand for fuel continues to grow at a fast rate. Apart from the possibility of depletion of these sources, pollution is a major disadvantage of fossil fuels. This is because they give off carbon dioxide when burned, thereby causing the greenhouse effect. This is also the main contributory factor to the global warming experienced by the Earth today. The current interest in the use of renewable is the need to reduce the high environmental impact of fossil-based fuels used in energy production. There are a number of renewable sources which are being investigated for harnessing energy on a large scale. Future energy sustainability depends heavily on how the renewable energy problem is addressed in the next few years and decades. The sources of renewable energy being considered for further development are solar, wind, biomass, geothermal, ocean, fuel cell and small hydro systems. The main problem with these resources is the cost and availability. Unlike conventional sources of electric power, the output of renewable sources is not controllable. Control is one of the key enabling technologies for the development of renewable energy systems. In this book, the generation of energy, mainly electrical, by using various renewable sources is addressed. Solar and wind power are the two major resources which are now in use in small- as well as large-scale power production. These sources also require effective use of advanced control techniques, which are discussed in this book. In addition, integration of renewable energy in the power grid and their working in the microgrid are also considered.

These contents are covered in 14 chapters. In Chapter 1, the types of renewable energy sources and the basic principles involving energy conversion are described. These include the theory of fluid mechanics which is required in hydro and ocean energy conversion and laws of thermodynamics required in understanding the thermal power conversion applied in biomass. The renewable energy systems extensively use power electronics for proper utilization of energy. Similarly, different types of generators are used in power generation which are more suitable for a particular type of energy source. Chapters 2 and 3, therefore, describe the theory of power electronics and various electric power generators. Since many renewable energy sources such as solar and wind are intermittent in nature, energy storage becomes essential for their utilization. Therefore grid-scale energy storage systems are described in Chapter 4.

Since solar energy has two main applications, one involving solar thermal and other photovoltaic, Chapter 5 deals with the solar thermal energy conversion technology and Chapter 6 covers the photovoltaic power generation. Wind energy conversion is presented in Chapter 7. Here, both horizontal and vertical wind turbines for power generation are described.

Biomass is another major source of renewable energy. Conversion of biomass into various forms of energy is described in Chapter 8. Chapter 9 deals with harnessing the geothermal energy for generation of electricity. Geothermal energy is heat from within the Earth. Geothermal power plants using hydrothermal resources are discussed in this chapter.

Technologies for generating electrical power from the ocean include tidal power, wave power, ocean thermal energy conversion, ocean currents, ocean winds and salinity gradients. All these methods are dealt with in Chapter 10.

A fuel cell converts stored chemical energy to electrical energy directly. Essentially, it takes the chemical energy that is stored within an energy source such as hydrogen, gasoline or methane, and then through two electrochemical reactions, it converts it directly to electricity. A number of fuel cell technologies are now available for various applications which are described in Chapter 11.

Hydropower is the most important and widely used renewable source of energy. Hydropower represents about 16% of total energy produced from renewable sources. In Chapter 12, only small hydropower is described as large power systems involving dams and reservoirs have already been in use for a long time and are generally described in many books on power systems.

With the availability of a large number of renewable energy sources distributed over a large area, the interconnection with distribution systems creates many new challenges. Wind energy and solar energy are not dispatchable sources, therefore, the power cannot be controlled. Solar and wind power plants exhibit changing dynamics, non-linearities and uncertainities. Effective control techniques are required to properly manage the distribution grid. Chapter 13 is devoted to the control and operation of grid-connected solar and wind energy systems.

As a large number of distributed sources of energy, mainly renewable, are available now, the secure and reliable power supply can be obtained with the help of the microgrid. Thus, Chapter 14 is devoted to integration of renewable sources in the microgrid and their operation and control. This type of material is not available in any renewable energy book and may be of interest to people working in the design of power supply with renewable sources.

This book can be used as a textbook for undergraduate and graduate courses. It can also be used by practicing engineers.

Mukhtar Ahmad

# 1

# Sources of Energy and Technologies

## 1.1 Energy Uses in Different Countries

As demand to meet social and economic development and improve human welfare and health is increasing, the demand for clean energy and associated services is also increasing. All societies require energy to meet basic human needs, for example, lighting, cooking, living comfort, mobility and communication also to run industries for various productive processes. Since around 1850, global use of fossil fuels (coal, oil and gas) has been the most dominant source of energy supply, leading to a rapid growth in $CO_2$ emissions. The per capita energy consumption which was about 200 W nearly 100 years ago has increased to more than 2000 W per capita now. The energy consumption has almost doubled during last 30 years. Globally, energy consumption grew most quickly in the transport and service sectors, because of rising passenger travel and freight transport and a rapid expansion in the service economy. In 2004, about 77.8% of total energy consumption was through fossil fuels, only 5.4% was nuclear and the rest 16.5% was from renewable resources which was mainly hydroelectric.

The energy consumption in the world is mainly from following six primary sources. These are (i) fossil fuels, ii) nuclear, iii) hydro, iv) wind, v) solar and vi) biomass.

According to Renewables 2010 Global Status Report, the renewable energy share of total energy consumption in 2008 was 19%, as shown in Fig. 1.1. Of this 19%, approximately 13% is used primarily for cooking and heating using traditional biomass which is growing slowly or even declining in some regions. The main reason for this is that now biomass is used more efficiently or is replaced by more modern energy forms. Hydropower represents 3.2% and is growing from a large base. Other renewables account for 2.6% and are growing very rapidly in developed as well as in some developing countries.

Three main sectors that account for approximately 70% of the total energy consumption in an industrialized country are as follows:

- Motors (approximately 40–45%)
- Lighting (15%)
- Home appliances (15%)

Energy consumption in a country is an indication of the level of development and quality of life [1]. The developing and developed countries have striking disparities in per capita annual energy consumption as shown Fig. 1.2. In India, energy use per capita in 2012 was 614 KWh. The electric power consumption kWh per capita was the highest

*Operation and Control of Renewable Energy Systems*, First Edition. Mukhtar Ahmad.
© 2018 John Wiley & Sons Ltd. Published 2018 by John Wiley & Sons Ltd.

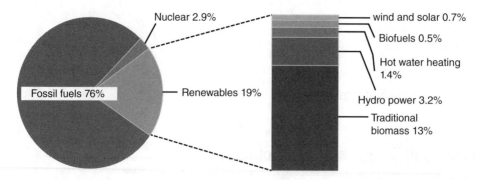

**Figure 1.1** Renewable energy share of global energy consumption in 2008. Source: REN21 (2010) [1]. Reproduced with permission from Renewable Energy Policy Network for the 21st Century.

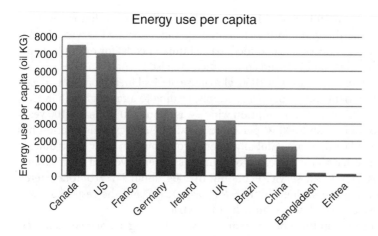

**Figure 1.2** Per capita energy consumption in countries.

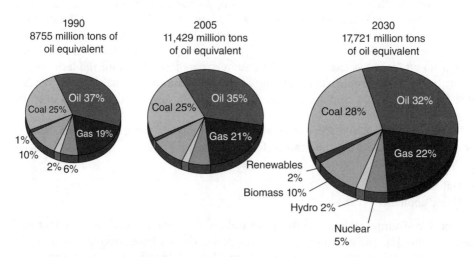

**Figure 1.3** World energy outlook and future prediction.

for Iceland 52,374 and minimum in Tanzania which was 92; for the United States, it was 11,919. The share of fossil fuels in primary sources of energy in 1995 and 2005 and the prediction for 2030 are shown in Fig. 1.3.

## 1.2    Energy Sources

All major sources of energy can be categorized as non-renewable and renewable. Non-renewable energy is mainly the fossil fuels energy obtained from coal, crude oil, natural gas and nuclear fuel. Renewable energy is obtained from hydropower, wind, solar, geothermal, ocean and biomass.

### 1.2.1    Non-Renewable Energy Resources

Coal, oil and gas are fossil fuels which are non-renewable sources of energy. These sources are non-renewable because they take millions of years to form. In addition, they are being used so extensively that the reserves are being depleted much faster than the new ones are being formed. One of the biggest benefits of using fossil fuels is their low cost. Another advantage is that these resources are available in abundance right now and relatively inexpensive to drill or mine. The estimate at current usage of coal suggests that its supply will last for 1500 years. However, if the consumption grows at 5% per year, the coal supply will last only for 86 years. It is expected that even greater usage of coal will be made in future as other fossil fuels become scarce. Other fuels such as oil and gas reserves are predicted on the basis of only proven reserve estimates. Total oil reserves in 2013 are now estimated at 1.64 trillion bbl, while the world's gas reserves are 7.02 quadrillion, up by 0.4% and 2%, respectively, from a forecast made a year earlier. It is estimated that oil production which will be economically viable will last at most by the end of 21st century. Similarly, the supply of natural gas may not last for more than 50 years. These fuels are being used extensively now because they are made of hydrocarbons. Hydrocarbons store energy in the form of the atomic bonds which can be released very easily – by simply burning them. The main drawback of using fossil fuels is the pollution due to the increase of toxic gases in the atmosphere. When fossil fuels are burnt, carbon and hydrogen react with oxygen in air to produce carbon monoxide (CO), carbon dioxide ($CO_2$), other pollutants such as $SO_x$ and water ($H_2O$). The increase of these greenhouse gases in the atmosphere has resulted in global warming.

### 1.2.2    Renewable Sources of Energy

Since the fossil fuels are fast depleting and also creating pollution, other sources of energy which are renewable must be explored for sustainable development. The term "renewable" is generally applied to those energy resources and technologies whose common characteristic is that they are non-depletable or naturally replenishable. Renewable energy technologies produce power, heat or mechanical energy by convert-ing those resources either to electricity or to motive power. At present, the renewable power generation provides about 18% of total electricity produced in the world. Most of this, that is, about 15%, is from hydroelectricity, and only 3% is produced from other renewable sources.

Major renewable resources include solar energy, wind, falling water, the heat of the earth (geothermal), plant materials (biomass), waves, ocean currents and the energy of the tides. In fact, all these energies are basically indirect form of solar energy except geothermal which is the heat generated deep inside the earth. Other sources of renewable energy are hydrogen energy, nuclear fusion and methane clathrate (methane ice). As will be seen in later chapters, the potential of renewable energy sources is enormous as they can in principle meet many times the world's energy demand for centuries.

The benefits of using renewable energy resources are many. Most of these benefits arise from their virtually inexhaustible nature and easy availability. Solar and wind resources are available as a continuous source of energy on a daily basis. Biomass can be grown through managed agricultural farming to provide continuous sources of fuel. In addition, these sources do not contribute to greenhouse effect and atmospheric pollution. Several of the renewable energy technologies, namely photovoltaics, solar thermal and wind, produce no emissions during power generation.

Sun is a big natural fusion reactor and retains its released gases due to gravitational forces. The surface of the Sun is maintained at a temperature of approximately $5800°K$. The Sun radiates energy uniformly in all directions in the form of electromagnetic waves which is about 120,000 TW. Solar energy has enormous potential as a clean, abundant and economical energy source but cannot be employed as such; it must be captured and converted into useful forms of energy such as electricity and heat. Currently, none of the technologies available to convert solar energy into heat, electricity and fuel is competitive with fossil fuels. However, because of environmental considerations and decreasing cost of solar energy systems, it is becoming attractive.

Wind energy has been used since early known history. It was used to propel boats, to pump water and grinding of grains and so on. Later, it was found that the energy contained in moving wind can be utilized to run turbines which can be connected to electric generators to produce electricity. Wind turbines work similarly to windmills but are used specifically to generate electricity. A wind turbine usually has fewer blades and is made of lighter materials, such as plastics, which allow the blades to turn more quickly and with less wind. The blades of the wind turbine capture the energy of the wind and send it down a shaft inside the nacelle. This shaft spins the turbines of a generator. A wind turbine can produce enough electricity to satisfy the needs of a home. Wind turbines can also be grouped together to create large quantities of electricity. This is referred to as a wind farm. Wind farms are becoming more widespread throughout the world.

Biomass is a plant-based material that stores light energy through photosynthesis. Biomass has unique property among renewable energy resources as it can be converted to carbon-based fuels and chemicals as well as electric power. Chemical or biological processes are used to transform biomass into carbon based fuels, such as methane, ethanol, producer gas and charcoal. Biomass plants, with a properly managed fuel cycle and modern emission controls, produce zero net carbon emissions and minimal amounts of other atmospheric effluents.

Another potential source of renewable energy is the heat contained inside the earth known as geothermal energy. Geothermal energy is the energy obtained from the earth (geo) from the hot rocks present inside the earth. It is produced due to the fission of radioactive materials in the earth's core because of which some places inside the earth become very hot. These are called hotspots. They cause the water deep inside the earth to form steam. As more steam is formed, it gets compressed at high pressure and comes out in the form of hot springs which produce geothermal power. Geothermal energy can be used to generate electricity or used as source of heat.

Energy from the oceans can be harnessed in the form of thermal energy, from the temperature difference of the warm surface waters and the cool deeper waters, as well as potential and kinetic energy from the tides, waves and currents.

Hydrogen is considered as an alternative fuel that can be produced from domestic resources. Hydrogen is locked up in enormous quantities in water ($H_2O$), hydrocarbons (such as methane, $CH_4$) and other organic matter. Hydrogen is a clean-burning fuel, and when combined with oxygen in a fuel cell, it produces heat and electricity with only water vapour as a by-product. But hydrogen does not exist freely in nature: it is only produced from other sources of energy, so it is often referred to as an *energy carrier*, that is, an efficient way to store and transport energy.

Hydropower engineering deals with mostly two forms of energy: of running water or from water stored in dams. Hydropower is still the largest source of renewable energy in electricity production. It is a proven, mature, predictable and economically viable technology. Hydropower has among the best conversion efficiencies of all known energy sources (about 90% efficiency). It requires relatively high initial investment but has a long lifespan with very low operation and maintenance.

Hydropower plants do not consume the water that drives the turbines. The water, after power generation, is available for various other essential uses. In fact, a significant proportion of hydropower projects are designed for multiple purposes. The dams help to prevent or mitigate floods and droughts, provide the possibility to irrigate agriculture, supply water for domestic, municipal and industrial use and can improve conditions for navigation, fishing and so on.

As there is more concern for greenhouse effect and global warming, the present trend in most of the countries is to gradually move away from fossil-fuel-based systems to renewable-energy-based technologies. As the costs of solar and wind power systems have dropped substantially in the past 30 years, and continue to decline, while the prices of oil and gas continue to fluctuate, a time will come soon when the price of energy due to renewables will be comparable with fossil fuel energy.

At present, the cost of power generation from renewables is high compared to fossil fuel energy. With development in renewable energy technologies, they may soon become more competitive. Since many renewable energy plants do not need to be built in large scale, they can be built in size increments proportionate to load growth patterns and local need. When constructed in smaller size, they can be located closer to the customer load, reducing infrastructure costs for transmission and distribution. Such "distributed" generations have a potentially high economic value more than just the value of the electricity generation as they help to guarantee local power reliability and quality.

In fact, fossil fuel and renewable energy prices, social and environmental costs are heading in the opposite directions. Thus, in a few years, the renewable energy systems may compete favourably with the present-day fossil fuel plants.

## 1.3  Energy and Environment

All fossil fuel energy sources have environmental impact during their life cycles. Climate change associated with greenhouse gas emissions is seen as the greatest environmental challenge facing humanity. At present, the power generation system is the largest contributor to overall emissions of greenhouse gases. The term greenhouse is used for a house made from transparent glass panes or sheets in which plants are grown. During the day, the Sun emits rays of short-wave infrared light. The short-wave infrared light is able to pass through glass. After hitting a surface, the waves turn into thermal energy. This energy is a long-wave infrared light that the glass roof does not allow it to pass. Hence, it maintains a controlled warmer environment inside the house suitable for growth of plants especially in harsh environments.

Similar effect as shown in Fig. 1.4 is observed naturally on Earth and has been given the name greenhouse effect. The greenhouse effect is a natural phenomenon by which certain gases in the atmosphere prevent re-emitting of solar radiation back into space. The burning of fossil fuels produces huge quantities of carbon dioxide, an important greenhouse gas. The Sun radiates vast amount of energy into space which is mostly in visible and near-visible parts of the spectrum. Various components of Earth's atmosphere absorb ultraviolet and infrared solar radiation before it penetrates to the surface, and the ozone layer in the Earth's stratosphere absorbs the harmful ultraviolet rays. However, the atmosphere is quite transparent to visible light, thus the Earth absorbs part of the Sun's radiation. This part of radiation which is absorbed by the Earth heats it which then radiates some of its energy out into space. Since the frequency at which any object emits radiation depends on its temperature and since the Earth is much cooler than the Sun, it emits energy at a lower frequency and therefore longer wavelength – in the IR region. Steady state is reached where the Earth is absorbing and radiating energy at the same rate, resulting in a fairly constant average temperature. If there were no greenhouse effect at all, then the surface temperature of the Earth would be about 256 K or −17°C (about the temperature of a domestic freezer), and life could not exist because

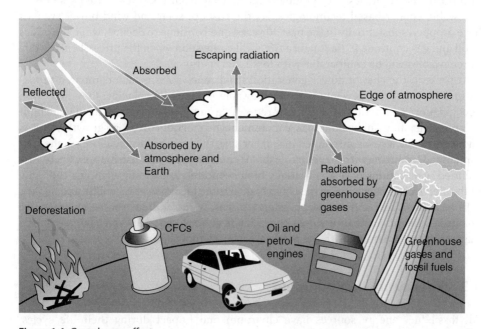

**Figure 1.4** Greenhouse effect.

water, which is fundamental to life, would be a solid. However, the IR radiation emitted by the Earth can be absorbed by gases in the troposphere and become trapped. The radiation is then re-emitted in all directions: some back towards the Earth, which is known as the "greenhouse effect". This leads to an increase in temperature and global warming, making the average surface temperature of the Earth about 286 K or 13°C. It is an essential part of keeping our planet hospitable and helps to sustain life. The gases which absorb IR radiation and then re-emit it are known as greenhouse gases.

The most significant greenhouse gas is $CO_2$. When fossil fuels such as oil, coal and gas are burnt, $CO_2$ is released into the Earth's atmosphere. The cutting down of forests also increases the concentration of $CO_2$ in the atmosphere. The trees absorb $CO_2$ from the atmosphere by wood, leaves and soil; this $CO_2$ is released back into the atmosphere when forests are burnt. Apart from $CO_2$, other greenhouse gases are methane, nitrous oxide, hydrofluorocarbons, sulfur hexafluoride and water vapour. The $CO_2$ emissions from developed countries account for 82% of the total greenhouse gas emissions of the world. China tops the countries that emit most $CO_2$ in the atmosphere followed by the United States and India. But only looking at carbon dioxide emissions doesn't give us the true picture of all greenhouse gases. Although the vast majority of emissions are carbon dioxide followed by methane and nitrous oxide, lesser amounts of CFC-12, HCFC-22, Perfluoroethane and Sulfur Hexafluoride are also emitted, and their contribution to global warming is magnified by their high global warming potential. The nitrous oxide is 310 times more absorptive than carbon dioxide and can remain in the atmosphere for over a hundred years.

### 1.3.1 Climate Change

The effect of increased greenhouse gases in the atmosphere due to burning of fossil fuels is to trap heat resulting in the rise of temperature of the planet. Earth's average temperature has risen by 0.9°C in the last century. Even a small amount of increase in global temperatures has resulted in changes in weather and climate of different parts of the world. The rising temperatures are leading to the melting of polar ice caps which in turn may result in rise in the levels of seawater. Other effects of climate change are visible in the form of frequent and severe hot and cold waves, tropical cyclones and heavy rains or draughts. Following are the impacts of global warming:

- Rise in the sea level resulting in inundation of low lying cities and islands.
- Changes in rainfall patterns resulting in draughts and fires in some areas and excess rainfall in other areas.
- Melting of ice caps may result in loss of habitat near the poles.
- Melting of glaciers.
- Bleaching of coral reefs due to warming of seas and carbonic acid formation.
- Extreme conditions such as hurricanes, floods.
- Spread of disease in the epidemic form of cholera, malaria and so on.

Changes in climate are visible in many parts of the world, where floods, heavy rains or draughts and intense heat waves are being observed. The scientists are forecasting an increase of about 1°C in another 100 years. In order to save the Earth from disaster, it is important that emission of greenhouse gases must be reduced. International concern for climate change led to Kyoto protocol negotiated in 1997. In order to satisfy the

Kyoto protocol, developed countries were required to cut back their emissions by a total of 5.2% between 2008 and 2012 from the levels in 1990. Specifically, the United States was supposed to reduce its presently projected 2010 annual emissions by 400 million tons of $CO_2$.

## 1.4 Review of Technologies for Renewable Energy System

The knowledge of the basic types of energy and the law of thermodynamics and fluid mechanics are required for understanding the technologies in renewable energy systems. The transfer of energy to and from a moving fluid is the basis of hydro, wind, wave and some solar power systems. In direct solar, geothermal and biomass, energy transfer is by heat rather than by mechanical or electrical process. A brief description of these theories is presented for the understanding of working of renewable energy systems. In addition, for operation and control of renewable energy systems, knowledge of types of generators used and power conversion through power electronic circuits is required. These technologies are discussed briefly in Chapters 2 and 3. Since most of the renewable energy sources are intermittent in nature, it is important to have knowledge of various energy storage techniques for large systems. It is described in Chapter 7.

### 1.4.1 Fluid Dynamics

In order to understand the operation of renewable energy systems requiring natural movement of air and water, the basic laws of mechanics must be understood. The basic equations of fluid dynamics are based on familiar laws of mechanics [2, 3].

- Conservation of mass
- Conservation of momentum
- Conservation of energy

The fluid considered here can be liquid or gas. If the flow also leads to compression of the fluid, then the law of thermodynamics can also be considered.

#### 1.4.1.1 Conservation of Mass
The principle of conservation of mass is based on the fundamental theory that the matter cannot be created or destroyed. This principle is applied to flowing fluids with fixed volumes known as control volumes or surfaces.

When studying the properties of moving fluid and the property of the objects that travel in a fluid, two types of fluid flow are considered: laminar flow and turbulent flow. Laminar flow occurs when a fluid flows in parallel layers, with no disruption between the layers. In laminar flow, the particles in the fluid follow in streamlines, and the motion of particles in the fluid is predictable. Laminar flow occurs when the velocity is low or the fluid is very viscous. If the flow rate is very large, or if objects obstruct the flow, the fluid starts to swirl in an erratic motion. This region of constantly changing flow lines is said to consist of turbulent flow. For studying the theory of conservation of mass, the simplest case of fluid flow – laminar flow with uniform velocity in a pipe – is considered. Fig. 1.5 shows a fluid of density d flowing through a uniform pipe of Area $A$ with velocity $v$.

**Figure 1.5** Flow of fluid in a pipe.

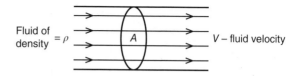

The rate at which mass $m$ flows through area $A$ over time $t$ is given as

$$r = \frac{m}{t} = \frac{\rho V}{t} \tag{1.1}$$

where $\rho$ is fluid density, $V$ is the volume and $A$ is the area. Since in a pipe no fluid can pass through the walls and there is no possibility of it being created or destroyed, the mass crossing each section of the pipe per unit time must be the same.

Or, $\rho AV = $ constant $\tag{1.2}$

This expression represents the law of conservation of mass for incompressible fluids. In the case of renewable energy systems using wind power, the air is treated as incompressible.

**Example 1.1**  Water flows through a 4 cm diameter hose with a speed of 2 m/s. Find the speed of water through the nozzle with the diameter being reduced to 1 cm. Using the principle of conservation of mass

$$\rho AV = \text{constant}$$

Reducing the diameter of the hose will reduce the area. Consequently, the velocity will increase by the same factor that the area is decreased.

$$\text{Area} = \pi r^2 = \pi(4)^2$$
$$\text{Area of the nozzle} = \pi(1)^2$$

Since $A_1 V_1 = A_2 V_2$,

$$V_2 = \frac{A_1}{A_2} V_1 = \frac{16}{1} \times 2 = 32 \, \text{m/s}$$

### 1.4.1.2  Conservation of Momentum

Newton's second law of motion for fluids may be defined as "At any instant in steady flow, the resultant force acting on the moving fluid within a fixed volume of space equals the net rate of change of momentum in that volume and is in the direction of force." To determine the rate of change of momentum in a moving fluid, let us consider a stream tube as shown in Fig. 1.6. Assuming that the fluid has steady and non-uniform flow, the volume of fluid entering the tube in time $dt$ is $A_1 V_1 dt$ and the momentum is equal to $\rho A_1 V_1 dt \, V_1$ and the momentum of fluid leaving the tube is $\rho A_2 V_2 dt \, V_2$. From Newton's second law, the force is equal to the rate of change of momentum.

$$F = \frac{\rho A_2 V_2 dt \, V_2 - \rho A_1 V_1 dt \, V_1}{dt} \tag{1.3}$$

Since from conservation of mass, $\rho A_1 V_1 = \rho A_2 V_2 = M$; hence,

$$F = M\rho(V_2 - V_1) \tag{1.4}$$

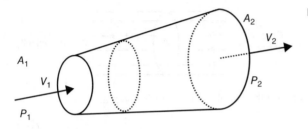

**Figure 1.6** Stream tube.

### 1.4.1.3 Conservation of Energy

Conservation of energy also known as Bernoulli's equation is stated as *The sum of the kinetic, potential and flow energies of a fluid particle is constant along a streamline during steady flow when compressibility and frictional effects are negligible.* Since the kinetic, potential and flow energies are the mechanical forms of energy, the Bernoulli equation can be viewed as the *conservation of mechanical energy.*

The Bernoulli equation states that during steady, incompressible flow with negligible viscosity, the various forms of mechanical energy are converted to each other, but their sum remains constant. In other words, there is no dissipation of mechanical energy during such flows since there is no friction that converts mechanical energy to thermal (internal) energy. The following two assumptions must be met for this Bernoulli equation to apply: the flow must be incompressible – even though pressure varies, the density must remain constant along a streamline; friction by viscous forces has to be negligible.

The total mechanical energy of a fluid exists in two forms: potential and kinetic and internal energy (flow energy). The law of conservation of energy can be written in terms of the quantities which are related to fluid flow (pressure, density, velocity, etc.). Bernoulli's equation states that at any point in the channel of a flowing fluid, the energy does not change.

Or, total energy = internal energy + potential energy + kinetic energy

The internal energy is required to overcome the pressure $P$ in the pipe. This pressure generates a force that resists the motion of the fluid. Thus, Bernoulli's equation is

$$P + \rho g h + \frac{1}{2}\rho V^2 = \text{constant} \tag{1.5}$$

Here $P$ is the pressure, $h$ is the height, $V$ is the velocity and $\rho$ is the density of fluid at any point in the flow channel. If the elevation of the fluid remains constant, or if the change in elevation is small enough to not change the gravitational potential energy of the fluid appreciably, then the potential energy term can be ignored. We then have

$$P + \frac{1}{2}\rho V^2 = \text{constant} \tag{1.6}$$

Despite the many approximations used in deriving the Bernoulli equation, it is commonly used in practice for solving practical fluid flow problems. This is because many flows of practical engineering interest are steady (or at least steady in the mean), compressibility effects are relatively small and net frictional forces are negligible in some regions of interest in the flow. Thus, Bernoulli's equation gives the result with reasonable accuracy.

**Example 1.2**

Water is flowing through a hose with a velocity of 1.5 m/s. The speed of the water when it leaves the nozzle is 30 m/s. The pressure on the water as it leaves the nozzle is 1 atm. Find the pressure of water inside the hose?

**Solution**

From Bernoulli's equation, $P_1 + \frac{1}{2}\rho V_1^2 = P_2 + \frac{1}{2}\rho V_2^2$.

Here $\rho$ = density of water = $1000\,\text{kg/m}^3$, $P_2 = 100{,}000\,\text{N/m}^2$, $V_2 = 30\,\text{m/s}$ and $V_1 = 15\,\text{m/s}$.

Therefore,

$$P_1 = 100{,}000 + \frac{1}{2} \times 1000 \times 30 \times 30 - \frac{1}{2} \times 1000 \times 1.5 \times 1.5$$

$$= 100{,}000 + 450{,}000 - 1125$$

$$= 548875\,\text{N/m}^2 = 5488\,\text{atm}$$

## 1.5  Thermodynamics

With direct solar, geothermal and biomass sources, most energy transfer is by heat rather than by mechanical or electrical processes. The knowledge of the basic types of energy and laws of thermodynamics is essential in understanding the conversion of heat into other forms of energy. Thermodynamics is the science of relations between heat, work and properties of the system or matter which are in equilibrium. In thermodynamics, the universe is arbitrarily divided into a *system* and its *surroundings*. The portion of the universe, which is chosen for thermodynamic consideration, is called a *system*. It usually consists of a definite amount of a specific substance as steam in turbine or CNG in a cylinder. The matter outside the system is called surroundings, and the separation between system and surroundings defines the boundary. A system may be homogeneous (gas or mixture of gases or liquid) or heterogeneous containing liquid and vapour or liquids which do not mix with each other. The system may be open, closed or isolated, depending on whether the matter and energy can flow in or out of the system. In a closed system, the energy can be exchanged but not matter with an external system. In an open system, both matter and energy can be exchanged with an external system. Most of the engineering systems are open systems. In an isolated system, neither energy nor matter can be exchanged with an external system.

The thermodynamic state of a system is defined by four properties: composition, pressure, volume and temperature. These properties define the system completely. The thermodynamic state of a system will not change from one state to another unless there is some interaction with the outside world. Whenever a thermodynamic system changes its state, it always involves addition or removal of heat or work. This is known as a process. A process occurs when a system undergoes change in state or an energy transfer at a steady state. A process has a certain path between the starting point and the end point and can be depicted as *P-V* (pressure–volume) or *T-S* (temperature–entropy) curve as shown in Fig. 1.7. The process may be reversible if it follows the same path in the reverse direction or irreversible if it follows a different path in the reverse direction. Both heat

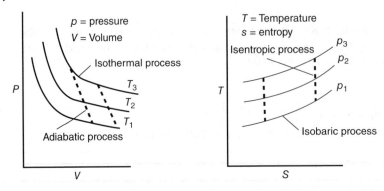

**Figure 1.7** *P-V* and *T-S* diagrams.

and work are functions of path, whereas internal energy is a function of state [4, 5]. Any process or a series of processes whose end states are identical is termed as a cycle. One of the thermodynamic properties of a system is its *internal energy*, E, which is the sum of the kinetic and potential energies of the particles that form the system. The other thermodynamic properties are as follows.

### 1.5.1 Enthalpy

Enthalpy H is defined as the sum of internal energy U and the pressure volume product PV as

$$H = U + PV \tag{1.7}$$

where P is the pressure in n/m$^2$, V is the volume in m$^3$ and H and U are in J. If pressure P is constant, then Eq. (1.7) can be written as

$$\Delta H = \Delta U + P\Delta V \tag{1.8}$$

$$\text{Or} \quad \Delta H = Q = mC_p\Delta T \tag{1.9}$$

Thus, the change in enthalpy is equal to the heat transferred Q at constant pressure. M = mass of fluid or gas, $c_p$ is the specific heat at a constant pressure and $\Delta T$ is the change in temperature.

### 1.5.2 Specific Heat

Specific heat is defined as the amount of heat required to raise the temperature of unit mass (mole) of any substance by 1°C. For gases, two different kinds of specific heat are used: specific heat at constant volume $C_v$ and specific heat at constant pressure $C_p$. For solids, there is one specific heat C only. For a solid of mass m and specific heat C, the heat dQ required to raise the temperature by dT degree is

$$dQ = mCdT \tag{1.10}$$

For gases in constant volume process, if the mass is m, the heat required to raise the temperature by dT is (here no work is done) hence dE = dQ.

$$dQ = mC_v dT \tag{1.11}$$

And for a gas of mass m at constant pressure, the work done is given by $PdV$; hence, $dE = dQ - PdV$, and the heat required is

$$dQ = mC_pdT \tag{1.12}$$

And for one mole of gas, $m = 1$ and $dE = dQ - PdV = C_pdT - PdV$. For ideal gases, $PV = RT$, and the expression for enthalpy change is

$$dH = d(E + PV) = dE + RdT$$

Or, $C_pdT = C_vdT + RdT$

Or, $C_p = C_v + R \tag{1.13}$

From these expressions, it is clear that for an ideal gas, internal energy and enthalpy are functions of temperature only.

Thermodynamics is governed by four laws, zero, first, second and third. The second law of thermodynamics was discovered first followed by the first law, then the third law and lastly the zeroth law. These laws are stated as follows.

### 1.5.3 Zeroth Law

The zeroth law of thermodynamics defines the concept of temperature and thermal equilibrium. This law of thermodynamics is actually an observation. For example, if two bodies A and B are at the same temperature and they are brought into contact, no heat will be exchanged between the two. In addition, if two bodies A and B are at the same temperature and B and a third body C are at the same temperature, then A and C are also at the same temperature (thermal equilibrium). If two bodies at different temperatures are in contact with each other for a long time, they will reach the same temperature. Thus, it can be stated that objects in thermodynamic equilibrium have the same temperature.

### 1.5.4 First Law

The first law, also known as the Law of Conservation of Energy, states that energy cannot be created or destroyed; it can only be redistributed or changed from one form to another. This law of thermodynamics is based on Joule's law which states that *the internal energy of a perfect gas is a function of temperature alone*. According to Joule's law. regardless of change in volume of an ideal gas, if the temperature does not change, the internal energy will remain constant. The internal energy is the sum of kinetic and potential energies of the particles that form a system. A way of expressing the first law of thermodynamics is that any change in the internal energy ($dU$) of a system is equal to the difference between heat added to the system and the work done by the system.

$$dU = Q - W \, J \tag{1.14}$$

here $Q$ is the heat input to the system and $W$ is work done by the system; $Q$ and $W$ are not functions of state, but $U$ depends only on the state of the system. This gives the result that $dU$ is independent of path, but $Q$ and $W$ depend on the path taken for change from one state to another. Sometimes, the first law is written in differential form as

$$dU = \delta Q - \delta W \tag{1.15}$$

In an isolated system, there is no interaction of the system with the surroundings; hence, $dQ = 0$, $dW = 0$ and $dU = 0$.

The first law of thermodynamics is not applicable to nuclear reactions.

### 1.5.4.1 Limitations of First law

The first law of thermodynamics has certain limitations which can be explained as follows. The first law does not give any indication of the directionality of a process, nor does it give any indication of the quality of energy. For example, if a current is passed through a resistor, it will heat up, but if the same resistor is heated to the same temperature, no current flows through it. That means that there are processes which can work in a certain direction only.

According to the first law, work and heat are directly related, and there is no quality associated. However, it will be shown from many examples that work is something which can rather easily be generated or which can be converted to heat, but on the other hand, heat cannot be directly converted into work, and it requires certain complicated devices which are called heat engines which will be described along with second law of thermodynamics.

### 1.5.5 Second Law of Thermodynamics

The second law of thermodynamics says that the entropy of any isolated system not in thermal equilibrium almost always increases. The entropy is a measure of disorder in the system, and for a reversible process between two equilibrium states, change in entropy is given by

$$dS = \left[\frac{dQ}{T}\right]$$

where $dS$ = change in entropy, $dQ$ = heat absorbed or expelled by the system in a reverse process and $T$ = absolute temperature. In other words, for a reversible process, the incremental change in entropy is a ratio of heat change and absolute temperature. When heat is absorbed by the system, $\Delta Q$ is positive, and hence, the entropy increases. When thermal energy is expelled by the system, $\Delta Q$ is negative and the entropy decreases. Entropy of the universe in all natural processes increases. All physical processes tend towards more probable states for the system and its surroundings. The more probable state is always one of higher disorder. The entropy is a measure of the disorder of a state.

*T-S* diagram is used to analyse energy transfer system cycles. This is because the work done by or on the system and the heat added to or removed from the system can be visualized on the *T-S* diagram. By the definition of entropy, the heat transferred to or from a system equals the area under the *T-S* curve of the process. Figure 1.8 is the *T-S* diagram for steam.

In order to understand the second law of thermodynamics, it is important to know the following terms:

*Thermal Reservoir*
  Thermal reservoir is a large body from which a finite quantity of energy can be extracted or to which a finite quantity of energy can be added as heat without changing its temperature.

*Heat Engine*

A heat engine is a device working cyclically which converts the energy it receives as heat into work. It receives energy as heat form a high-temperature body and converts part of it into work, and the rest is rejected to a low-temperature body.

*Heat Pump*

Figure 1.9 shows a heat engine and a heat pump. Heat pump is a cyclically operating device which absorbs energy form a low-temperature reservoir and delivers energy as heat to a high-temperature reservoir when work is performed on the device. Refrigerator is an example of heat pump.

*Coefficient of Performance (COP)*

The ratio of heat transfer to work input is not called the efficiency, but the coefficient of performance

$$\text{COP} = \frac{Q_2}{W} \tag{1.16}$$

There are two coefficients of performance for such a cycle: one for the refrigeration effect and one for the heat pump effect. There are two classical statements of second law of thermodynamics: Kelvin–Planck statement and Clausius statement.

### 1.5.5.1 Kelvin–Planck Statement

It is impossible to construct an operating device working in a cycle such that it produces no other effect than the absorption of energy as heat from a single thermal reservoir and performs an equivalent amount of work. It means that some of the energy received must be rejected to a lower temperature sink. Thus, the Kelvin–Planck statement further implies that no heat engine can have a thermal efficiency of 1 (100%). This does not violate the first law of thermodynamics. For example, a fluid can flow from high pressure or potential to low pressure or potential, but to send back the fluid to higher potential, a pump (energy) is required. Current can flow from a point of high potential to low potential, but energy must be spent if the reverse is required. Battery can discharge through higher potential to lower one through a resistor, but to charge it, electrical energy is required.

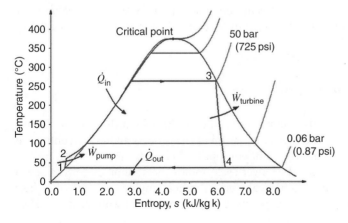

**Figure 1.8** *T-S* diagram for steam.

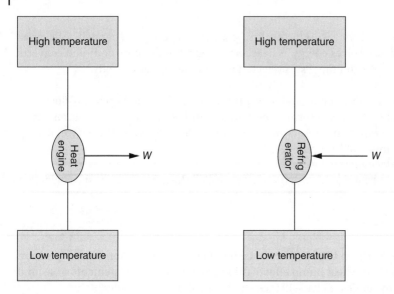

**Figure 1.9** Schematic diagram of a heat engine and a refrigerator.

### 1.5.5.2 Clausius Statement

Heat always flows from a body at higher temperature to a body at a lower temperature. The reverse process never occurs spontaneously. Clausius statement of the second law says: *It is impossible to construct a device which, operating in a cycle, will produce no effect other than the transfer of heat from a low-temperature body to a high temperature body.*

### 1.5.6 Third Law of Thermodynamics

The third law of thermodynamics states that the *entropy of a system approaches a constant value (zero for perfect crystal) as the temperature approaches zero.* The crystal must be perfect for entropy to become zero; else, there will be some inherent disorder. It must also be at 0 K; otherwise, there will be thermal motion within the crystal which leads to disorder. The third law also says that it is not possible for any system to reach absolute zero in a finite number of steps. This effectively makes it impossible to ever attain a temperature of exactly 0 K.

## 1.6 Thermodynamic Power Cycles

According to the first law of thermodynamics, the net heat input is equal to the net work output over any cycle. The repeating nature of the process allows for continuous operation, making the cycle an important concept in thermodynamics. There are two types of thermodynamic cycles – the power cycle and the heat pump cycle. The power cycles convert heat input into mechanical work. The heat engine cycle transfers heat from lower temperature to higher temperature using mechanical work as input. Here only power cycles are discussed.

**Figure 1.10** Carnot cycle.

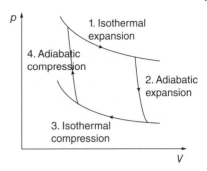

There are four power cycles that are generally used in the generation of electricity: the Rankine cycle used in steam turbines, the Brayton cycle for gas turbine, the Otto cycle for petrol engines and the Diesel cycle for diesel engines. These power cycles are the basis for the operation of heat engines, which supply most of the world's electric power and run almost all motor vehicles. Before discussing these power cycles, it will be useful to study the ideal cycle or Carnot cycle which describes the maximum efficiency of a thermodynamic cycle. In addition, only Rankine and Brayton cycles are described as these are used in renewable energy systems.

### 1.6.1 Ideal Cycle (Carnot Cycle)

The Carnot cycle shown in Fig. 1.10 is a cycle composed of totally reversible processes of isentropic compression and expansion and isothermal heat addition and rejection. Any fluid can be used in the operation of the Carnot cycle. There are four stages of a Carnot cycle.

In stage 1, (process 1–2) known as isothermal expansion, the heat is applied, and the fluid expands isothermally at constant temperature $T_1 = T_H$.

In stage II (process 2–3) known as adiabatic expansion, no heat is applied or taken away. The fluid expands adiabatically, and temperature falls from $T_1 = T_H$ to $T_2 = T_L$.

In stage III (process 3–4) known as isothermal compression, the heat is rejected by the fluid as it is compressed isothermally at constant temperature $T_L$.

In stage IV (process 4–1) known as adiabatic compression, no heat is allowed to enter or leave the fluid. The temperature of the fluid is raised due to compression to $T_H$.

The thermal efficiency of a Carnot cycle depends only on the absolute temperatures of the two reservoirs in which heat transfer takes place, and for a power cycle, it is given by

$$\eta = 1 - \frac{T_L}{T_H} \tag{1.17}$$

**Example 1.3**
300 kg of fruits are preserved in a cold storage at 30°C. The cold storage maintains a temperature of −5°C, and the fruits take 8 h to be at the storage temperature. The latent heat of freezing is 105 kJ/kg, and the specific heat of the fruits is 1.256 kJ/kg K. Find the refrigeration capacity of the plant.

**Solution**

Mass of fruits $= 300\,\text{kg}$; temperature $= 273 + 30 = 303\,\text{K}$.

It is to be cooled to $273 - 5 = 268$ K.

The heat removed from the fruits in $8$ h $= 300 \times 1.256 \times (303 - 268)$

$\qquad = 13,188$ kJ.

Total latent heat of freezing $= 300 \times 105 = 31,500$ kJ.

Total heat removed in 8 h $= 13,188 + 31,500 = 44,688$ kJ.

Heat removed per minute $= 93.1$ kJ/min $= 93.1/210 = 0.423$ TR

$\qquad (1\,\text{TR} = 210\,\text{kJ/min})$.

### 1.6.2 Rankine Cycle

Rankine cycle is an ideal cycle for steam power plants. The shortcomings inherent in a steam power plant realizing the Carnot cycle with wet steam can be partly remedied if heat is rejected from the wet steam in the condenser before the entire steam condenses. Thus, instead of wet steam with a small density, water is compressed from pressure $P_2$ to pressure $P_1$. The cycle consists of four processes similar to Carnot cycle as shown in Fig. 1.11(a) and (b).

In process 1–2, heat $Q_1$ is added to increase the temperature of the high-pressure water up to its saturation value.
In process 2–3, the work delivered by turbine, $W_T = h_3 - h_4$.
In process 3–4, the energy rejected in the condenser, $q_2 = h_4 - h_1$.
In process 4–1, the work is done in pumping water (negligible).

The thermal efficiency of the Rankine cycle is given by the area under the closed loop 1–2–3–4 shown in the *T-S* diagram:

$$\eta = \frac{Q_1 - Q_2}{Q_1} = \frac{(h_3 - h_2) - (h_4 - h_1)}{(h_3 - h_2)} \tag{1.18}$$

Normally, the water-pumping power is very small compared to the power produced by the turbine.

$$\eta = \frac{(h_3 - h_4)}{(h_3 - h_2)} \tag{1.19}$$

Rankine cycle efficiency can be improved by the following.

- Superheat: If the steam is superheated before entering the turbine, the efficiency is improved.
- Increased boiler pressure. By increasing the boiler pressure, the efficiency is increased, reaching a maximum value. Further increase in pressure decreases the efficiency.
- Reducing the condenser pressure: By reducing the condensing pressure, the temperature at which heat is rejected can be decreased, thereby improving the efficiency.

### 1.6.3 Brayton Cycle

The Brayton cycle was proposed by George Brayton in 1870 for use in reciprocating engines. Now it is used in gas turbines with rotating machines. This cycle consists of

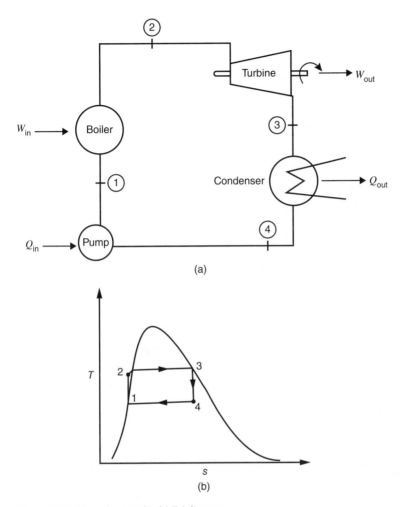

**Figure 1.11** (a) Rankine cycle; (b) *T-S* diagram.

two isentropic and two constant-pressure processes. Figure 1.12 shows the Brayton cycle with *P-V* and *T-S* diagrams. Brayton cycle basically consists of four internally reversible processes. Process 1–2 is isentropic compression. It is basically compression in a compressor. Here work is accepted as input. Process 2–3 is constant-pressure transfer of heat in combustor. Process 3–4 is isentropic expansion. In the case of a gas turbine engine, it is in a turbine. A work is produced. Process 4–1 is constant-pressure heat rejection by the condenser. All these processes are internally reversible. The thermal efficiency of Brayton cycle can be obtained as

$$\eta = \frac{\text{net work}}{\text{Heat input}} \tag{1.20}$$

In an ideal Brayton cycle, heat is added to the cycle at a constant-pressure process (process 2–3).

$$q_{in} = h_3 - h_2 = C_p \, (T_3 - T_2)C_p \tag{1.21}$$

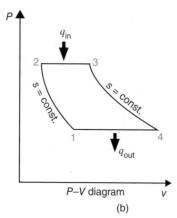

**Figure 1.12** (a) Brayton cycle; (b) *P-V* diagram.

Heat is rejected at a constant-pressure process (process 4–1).

$$q_{out} = h_4 - h_1 = C_p (T_4 - T_1) \tag{1.22}$$

Then the thermal efficiency of the ideal Brayton cycle under the cold-air standard assumption is given as

$$\eta = \frac{c_p(T_4 - T_1)}{c_p(T_3 - T_2)} \tag{1.23}$$

Process 1–2 and process 3–4 are isentropic processes; thus,

$$\eta = 1 - \left\{ \frac{p_1}{p_2} \right\}^{\frac{(\gamma-1)}{\gamma}} \tag{1.24}$$

$$\eta = 1 - \frac{1}{r_p^{(\gamma-1)/\gamma}} \tag{1.25}$$

where $r_p = \frac{p_2}{p_1}$ is the pressure ratio and $\gamma$ is the specific heat ratio. In most designs, the pressure ratio of gas turbines ranges from about 11 to 16.

## 1.7 Summary

As the demand for renewable energy sources to generate power is growing, different sources of renewable energy are being exploited. The greenhouse effect and climate change are deriving most of the countries to apply power sources that are pollution free. The technologies involved in various renewable energy sources are based on theories of fluid dynamics and heat transfer (thermodynamics). The transfer of energy to and from a moving fluid is the basis of hydro, wind, wave and some solar power systems. This chapter therefore describes some of the basic properties of fluid dynamics and thermodynamics.

Since similarly power electronics is finding applications in RES to improve performance and efficiency of these systems, these systems require greater attention and will be described in Chapter 9.

## References

1 REN 21 (2010) *Renewable Energy Policy Network for the 21st Century*, p. 15.
2 Boyle, G. (2012) *Renewable Energy: Power for a Sustainable*, Future.
3 Dixon, S.L. and Hall, C. (2010) *Fluid Mechanics and Thermodynamics of Turbomachinery*, Butterworth-Heinemann.
4 Alexander, O. (1969) *Treatise on Thermodynamics*, 3rd edn, Dover Publications.
5 Wolfgang, P. (2003) *Thermodynamics and the Kinetic Theory of Gases*, Dover Publications.

# 2

# Power Electronic Converters

## 2.1 Types of Power Electronic Converters

In order to utilize the power generated by various renewable energy sources, power electronic converters are essentially required. Many renewable resources such as wind are intermittent in nature and require power electronic circuits to maintain frequency and voltage at the desired level. In addition, the present trend is to use variable-speed wind turbine with pitch control. Similarly, the power generated by photovoltaic conversion is dc and must be converted to ac before it can be utilized in the existing system. Moreover, due to the intermittent nature of power generation in renewable energy systems, energy storage is required to maintain balance between generation and load. Energy is generally stored in batteries, which requires PE for control. The function of power electronics in renewable energy systems is to convert energy generated by renewable energy sources to be used in power grid with maximum possible efficiency and minimum cost. Thus, power electronic converters in the form of ac-dc-ac, dc-dc-ac or dc to ac are commonly used in renewable energy systems and are briefly described in this chapter. Power converters for renewable energy systems integration present a more complex system compared to stand-alone system. It is therefore important to study PE converters from the point of view of their application in renewable energy systems.

## 2.2 Power Semiconductor Devices

Power electronic converters utilize power semiconductor devices that were first introduced in the form of thyristor or SCR in 1975 by the General Electric Company. In last 40 years, many new power electronic devices have been developed, and now power electronic control is used wherever large power control is required [1–4]. In the most simple form, a power electronic device may be considered as a switch which may be controlled or uncontrolled. Simple p–n diode is an uncontrolled switch. It is a two-terminal device shown in Fig. 2.1(a). In an ideal diode, current can only flow in one direction from anode to cathode. However, in an actual diode, small current can flow in the reverse direction as well. The $i$-$v$ characteristic of an actual diode is shown in Fig. 2.1(b). As can be seen from the figure, the $i$-$v$ characteristic of the diode is not linear, and it has exponential current–voltage relationship. When a diode is connected in a *Forward Bias* condition, a positive voltage is applied to the p-type material and negative voltage is applied to

*Operation and Control of Renewable Energy Systems*, First Edition. Mukhtar Ahmad.
© 2018 John Wiley & Sons Ltd. Published 2018 by John Wiley & Sons Ltd.

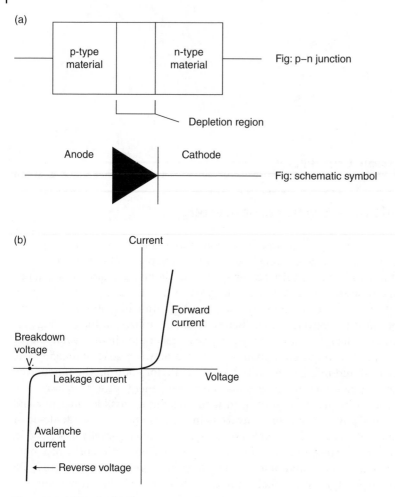

**Figure 2.1** (a) Diode. (b) Characteristic of a real diode.

the n-type material. If this external voltage becomes greater than the potential barrier, which is approximately 0.7 V for silicon and 0.3 V for germanium, the opposition due to potential barriers will be overcome and current will start to flow.

In a *Reverse Bias* condition, a positive voltage is applied to the n-type material and a negative voltage is applied to the p-type material. The positive voltage applied to the n-type material attracts electrons towards the positive electrode and away from the junction, while the holes in the p-type end are also attracted away from the junction and towards the negative electrode. This results in a high-potential barrier, thus preventing the current from flowing through the semiconductor material. In a practical diode, a small reverse current does flow, known as leakage current, but it can be ignored for most purposes. As can be seen from the *i-v* characteristic, if the reverse applied voltage is increased to a large value, breakdown occurs. The maximum reverse bias voltage that a diode can withstand without "breaking down" is called the *Peak Inverse Voltage* or *PIV* rating of the diode.

### 2.2.1 Thyristor

The thyristor, or silicon controlled rectifier (SCR), introduced by GEC in 1958 is responsible for modern power electronic age. A thyristor consists of p–n–p–n layers with three junctions and has three terminals: anode, cathode and gate. Its diagrammatical representation and circuit symbol are shown in Fig. 2.2. The static *V-I* characteristics of thyristor are shown in Fig. 2.3. As can be seen from the characteristic, the thyristor has three modes of operation. When thyristor is reverse biased, that is, the anode is negative with respect to the cathode, junctions $j_1$ and $j_3$ are reverse biased and do not allow the flow of current through them; this condition is called reverse-blocking state of thyristor. This condition is similar to p–n diode under reverse bias condition. When thyristor is forward biased, that is, the anode is positive with respect to the cathode, the junctions $j_1$ and $j_3$ are forward biased but junction $j_2$ is reverse biased which blocks the flow of current through the device. This state is known as forward-blocking state. In forward-blocking state, the thyristor can be made to conduct if the gate-to-cathode positive voltage is applied. For different values of gate current, the thyristor will conduct at different values of anode-to-cathode voltage as shown in Fig. 2.3. However, once the thyristor is turned on, gate current is not required to keep it in the *on* state. Thus, it is possible to turn on the thyristor by simply applying a pulse, and constant voltage at the gate is not required. Steep-fronted gate pulses are therefore desirable for turning on the thyristor. The gate pulse should be sufficient to allow the gate current to reach a minimum level known as latching current; otherwise, the thyristor will turn off. The thyristor can also be turned on by applying a beam of light directed at the gate-to-cathode junction. For light-triggered SCRs, a special terminal niche is made inside the inner P layer instead of the gate terminal. When light is allowed to strike this terminal, free charge carriers are generated. When the intensity of light becomes more than a normal value, the thyristor starts conducting. This type of SCRs are called as Light-Activated SCR (LASCR)

Turning off a thyristor known as commutation can be achieved by reducing the anode current below the holding current by providing a reverse bias across it. Thyristor commutation schemes are broadly classified as follows:

- Line commutation or natural commutation
- Load commutation
- Forced commutation

#### 2.2.1.1 Line Commutation
If the thyristor is connected to ac supply, the natural reversal of ac supply voltage can commutate the thyristor without any external commutation circuit. The line

**Figure 2.2** Thyristor.

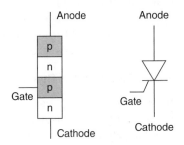

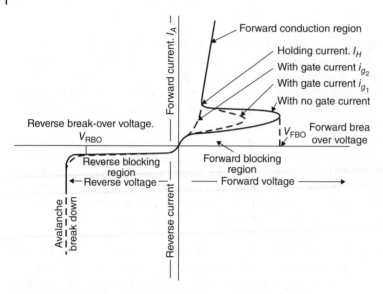

**Figure 2.3** Conduction of thyristor for different gate currents.

commutation is simple and is used in ac-to-dc converters, ac regulators and cyclocon-verters to be described later.

#### 2.2.1.2 Load Commutation

If the thyristor is connected to dc supply and the load contains a capacitor, the current will automatically fall to zero when the capacitor is fully charged. The thyristor will turn off automatically, and in order to turn it on again, the capacitor must be discharged. This type of commutation is called load commutation.

#### 2.2.1.3 Forced Commutation

When a thyristor is connected to a dc source and load commutation is not possible, it can be turned off by switching a previously charged capacitor across it to provide reverse bias. This turn-off process is known as forced commutation. The forced commutation process can also be classified as voltage commutation or current commutation, depend-ing on whether the thyristor is turned off by applying a pulse of large reverse voltage or by passing an external pulse of current. A voltage commutation circuit is shown in Fig. 2.4. Here $T_1$ is the main thyristor, and $T_A$ is the auxiliary thyristor used to turn off $T_1$. When $T_1$ is "on", the commutating capacitor $C$ is charged to slightly higher than $-V$ V due to inductor $L$. To turn off $T_1$, auxiliary thyristor $T_A$ is turned on by applying voltage pulse to its gate. This connects the charged capacitor in parallel with the main thyristor and applies a reverse voltage across it. Since its voltage is slightly higher than the applied voltage, the current through $T_1$ tries to reverse and turns it off. The load current during this period freewheels through diode D.

### 2.2.2 Gate Turn-Off Thyristor (GTO)

In order to avoid the use of bulky commutation circuits for turning off a thyristor, modification in the structure of the thyristor is made so that it can be turned off through gate current. This thyristor is called GTO. Similarly to thyristor, the GTO is a

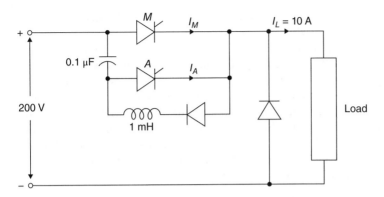

**Figure 2.4** Turning off a thyristor (voltage commutation).

**Figure 2.5** Gate turn-off thyristor.

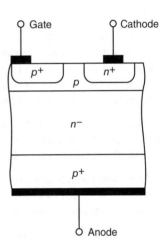

current-controlled minority carrier (i.e. bipolar) device. GTOs differ from conventional thyristor in that they are designed to turn off when a negative current is sent through the gate, thereby causing a reversal of the gate current. A relatively high gate current is needed to turn off the device with typical turn off gains in the range of 4–5.

It can be used in power inverter circuits as it can be switched on and off at a higher speed compared to a normal thyristor. The GTO circuit symbol is shown in Fig. 2.5. The two arrows at the gate indicate the bidirectional gate current capability. GTO is also a four-layer (p–n–p–n), three-terminal device having an anode, a cathode and gate terminals. The four layers are modelled as two transistors p–n–p and n–p–n. To turn on the GTO, a gate current is injected as in a normal thyristor. Once the device is turned on, external gate current is not necessary to keep it in on condition. To turn off a conducting GTO, the gate terminal is biased negative with respect to the cathode. However, the current required to turn off GTO is much larger than the current required for turning it on.

### 2.2.3  Power Bipolar Junction Transistor

Power transistors like GTO are controlled turn-on and turn-off devices. However, a GTO requires a large reverse current through its gate to turn it off. In addition, the switching speed of GTO is very low. These disadvantages are eliminated in power transistors. They are turned on when a current signal is applied to the base or control terminal. The transistor remains in the on state so long as the control signal is present. Bipolar junction transistor (BJT) was the first power transistor developed which simplified the design of a large number of power electronic circuits that earlier used forced commutated thyristors or GTOs. Subsequently, many other devices that can broadly be classified as "Power Transistors" have been developed. Many of them have superior performance compared to the BJT in some respects. They have, by now, almost completely replaced BJTs. However, it should be emphasized that the BJT was the first semiconductor device to closely approximate an ideal fully controlled power switch. Power transistors are classified as follows:

- Bipolar junction transistors (BJTs)
- Metal–oxide–semiconductor field-effect transistors (MOSFETs)
- Insulated-gate bipolar transistors (IGBTs)

A transistor is a three-layer and three-terminal device which is either p–n–p or n–p–n with terminals marked as collector, emitter and base as shown in Fig. 2.6(a). From the point of view of construction and operation, BJT is a bipolar (i.e. minority carrier) current-controlled device. The bipolar transistors have the ability to operate within three different regions:

- Active: In this region, the transistor works as an amplifier.
- Saturation: The transistor is in the on state, offering very low forward resistance.
- Cutoff: The transistor is in the blocking state or off state.

In a power electronic circuit, the power transistor is usually employed as a switch; that is, it operates in either "cutoff" (switch OFF) or saturation (switch ON) regions. However, similarly to any other switching device, it has the following limitations:

- It can conduct only a finite amount of current in the on state.
- It can have limited blocking voltage in the off state.
- It has a small forward resistance and voltage drop during "ON" condition.
- It carries a small leakage current during OFF condition.
- Can not switch on or off instantaneously.

As can be seen from its *i-v* characteristics in Fig. 2.6(b), a significantly large base current results in switching the device to saturation (on) state. The on-state voltage of the

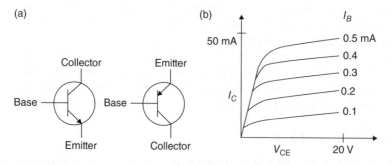

**Figure 2.6** (a) n–p–n and p–n–p transistor; (b) *i-v* characteristic of a bipolar transistor.

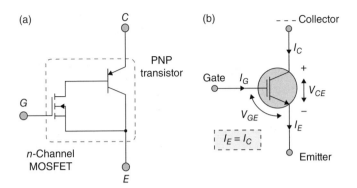

**Figure 2.7** (a) Simplified equivalent circuit n-channel power MOSFET and (b) circuit symbol.

power transistors is about 1–2 V. BJTs are current-controlled devices, and base current must be supplied continuously to keep them in the on state. However, to turn off BJTs, simply the base current is removed.

### 2.2.4 Power MOSFET

A power MOSFET (Metal–Oxide–Semiconductor Field-Effect Transistor) is a voltage-controlled device and requires only a small input current to turn on. The circuit symbol of an n-channel MOSFET is shown in Fig. 2.7(a), and circuit symbol shown in Fig. 2.7(b). It has three external terminals designated as drain, source and gate. MOSFETs have very high input impedance. The current gain which is the ratio of drain current to gate current is also very high of the order of $10^9$. They require low gate energy and have very fast switching speed and low switching losses.

MOSFETs require continuous application of gate source voltage of appropriate magnitude to keep them in the on state. The MOSFET is turned off when the gate-to-source voltage is less than the threshold value which is typically 2–3 V for most of the devices. Gate current flows only during transitions from on to off or off to on conditions only. The main drawback of MOSFET is its high on-state resistance which increases with increased voltage rating. These devices are therefore more suitable for high-speed switching.

### 2.2.5 Insulated Gate Bipolar Transistor (IGBT)

In IGBT, the qualities of BJT and MOSFET are combined so that high switching speed of MOSFET with low conduction losses of BJT is obtained. The equivalent circuit of an n-channel IGBT is shown in Fig. 2.8(a) For p-channel IGBT, the arrow direction is reversed. The arrow basically indicates the injection of current. It also has three terminals: collector, emitter and gate. The circuit symbol of an n-channel IGBT is shown in Fig. 2.8(b). An insulated gate bipolar transistor is simply turned "ON" by applying a positive input voltage signal between the gate and the emitter, while it can be turned off by making the input gate signal zero or slightly negative with respect to the emitter. The characteristic of p-channel IGBT will be the same except that the polarities of the voltages and currents will be reversed.

When the gate source voltage is enough, electrons are drawn towards the gate, allowing the flow of current from source to drain. In this mode, the IGBT enters into the active (a)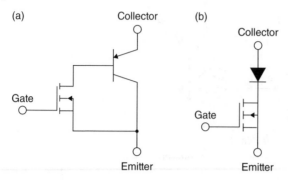

(b)

**Figure 2.8** (a) n-channel IGBT equivalent circuit and (b) circuit symbol.

region of operation, and the drain current is determined by the transfer characteristics of the device. This characteristic is qualitatively similar to that of a power MOSFET and is reasonably linear over most of the collector current range. As the gate emitter voltage is increased further, the collector current also increases, and for a given load resistance, the collector–emitter voltage decreases. When collector–emitter voltage becomes less than the gate source voltage, the IGBT enters into the saturation region. If the source gate voltage is below the threshold value, only a very small leakage current flows though the device while the collector–emitter voltage almost equals the supply voltage. In this condition, the IGBT is said to be operating in the cutoff mode.

The main advantages of using the IGBT over other types of transistor devices are its high voltage capability and low ON resistance. It has fast switching speeds compared to BJT but less than MOSFET combined with zero gate drive current, making it a good choice for moderate-speed, high-voltage applications. IGBT is most suitable for high-power applications such as the following:

- Electric vehicle motor drives
- Power factor correction converters
- Solar inverters
- Uninterruptable power supplies (UPS)
- Inductive heating cookers

## 2.3 ac-to-dc Converters

Wind turbines with a PM synchronous generator produce electricity at variable frequency and voltage. It therefore is first converted into dc voltage in an uncontrolled manner. The output of this converter is connected to an inverter which controls the frequency and voltage suitable for the ac system [5, 6]. The most simple ac-dc converter is single-phase uncontrolled half-wave converter which can provide a fixed output voltage. If variable voltage is required, input to this converter may be made through autotransformer. But it suffers from poor output voltage and/or input current ripple factor. The ripples can be reduced, and dc supply smoothened by connecting a capacitor at the output terminal. In addition, the input current contains a dc component which may cause problem (e.g. transformer saturation) in the power supply system. The output dc voltage is also relatively less. Some of these problems can be addressed using a full-wave rectifier described here.

### 2.3.1 Single-Phase Diode Bridge Rectifiers

This is one of the most popular rectifier configurations and is used widely for applications requiring dc. power output from a few hundred watts to several kilowatts. Figure 2.9(a) shows the rectifier supplying a resistive load. These rectifiers are also very widely used with capacitive loads particularly as the front end of a variable-frequency voltage source inverter.

During the positive half cycle of the supply, diodes $D_1$ and $D_2$ are forward biased and conduct in series while diodes $D_3$ and $D_4$ are reverse biased. The current flows through the load as shown in Fig. 2.9. During the negative half cycle of the supply, diodes $D_3$ and $D_4$ are forward biased and diodes $D_1$ and $D_2$ are reverse biased. Now the current flows through the diodes $D_3$ and $D_4$ and through the load. As can be seen from Fig. 2.9, the direction of the current is the same as before. As the current flowing through the load is unidirectional, the voltage developed across the load is also unidirectional. The ripple frequency is now twice the supply frequency (e.g. 100 Hz for a 50 Hz supply or 120 Hz for a 60 Hz supply). If a smoothing capacitor C is connected, it converts the full-wave rippled output of the rectifier into a smooth dc output voltage. The output voltage with and without a capacitor is shown in Fig. 2.9(b). Generally for dc power supply circuits, the smoothing capacitor is an aluminium electrolytic type that has a capacitance value of 100 μF or more with repeated dc voltage pulses from the rectifier charging up the capacitor to peak voltage.

The main advantages of a full-wave bridge rectifier is that it has a smaller ac ripple value for a given load and a smaller smoothing capacitor compared to an equivalent half-wave rectifier. Therefore, the fundamental frequency of the ripple voltage is twice

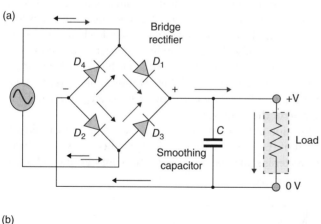

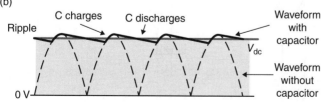

**Figure 2.9** (a) Single-phase diode bridge rectifier. (b) Resultant output waveform.

that of the ac supply, whereas for the half-wave rectifier, it is exactly equal to the supply frequency.

### 2.3.2 Three-Phase Full-Wave Bridge Diode Rectifiers

In large-capacity wind turbines using three-phase variable-speed generators, three-phase full-wave bridge rectifiers are used. A three-phase, six-pulse full-wave bridge diode rectifier is shown in Fig. 2.10. As shown in the figure, the anodes of three diodes are connected together in one group and the cathodes of other three diodes are connected together in another group. The three phase balances ac supply is connected as shown. The device connected to the most positive voltage will conduct in the cathode group; the other two will be reverse biased. In addition, the device connected to the most negative voltage will conduct in the anode group; the other two in this group will be reverse biased. At least two devices from positive and negative groups are simultaneously in the open state here, and at least one device from each group must conduct to facilitate the flow of the current. The voltage ripple is low because the output voltage consists of six pulses per voltage period. This circuit does not require the neutral connection of the three-phase source; therefore, a delta-connected source can also be used.

The positive output terminal voltage is equal to the maximum of the phase voltages, and the voltage of the negative output terminal voltage is equal to the minimum of the phase voltages. In the figure, only the positive waveform of the output voltage is shown.

### 2.3.3 Single-Phase Fully Controlled Rectifiers

A single-phase fully controlled rectifier is shown in Figure. 2.11. It is similar to diode bridge rectifier with the difference that four diodes are replaced by four thyristors or transistors. These switches can be turned on by applying controlling pulse at any instant of time during which the anode is at a higher voltage. For $R\_L$ load and continuous load current, switches $Q_1$ and $Q_4$ remain in the on state beyond the positive half-wave of the source voltage $E_s$. For this reason, the load voltage $E_0$ can have a negative instantaneous value. The turning on of switches $Q_2$ and $Q_4$ has two effects: in the case of thyristors, this will turn off $Q_1$ and $Q_4$ and the load current will flow through them. This type of converter is known as line-commutated converter. The output voltage is controlled by controlling the turning-on period of the switches.

Under normal operating condition of the converter, the load current may or may not remain zero over some interval of the input voltage cycle. If load current $I_0$ is always greater than zero, then the converter is said to be operating in the continuous conduction

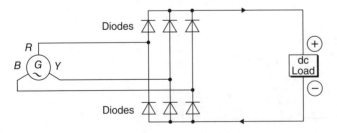

**Figure 2.10** Three-phase six-pulse diode bridge rectifier.

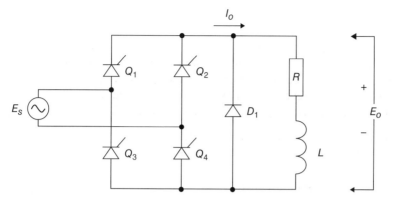

**Figure 2.11** Single-phase controlled bridge rectifier.

mode. In the discontinuous conduction mode, none of the switches carry the current over some portion of the input cycle. The load current remains zero during that period. The disadvantage of phase-controlled converters is that for a low voltage output, their power factor is low.

### 2.3.4 Three-Phase Fully Controlled Bridge Converter

In three-phase fully controlled bridge converter, the diodes as shown in Fig. 2.10 are replaced by thyristors (transistor) as shown in Fig. 2.12. This circuit is also known as a three-phase full-wave bridge or as a six-pulse converter. The thyristors are triggered at an interval of $\pi/6$ radians (i.e. at an interval of 30°). The frequency of the output ripple voltage is 6 fs.

Three-phase converters are used when large power is to be controlled. Control over the output dc voltage is obtained by controlling the conduction interval of each thyristor. This method is known as phase control, and the converters are also called "phase-controlled converters".

The three-phase bridge rectifiers have following advantages:

- Low ripple
- High power factor
- Simple construction

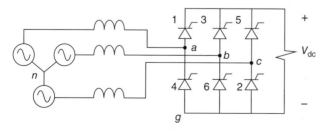

**Figure 2.12** Three-phase controlled bridge rectifier.

## 2.4 dc-to-ac Converters (Inverters)

In renewable energy systems, power electronic interface is required to convert dc power generated by PV solar system or fuel cell into ac power which may be directly used in isolated systems [7, 8] or for connection to grids. When connected to a power grid, these are used to inject sinusoidal current into the power system with required THD and high power factor. The inverters available for power conversion are as follows:

- Single-phase and three-phase voltage source inverters
- Single-phase and three-phase current source inverters

### 2.4.1 Single-Phase Voltage Source Inverters

Figure 2.13 shows the power circuit diagram for single-phase voltage source H-bridge inverter. In this, four switches (in two legs) are used to generate the ac waveform at the output. Any semiconductor switch such as GTO, IGBT, MOSFET or BJT can be used. For inductive load, the stored energy is fed back through diodes connected across the switches. These diodes are called as *Feedback Diodes*.

Anti-parallel diodes are not required for resistive load because load current $i_o$ is in phase with output voltage $v_o$.

### 2.4.2 Square-Wave PWM Inverter

In full-wave bridge inverter, shown in Fig. 2.13, there are two arms of the bridge with two switches in each arm. The switches in each branch are operated alternatively so that they are not in the same mode (ON/OFF) simultaneously. When $S_1$ and $S_4$ are on and $S_2$ and $S_3$ are off, the output voltage across the load is $V_s$, and when $S_3$ and $S_2$ conduct, the output voltage is $-V_s$. The switches $S_1$ and $S_4$ conduct for a period of $0 < t \leq T/2$, and the switches $S_3$ and $S_2$ conduct for a period of $T/2 < t \leq T$, where $T$ is the time period of the conduction of switches. The frequency of output ac voltage is given by $1/T$. The frequency can be varied by varying the $T$ or conduction period of the switch with the help of the gate (base) signal. The root mean square (rms) value of the output ac voltage is

$$V_0 = \left( \frac{1}{T} \int_0^T V_s^2 \, dt \right)^{\frac{1}{2}} \tag{2.1}$$

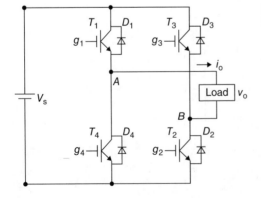

**Figure 2.13** Single-phase voltage source bridge inverter.

The following PWM techniques are used for controlling the output ac rms voltage and frequency control in an inverter.

- Single-pulse-width modulation
- Multiple-pulse-width modulation
- Sinusoidal-pulse-width modulation

### 2.4.3 Single-Pulse-Width Modulation

In single-pulse-width modulation control, there is only one pulse per half cycle and the output rms voltage is changed by varying the width of the pulse. The gating signals for the switches are generated by comparing a rectangular reference signal of amplitude $V_c$ with a triangular reference carrier signal $V_{car}$. The time period of the two signals is nearly equal. The gating signals and output voltages of single-pulse-width modulation are shown in Fig 2.14(a) and (b). The frequency of the control signal determines the frequency of the output voltage. The instantaneous output voltage is $v_0 = V_s$ corresponding to $T_1$ and $T_2$ conducting and equal to $-V_s$ corresponding to $T_3$ and $T_4$ conducting. If $\delta$ is the time period for which $T_1$ and $T_2$ conduct and also $T_3$ and $T_4$, then the rms value

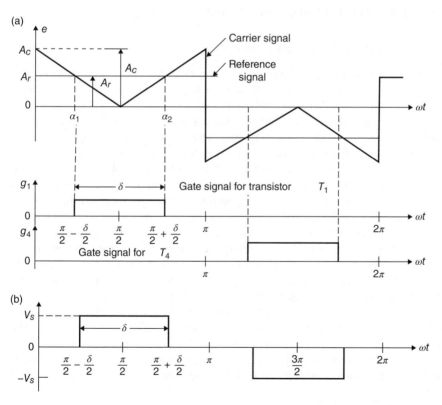

**Figure 2.14** (a) Gating signal for single-pulse-width modulated single-phase inverter. (b) Output voltage.

of output voltage is

$$V_0 = \left[ \frac{2}{2\pi} \int_{(\pi-\delta)/2}^{(\pi+\delta)/2} V_s^2 d(\omega t) \right]^{1/2} \tag{2.2}$$

By varying $V_c$ from 0 to $V_{car}$, the pulse width $\delta$ can be varied from 0 to $\pi$, and the rms output voltage from 0 to $V_s$.

The output voltage consists of a rectangular pulse in each half cycle which can be expressed as

$$v_0(t) = \sum_{n=1,2,3,\ldots}^{\infty} \frac{4V_s}{n\pi} \sin \frac{n\delta}{2} \sin n\omega t \tag{2.3}$$

### 2.4.4 Multiple-Pulse-Width Modulation

In this modulation, there are multiple numbers of output pulses per half cycle, and all pulses are of equal width. The gating signals are generated by comparing a rectangular reference with a triangular reference. The frequency of the reference signal $f_{carr}$ determines the number of pulses per half cycle $m$, whereas the frequency of reference signal $V_c$ sets the output frequency $f$. The modulation index controls the output voltage. This type of modulation is also known as symmetrical pulse-width modulation. The number of pulses $N_p$ per half cycle is found from the expression $N_p = \dfrac{f_{carr}}{2f_c}$.

### 2.4.5 Sinusoidal-Pulse-Width Modulation

In sinusoidal pulse-width modulation, there are multiple pulses per half-cycle as described earlier, but the width of each pulse is varied over the output cycle, in a sinusoidal manner. The scheme, in its simplified form, involves comparison of a high-frequency triangular carrier voltage with a sinusoidal modulating signal that represents the desired fundamental frequency component of the voltage waveform. The peak magnitude of the modulating signal is limited to the peak magnitude of the carrier signal. The comparator output is then used to control the high-side and low-side switches. The two types of pulse-width modulation inverters are bipolar and unipolar switching. Each unique switching technique creates either a unipolar or a bipolar output at the load. Figure 2.15 shows the gating signals and output voltage of SPWM with unipolar switching. In this scheme, the switches in the two legs of the full-bridge inverter are not switched simultaneously, as in the bipolar scheme. In this unipolar scheme, the legs A and B of the full-bridge inverter are controlled separately by comparing the carrier triangular wave $v_{car}$ with the control sinusoidal signals $v_c$ and $-v_c$, respectively. This SPWM is generally used in industrial applications. The number of pulses per half cycle depends upon the ratio of the frequency of the carrier signal ($f_c$) to the modulating sinusoidal signal. The frequency of the control signal or the modulating signal sets the inverter output frequency ($f_o$) and the peak magnitude of the control signal controls the modulation index $m_a$ which in turn controls the rms output voltage. If $t_{on}$ is the width of nth pulse, the rms output voltage can be determined by

$$V_0 = V_s \left\{ \sum_{n=1}^{2p} 2t_{on}/T \right\}^{\frac{1}{2}} \tag{2.4}$$

(a)

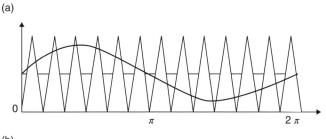

(b)

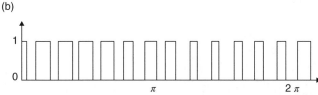

**Figure 2.15** (a) Gating signal and (b) output voltage for sinusoidal PWM of single phase.

The *amplitude modulation index* is defined as

$$m_a = \frac{\widehat{V}_c}{\widehat{V}_{\text{car}}} \tag{2.5}$$

where,

$\widehat{V}_c$ = peak magnitude of the control signal (modulating sine wave)

$\widehat{V}_{\text{car}}$ = peak magnitude of peak magnitude of the carrier signal (triangular signal).

The *frequency modulation ratio* is defined as

$$m_f = \frac{f_{\text{car}}}{f_c} \tag{2.6}$$

### 2.4.6   Three-Phase Voltage Source Inverters

Figure 2.16 shows a three-phase voltage source inverter circuit. The inverter is fed by a dc voltage and has three phase legs, each consisting of two transistors and two diodes (labelled) with subscripts (*a*, *b*, *c*). A common inverter control method is sine–triangle pulse-width modulation (STPWM) control. With STPWM control, the switches of the inverter are controlled based on a comparison of a sinusoidal control signal and a triangular switching signal. There are three sinusoidal reference waves, each displaced by 120°. A triangular carrier wave is compared with the reference signal of each phase to generate a control signal for that phase.

The sinusoidal control waveform establishes the desired fundamental frequency of the inverter output, while the triangular waveform establishes the switching frequency of the inverter. The ratio between the frequencies of the triangle wave and the sinusoid is referred to as the modulation frequency ratio.

The sinusoidal PWM control signal generation is shown in Fig. 2.17(a). In theory, the switches in each leg are never both on or off simultaneously; therefore, the voltages $V_{ag}$, $V_{bg}$ and $V_{cg}$ fluctuate between the input voltage ($V_{\text{dc}}$) and zero. By controlling the switches in this manner, the output voltages are ac, with the fundamental frequency

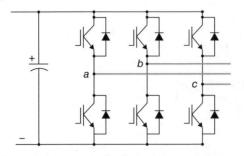

**Figure 2.16** Three-phase voltage source inverter.

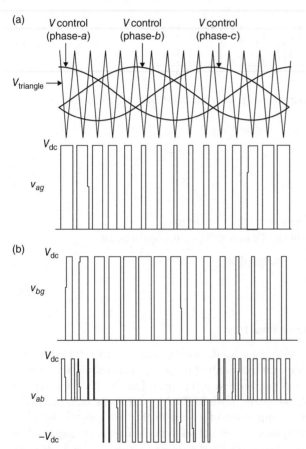

**Figure 2.17** (a) Generation of control signal for sinusoidal PWM three-phase inverter. (b) Output voltage waveform.

corresponding to the frequency of the sinusoidal control voltage as shown in Fig. 2.17(b). In most instances, the magnitude of the triangle wave is held fixed. The amplitude of the inverter output voltages is therefore controlled by adjusting the amplitude of the sinusoidal control voltages. The ratio of the amplitude of the sinusoidal waveforms relative to the amplitude of the triangle wave is the amplitude modulation index similarly to a single-phase inverter. The diodes provide paths for current when a transistor is turned on but cannot conduct the polarity of the load current. For example, if the load current is negative at the instant, the upper transistor is gated on, the diode in parallel with the upper transistor will conduct until the load current becomes positive at which time the upper transistor will begin to conduct.

### 2.4.7 Single-Phase Current Source Inverters

The voltage source inverters are fed from a voltage source, and the load current alternates from positive to negative value. In a current source inverter, the output current is maintained constant irrespective of the load, and the voltage is changed between positive and negative values. A single-phase current source inverter with transistor switches is shown in Fig. 2.18. The output of this inverter is near-square-wave current. In CSI, no feedback diodes are required, and the switching devices must withstand peak reverse voltages. The transistors conduct in the sequence $Q_1$, $Q_2$; $Q_2$, $Q_3$; $Q_3$, $Q_4$; $Q_4$, $Q_1$, and so on.

The load current can be expressed as

$$i_0(t) = \sum_{n=1,2,3,\ldots}^{\infty} \frac{4I_s}{n\pi} \sin \frac{n\delta}{2} \sin(n\omega t) \tag{2.7}$$

#### 2.4.7.1 Three-Phase Current Source Inverter

The circuit of a three-phase current source inverter (CSI) is shown in Fig. 2.19. As in single-phase CSI, the three–phase inverter is also operated from a current source. There are six switches, two in each arm as in the case of voltage source inverters. There is a 60° phase displacement between commutation of the upper device followed by commutation of the lower device.

If thyristors are used as switches then, six diodes, each one in series with the respective thyristor, are needed here, as used for single-phase CSI. Six capacitors, three each in two (top and bottom) halves, are used for commutation. If transistors are used instead of thyristors, capacitors for commutation are not required.

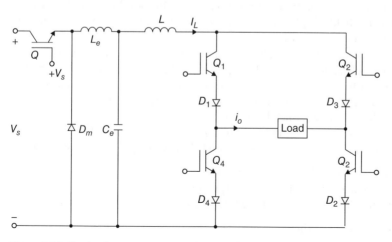

**Figure 2.18** Single-phase current source inverter.

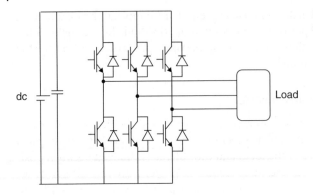

**Figure 2.19** Three-phase current source inverter.

For three-phase current source inverter and star-connected load, the output current in Fourier series can be written as

$$i_0(t) = \sum_{n=1,2,3,\ldots}^{\infty} \frac{4I_s}{n\pi} \sin\frac{n\pi}{3} \sin n\,(\omega t + \pi/6) \tag{2.8}$$

## 2.5  Multilevel Inverters

Multilevel converters have been developed to provide high power with medium-voltage semiconductor switches. Multilevel inverters achieve higher power by using a series of power semiconductor switches along with several lower voltage dc sources to perform the power conversion by synthesizing a staircase voltage waveform. The term multilevel was used initially for the three-level converter. Subsequently, several multilevel converter topologies have been developed. To achieve a three-level waveform, a single full-bridge inverter is employed. Basically, a full-bridge inverter is known as an H-bridge cell. This inverter produces three output voltage levels, $+V$, $-V$ and 0. In order to produce high power and high voltage, multilevel inverters are used in which the number of voltage levels is increased. As the number of voltage level increases, the harmonic contents of the output voltage decreases. Multilevel inverters are widely utilized in industrial field employing motor drives, static VAR compensators and renewable energy systems and so on.

The general structure of a multilevel inverter is to synthesize a near-sinusoidal waveform from several levels of dc voltages. The capacitor voltages can be added by connecting them in series with the help of power electronic switches to provide multilevel of voltages. In this way, a high voltage is reached at the output, while the power semiconductor devices withstand only reduced voltages.

The most attractive features of multilevel inverters are as follows [9–11].

They can generate output voltages with extremely low distortion and lower $dv/dt$.

1) They draw input current with very low distortion.
2) They generate smaller common-mode (CM) voltages.
3) They can operate with a lower switching frequency.

There are three types of multilevel inverters in use. These are as follows:

- Diode-clamped multilevel inverter

- Flying-capacitor multilevel inverter
- Cascaded multicell inverter

## 2.5.1 Diode-Clamped Multilevel Inverter

The diode-clamped inverter provides multiple voltage levels through connection of the phases to a series bank of capacitors. It typically consists of $(m-1)$ capacitors on the dc bus and produces m levels of phase voltage. A three-level diode-clamped inverter is shown in Fig. 2.20(a). In this circuit, the dc bus voltage is split into three levels by two series-connected bulk capacitors, $C_1$ and $C_2$. The middle point of the two capacitors $n$ can be defined as the neutral point. It is therefore also known as *neutral-point clamped multilevel inverter.* The output voltage $V_{an}$ is ac and has three states: $V_{dc}/2$, 0 and $-V_{dc}/2$. To obtain the voltage level of $V_{dc}/2$, switches $S_1$ and $S_2$ need to be turned on; for $-V_{dc}/2$, switches $S_1'$ and $S_2'$ need to be turned on; and for the 0 level, $S_2$ and $S_1'$ need to be turned on.

The key components that distinguish this circuit from a conventional two-level inverter are two diodes $D_1$ and $D_1'$. These two diodes clamp the switch voltage to half the level of the dc bus voltage. If the output is taken between $a$ and 0, then the circuit becomes a dc/dc converter, which has three output voltage levels: $V_{dc}$, $V_{dc}/2$ and 0. It is the most commonly used topology in the industry for a number of levels equal to 3.

Figure 2.20(b) shows a five-level diode-clamped converter in which the dc bus consists of four capacitors, $C_1$, $C_2$, $C_3$ and $C_4$. For input voltage $V_{dc}$, the voltage across each capacitor is $V_{dc}/4$, and each device will be stressed to one capacitor voltage level through clamping diodes. The staircase voltage is synthesized through five switch combinations

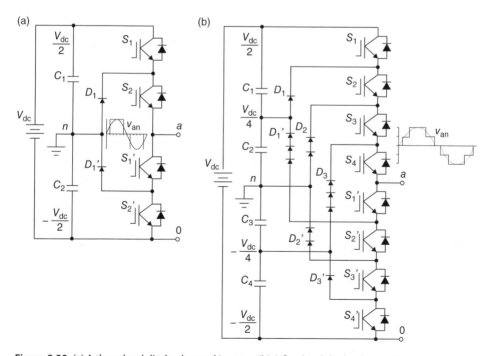

**Figure 2.20** (a) A three-level diode-clamped inverter. (b) A five-level diode-clamped inverter.

of five voltage levels across $a$ and $n$. The five voltage levels are $-V_{dc}/2, -V_{dc}/4, 0, V_{dc}/4$ and $V_{dc}/2$.

### 2.5.2 Flying-Capacitor Multilevel Inverter

Flying-capacitor inverter is quite similar to diode-clamped multilevel inverter. This type of multilevel inverter requires the capacitor to be pre-charged to a varying voltage level. By changing the switching states, the capacitors and dc source are connected in different connection configurations and produce various line-to-ground output voltages. Figure 2.21 illustrates the fundamental building block of a phase-leg capacitor-clamped inverter. The circuit is also called the flying-capacitor inverter with independent capacitors clamping the device voltage to one capacitor voltage level. The inverter in Fig. 2.21(a) provides a three-level output across $a$ and $n$, $V_{an}$, which is $V_{dc}/2$, 0 and $-V_{dc}/2$. Every cell has a single capacitor and two power switches. Power switch is a combination of a transistor connected with an anti-parallel diode. The switching of devices is similar to the one used for diode-clamped inverter.

Five-level capacitor-clamped inverter is shown in Fig. 2.21(b). The voltage synthesis in a five-level capacitor-clamped converter has more flexibility compared to a diode-clamped converter. The voltage of the five-level phase leg $a$ output with respect to the neutral point $n$ can be synthesized with many combinations of switch positions. As the number of levels is increased, many capacitors are required which makes this topology heavy and cumbersome. Since large numbers of capacitors are required, it can provide storage capability during power outages.

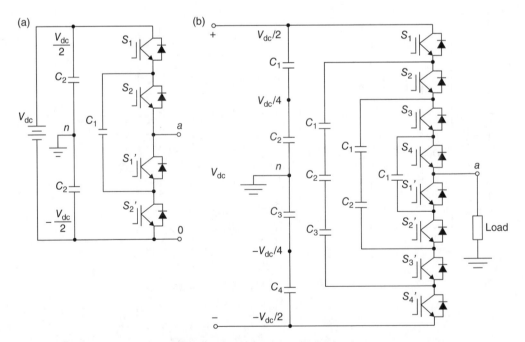

**Figure 2.21** (a) Three-level capacitor-clamped inverter; (b) five-level.

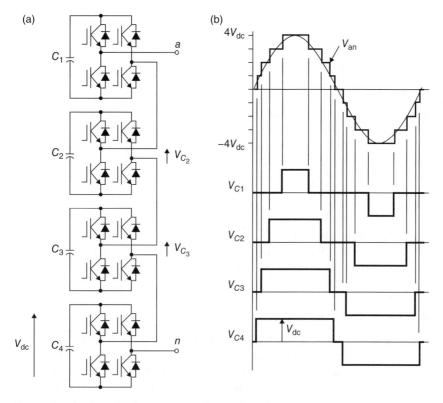

**Figure 2.22** (a) Cascaded inverter circuit; (b) waveforms.

### 2.5.3   Cascaded Multicell with Different dc Source Inverter

Cascaded multicell inverter is based on the series connection of single-phase inverters with separate dc sources. A cascaded multicell inverter (CMCI) differs in several ways from diode-clamped MLI and CCMLI in the manner in which the multilevel voltage waveform is obtained. It uses cascaded full-bridge inverters with separate dc sources, in a modular setup, to create the stepped waveform. Figure 2.22 shows the power circuit of one phase leg of a nine-level cascaded multicell inverter with four cells in each phase. It basically consists of a series of H-bridge inverter units. The resulting phase voltage is synthesized by the addition of the voltages generated by the different cells, each supplied by a different dc source. Each single-phase full-bridge inverter generates three voltages at the output: $V_{dc}$, 0 and $-V_{dc}$. This is made possible by connecting the capacitors sequentially to the ac side via the four power switches. The resulting output ac voltage swings from 4 to $-4$ with nine levels, and the staircase waveform is nearly sinusoidal, even without filtering.

## 2.6   Resonant Converters

In the PWM and other inverters described earlier, the switches are operated in the switch mode where they are subjected to full load current during each switching.

Thus, these switches are subjected to high switching stresses and losses. They are also responsible for electromagnetic interference (EMI). Normally, a higher switching frequency of operation of power converters is desirable, as it allows the size of inductors and capacitors to be reduced, leading to cheaper and more compact circuits. However, with a higher switching frequency, the losses and also EMI are increased.

Resonant power converters contain resonant *L-C* circuits whose voltage and current waveforms have sinusoidal variations during one or more subintervals of each switching period. Since a sinusoidal waveform is produced, the current and voltage pass through zero at each half cycle. The switches can be turned on or off when either the current though them or the voltage across them is zero.

The following types of resonant converters topologies are commonly used.

- Series resonant converters
- Parallel resonant converters
- Series–parallel resonant converters
- Zero-voltage switching (ZVS) resonant converters
- Zero-current switching (ZCS) resonant converters
- Resonant dc-link inverters

### 2.6.1 Series Resonant Converter

In series resonant converters (SRCs), the resonating circuit and the switching device are connected in series with the load to form an underdamped circuit. These circuits are based on resonant current oscillations. The series resonant inverters can be implemented by employing either unidirectional or bidirectional switches. The unidirectional switch can be a thyristor, GTO, bipolar transistor, IGBT and so on. In a bidirectional switch, an anti-parallel diode connected with the switching device or RCT (reverse-conducting thyristor) can be used. The circuit diagram of a full-bridge SRC with bidirectional switches is shown in Fig. 2.23. One method of controlling the output voltage in resonant converters is to control the switching frequency of the full-bridge inverter. This method is called "variable frequency control".

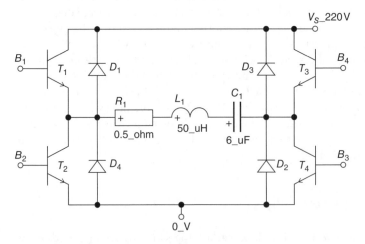

**Figure 2.23** Series resonant converter.

There are three possible modes of operation depending on whether switching frequency $f_s$ is more than, less than or equal to resonance frequency $f_0$. There are no switching loss $d$ in the switches at $f_s = f_0$ since the load current will be passing through zero exactly at the time when the switches change state. However, when $f_s < f_0$ or $f_s > f_0$, the switches are subjected to losses. For instance, if $f_s < f_0$, the load current will flow through the switch at the beginning of each half cycle and then commutate to the diode when the current changes polarity. These transitions are lossless. However, when the switch turns on or when the diode turns off, they are subjected to simultaneous step changes in voltage and current. These transitions therefore produce losses. As a result, each of the four devices is subjected to only one lossy transition per cycle.

### 2.6.1.1 Discontinuous Conduction Mode

In this mode, the resonant current is interrupted in every cycle. This control mode theoretically avoids switching losses because whenever a switch turns on or off, its current is zero. The shortcoming of this control mode is the distorted current waveform which may not be suitable for many applications including PV system.

### 2.6.2 Parallel Resonant Inverter

A parallel resonant inverter is supplied from a current source. This circuit offers higher impedance to switch current. Since here the current is continuously controlled, it has better short-circuit response. It is used in low-power applications. The circuit is shown in Fig. 2.24.

### 2.6.3 ZCS Resonant Converters

The switches in a zero-current resonant switch converter are turned on and off at zero current. A step-down dc-dc converter using the ZCS configuration is shown in Fig. 2.25. Prior to turning the switch on, the output current freewheels through the diode, both the current $i_L$ in $L$ and the voltage $v_C$ across $C$ are zero. After turning the switch on, the diode conducts and the total input voltage develops across $L$, the current $i_L$ rises linearly, *ensuring ZCS* and soft current change. When $i$ reaches $i_0$, the current conduction stops in diode $D_m$. At this time, the $L$-$C$ resonant circuit starts resonating and the change in $i_L$ and $v_C$ will be sinusoidal. The peak current $i = i_0 + v_s/Z_0$ where $Z_0$ is $\sqrt{L/C}$ and peak voltage $v_c = 2/v_s$ at $v_s/Z_0$ must be larger than $I$ $i_0$; otherwise, $i_L$ will not swing back to zero. When the current $i_L$ becomes zero, the switch is turned off automatically. Now the

**Figure 2.24** Parallel resonant converter.

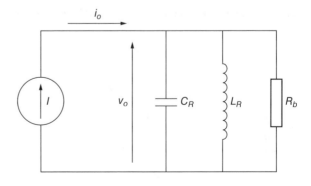

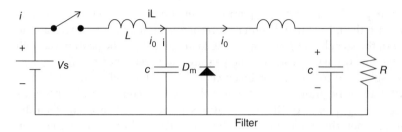

**Figure 2.25** ZCS resonant converter.

capacitor supplies the load current and therefore its voltage falls linearly till it becomes zero. The load current during this time freewheels through the diode. Now the switch can be turned on after this time, and the cycle is repeated. The output voltage will be equal to the average value of voltage $v_C$ which can be varied by changing the switching frequency.

### 2.6.4 ZVS Resonant Converter

The switches of ZVS resonant converter are turned on and off at zero voltage. The capacitor of ZVS is connected in parallel with the switch as shown in Fig. 2.26. Initially, the switch and the diode $D_b$ are off, and the capacitor charges at a constant rate of load current $I$. When the capacitor is fully charged, the current becomes zero and then starts to flow through diode $D_b$ because of inductance in the circuit. The capacitor voltage reaches a peak value and then starts decreasing till it is equal to the source voltage. After this, the capacitor voltage continues to fall till it becomes zero. Now the switchs is turned on at zero voltage. The diode will still be conducting till the current through the inductor equals the load current. At this time, the diode stops conducting and the load current conducts through the switch till it is turned off again.

### 2.6.5 Resonant dc-Link Inverters

In a resonant dc-link inverter, a simple $LC$ resonant circuit is placed between the dc bus and a PWM inverter. A three-phase resonant dc-link inverter is shown in Fig. 2.27. The six switches $Q_1 \ldots Q_6$ are switched in such a manner that it sets up periodic oscillations in the dc-link $LC$ circuit. The switches are turned on and off at zero link voltages to reduce the losses. The turning on of the switch as zero voltage is no problem as the device is turned on when its anti-parallel diode is conducting, presenting zero voltage.

**Figure 2.26** ZVS resonant converter.

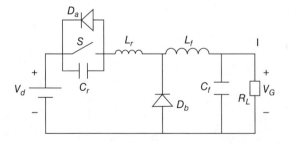

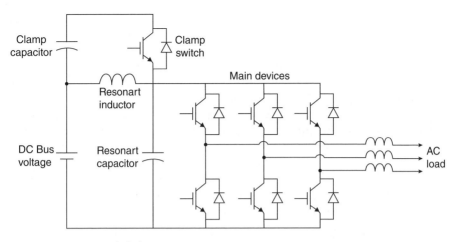

**Figure 2.27** Resonant dc-link inverter.

The dc-link resonant cycle is normally started with a fixed value of the initial capacitor current. This may result in the voltage across the dc link to be more than twice the input voltage. This voltage will appear across the switches which may require high rating devices. An active clamp shown in Fig. 2.27 is connected to limited link voltage.

## 2.7 Matrix Converters

The matrix converter is a direct frequency converter, which can perform ac/ac, dc/ac and dc/dc conversions [12–14]. It is single-stage converter which has an array (matrix) of $m \times n$ bidirectional switches connecting $m$-phase voltage source to $n$-phase load. It uses bidirectional switches, with a switch connected between each input terminal and each output terminal as shown in Fig. 2.28. The matrix converter shown in Fig. 2.28 is a $3 \times 3$ matrix converter which consists of nine bidirectional switches that allow any output phase to be connected to any input phase. A $3 \times 3$ matrix has 27 possible switching states. The source voltage and load voltage are related by the following equation:

$$\begin{bmatrix} v_A \\ v_B \\ v_C \end{bmatrix} = \begin{bmatrix} S_{Aa} & S_{Ba} & S_{Ca} \\ S_{Ab} & S_{Bb} & S_{Cb} \\ S_{Ac} & S_{Bc} & S_{Cc} \end{bmatrix} \begin{bmatrix} v_a \\ v_b \\ v_c \end{bmatrix} \tag{2.9}$$

**Figure 2.28** Matrix converter.

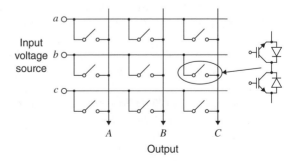

Equation (2.9) gives the instantaneous relationship between the input and output voltages. The switches can have a value of either 0 or 1, depending on whether they are off or on. The operating conditions of this type of converter demands that the input phases of the converter should never be short-circuited and the output currents should not be interrupted. Basically, this means that one and only one bidirectional switch per output phase must be switched on at any instant.

Since no energy storage components are present between the input and output sides of the matrix converter, the output voltages have to be generated directly from the input voltages. Each output voltage waveform is synthesized by sequential piece-wise sampling of the input voltage waveforms. The sampling rate has to be set much higher than both input and output frequencies, and the duration of each sample is controlled in such a way that the average value of the output waveform within each sample period tracks the desired output waveform. The maximum output voltage the matrix converter can generate without entering the over-modulation range is equal to $\frac{\sqrt{3}}{2}$ of the maximum input voltage; this is an intrinsic limit of the matrix converter, and it holds for any method of control.

The control methods for a matrix converter are quite complex and will not be discussed here.

## 2.8 Summary

All the renewable energy systems require specific power electronic converters to convert the power generated into useful power that can be directly interconnected with the utility grid and/or can be used for specific consumer applications locally. In this chapter, power electronic converters to convert ac to dc, dc to ac, dc to dc and ac-dc-ac conversion are presented. First, various types of power electronic devices that can be used as switches are discussed. Finally, the converter topologies which are suitable for wind, solar and energy storage are discussed.

## References

1 Baliga, B.J. (1987) *Modern Power Electronic Devices*, Wiley.
2 Rashid, M.H. (2004) *Power Electronic Circuits Devices and Applications*, 3rd edn, Pearson.
3 Ahmad, M. (1996) *Industrial Drives*, Chapter 10, Macmillan, India.
4 Bacha, S. *et al.* (2014) *Power Electronic Converters Modeling and Control with Case Studies*, Springer Verlag, London.
5 Erickson, R.W. (1997) *Fundamental of Power Electronics*, Springer Science.
6 Mohan, N. *et al.* (2002) *Power Electronics: Converters, Applications, and Design*, 3rd edn, Wiley.
7 Blaabjerg, F. *et al.* (2004) Power electronics as efficient interface in dispersed power generation systems. *IEEE Transactions on Power Electronics*, **19** (5), 1184–1194.
8 Kramer W *et al.* (2008) Advanced Power Electronic Interfaces for Distributed Energy Systems Technical Report NREL

**9** Rodriguez, J. *et al.* (2002) Multilevel inverters: a survey of topologies, controls, and applications. *IEEE Transactions on Industrial Electronics*, **49** (4), 724–738.

**10** Lai, J.S. and Peng, F.Z. (1996) Multilevel converters-a new breed of power converters. *IEEE Transactions on Industry Applications*, **32** (3), 509–517.

**11** Mei, J. *et al.* (2013) Modular multilevel inverter with new modulation method and its application to photovoltaic grid connected generator. *IEEE Transactions on Power Electronics*, **28** (11), 5063–5073.

**12** Wheeler, P. *et al.* (2002) Matrix converters: a technology review. *IEEE Transactions on Power Electronics*, **49** (2), 276–288.

**13** Empringham, L. *et al.* (2013) Technological issues and industrial application of matrix converters: a review. *IEEE Transactions on Industrial Electronics*, **60** (10), 4260–4271.

**14** Wheeler, P.W. *et al.* (2002) Matrix converters: a technology review. *IEEE Transaction Industrial Electronics*, **49** (2), 276–288.

# 3

# Renewable Energy Generator Technology

## 3.1 Energy Conversion

Energy conversion uses the primary energy present in raw natural form and converts it into a useful and efficient form of electric energy. In the renewable sources of energy, the conversion to electricity in the form suitable for domestic and industrial use is either by a rotating generator or static power electronic converters. Solar and fuel cells use static power electronic converters to generate electricity for practical use. The static power converters have already been described in detail in Chapter 2. In this chapter, rotating types of generators suitable for renewable energy sources, mainly hydro, wind and ocean, are discussed.

The large majority of sources of energy use rotating generators for example, wind, diesel, steam and hydro turbines, ocean waves and microturbines. A variety of electric generators are available for power conversion, and selection of proper generator is important factor in designing a renewable energy systems. It is, therefore, important to have knowledge of the characteristics of the available generator so that it may be selected for a specific application. Energy storage systems may be static or rotating as discussed in Chapter 4.

## 3.2 Power Conversion and Control of Wind Energy Systems

The major components of a typical wind energy conversion system include a wind turbine, generator, interconnection apparatus and control systems, as shown in Fig. 3.1. The generating systems used in wind energy conversion are different from the synchronous generator used in conventional power plants mainly because of the following:

- The speed of wind turbines fluctuates randomly because the input to it is wind, which cannot be controlled.
- The power generated is completely determined by the wind speed, and there is limited control over it.
- The capacity of wind turbines is much lower than that of conventional turbines of a power plant.

The type of generator that is most suitable for conversion of wind power depends on the range of speed of the wind turbine. Generators for wind turbines which

*Operation and Control of Renewable Energy Systems,* First Edition. Mukhtar Ahmad.
© 2018 John Wiley & Sons Ltd. Published 2018 by John Wiley & Sons Ltd.

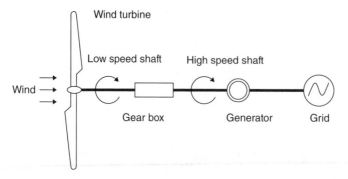

**Figure 3.1** Wind energy conversion system.

are commonly used are synchronous generators, permanent magnet synchronous generators and induction generators, including the squirrel cage type and wound rotor type. For small- to medium-power wind turbines, permanent magnet generators and squirrel-cage induction generators are often used because of their reliability and cost advantages. Induction generators, permanent magnet synchronous generators and wound-field synchronous generators are currently used in various high-power wind energy systems.

In the application of conversion of wave energy to electrical energy, linear generators are used as they are more suitable because of translational movement of body by the waves. Linear generators can extract the power of the waves in the form of a reciprocating motion at a low speed. High torque energy is possible with linear generators because of direct extraction of low speed. However, the electrical output voltage obtained is not suitable for direct connection to grid and needs a power electronic circuit for processing of the waveform. The following linear generators are considered for wave-energy-converting power plants:

i)  Linear permanent magnet synchronous machines, both with surface and buried permanent magnets
ii) Vernier hybrid linear machines
iii) Air-cored permanent magnet tubular linear machines

The working of these generators is discussed later in this chapter.

### 3.2.1  Induction Generator

An important step for installation of wind energy system is to select the turbine rating, the type of generator, and the distribution system. In general, the output characteristics of the wind turbine power do not follow exactly those of the generator power; so they have to be matched in the best possible way. Based on the range of speed expected for the turbine and taking into account the cubic relationship between the wind speed and the generated power, the generator and the gearbox are selected which match the speeds. For small renewable energy power plants, mostly induction machines are used, because they are widely and commercially available and very inexpensive. It is also very easy to operate them in parallel with large power systems.

Induction generators are similar in construction to induction motors. They produce electrical power when their shaft is rotated by a prime mover above the synchronous speed. Induction generators are simpler in construction compared to other generator types. They are also more rugged, requiring no brushes or commutator. The types of induction generators that can be used in wind energy conversion depend on the type of wind turbine. For limited variable-speed turbine, a squirrel-cage or wound-rotor induction generator with rotor resistance variation is used. Wind turbines with variable speed in a long range use a doubly fed induction generator (DFIG) consisting of a wound-rotor induction generator and an ac/dc/ac PWM converter. The stator winding is connected directly to the 50 Hz grid while the rotor is fed at a variable frequency through the ac/dc/ac converter. The DFIG technology is now the most preferred as it allows extracting maximum energy from the wind for low wind speeds by optimizing the turbine speed. There are many advantages of adjustable speed generators (ASGs) compared to fixed-speed generators (FSGs) which will be described later.

### 3.2.2 Permanent Magnet Synchronous Generator

Apart from DFIG, permanent magnet generators are now preferred because they are highly stable and do not require additional dc supply. A permanent magnet synchronous generator is a generator where the excitation field is provided by a permanent magnet. Typically, the permanent magnets will be mounted on the rotor with the three-phase winding on the stator. Permanent magnet motor and generator technology has advanced greatly in the past few years with the creation of rare earth magnets (neodymium, samarium–cobalt and alnico).

Permanent magnet alternators can be very efficient, in the range of 60%–95%, (typically around 70). They were initially used only for small and medium powers but are now used for higher powers also because of their advantages.

### 3.2.3 Linear PM Synchronous Machine

In order to convert the slow linear reciprocating motion of sea waves into high rotating speed compatible with rotating generators, mechanical or hydraulic systems were developed. However, a major drawback of such a system is that the overall efficiency is greatly reduced by the gearboxes, turbines or hydraulic systems used. In addition, it has lower reliability because of a large number of moving parts employed. In order to overcome these problems, linear generators were developed which can be directly connected to the wave energy absorber.

## 3.3 Operation and Control of Induction Generators for WES

If an induction motor is driven by a prime mover such as wind turbine so that its speed is more than the synchronous speed, then it works as an induction generator. Depending on the method by which the machine receives its excitation current, the induction generator can be operated in two different conditions. It can be connected to an already available power supply or in an isolated mode. When the induction generator is connected to an already available supply, it gets its excitation from the external source. In isolated mode or self-excited mode, capacitors are connected at the terminal of the generator to provide excitation. In order to understand the operation of an induction generator, a brief description of its working based on the equivalent circuit of an induction motor is described [1–3].

### 3.3.1 Equivalent Circuit

An equivalent circuit of the induction motor is shown in Fig. 3.2. Here $R_1$ and $X_1$ are the resistance and leakage reactance, respectively, of the stator winding; $R_0$ and $X_0$ are the loss resistance and magnetizing reactance; and $R'_2$ and $X'_2$ are the rotor resistance and leakage reactance, respectively, referred to the stator side; and s is the slip. The slip is given by

$$s = \frac{n_s - n_r}{n_s} \tag{3.1}$$

where $n_s$ is the synchronous speed of the machine and is related to the number of poles P of the machine and the frequency of stator supply f, and $n_r$ is the speed of the rotor.

$$n_s = \frac{120f}{P} \tag{3.2}$$

From the equivalent circuit, it can be seen that the dissipation in $R_1$ represents the stator copper loss, and the dissipation in $R_0$ represents the iron loss. Therefore, the power absorption indicated by the rotor part of the circuit must represent the actual mechanical output, friction and windage loss components and the rotor copper loss components. If losses in $R_0$ are neglected, the power delivered to the air gap from the equivalent circuit is

$$P_{airgap} = \frac{3I_2^2 R_2}{s} \tag{3.3}$$

The torque is given by

$$\frac{P_{airgap}}{\omega_s} = \frac{3I_2^2 R_2}{s\omega_s} \tag{3.4}$$

Since the dissipation in $R_2$ is rotor copper loss, the power dissipation in $R'_2\left(\frac{1-s}{s}\right)$ represents the mechanical power output in the induction motor. The induction generator does not differ much in construction from the induction motor. The mechanical power converted into electrical power at negative slip (when the rotor speed is more than the synchronous speed) is given by

Air gap power − copper losses = $\frac{3I_2^2 R_2}{s} - 3I_2^2 R_2$, where s is negative.
Or, the electrical power output is $= (1 - s)P_{airgap}$.

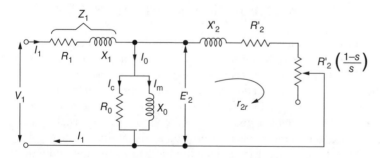

**Figure 3.2** Equivalent circuit of induction machine. Source: Reproduced from electronic hub tutorials on thyristor. (http://www.electronicshub.org/thyristor-basics/#Types_of_Thyristors).

The torque–slip characteristic of the induction machine is shown in Fig. 3.3. As can be seen from Fig. 3.3, there is no torque at s = 0 (synchronous speed) and the torque slip curves are almost linear from no load speed (s = 0) to full load speed (s = 1–2%). The peak power supplied by the induction generator happens at a speed slightly above the synchronous speed (approximately 1 to 2 %). Thus, a cage-type induction generator is suitable for a constant-speed or fixed-speed wind turbine.

A cage-type induction generator always consumes reactive power. The reactive power is therefore partly compensated by connecting capacitors to it as shown in Fig. 3.4. The capacitors connected may provide part or full reactive power compensation.

### 3.3.2 Wound-Rotor Induction Machine

A three-phase wound-rotor induction machine is shown in Fig. 3.5. The wound-rotor generator is not as rugged as the squirrel cage type, but the brushes have little wearing and sparking when compared to dc machines. In wound-rotor machine, the rotor has a three-phase winding similar to the stator windings. The rotor winding terminals are connected to three slip rings which turn with the rotor. The slip rings/brushes allow external resistors to be connected in series with the winding. The effect of

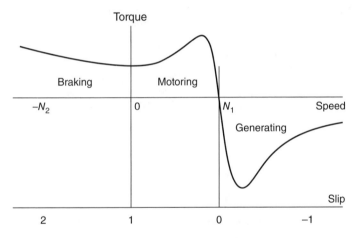

**Figure 3.3** Torque–slip characteristic of induction machine.

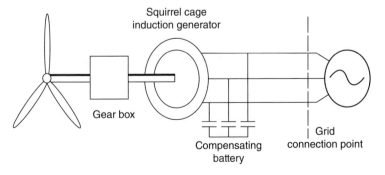

**Figure 3.4** Capacitor connected for reactive power compensation.

rotor resistance variation on the torque of the induction machine is presented in the torque–slip characteristic shown in Fig. 3.7. As can be seen from the figure, the maximum torque occurs at a higher value of slip if the rotor resistance is increased. With rotor resistance control, the generator can be run up to 10% above the synchronous speed. By adding a variable external resistance to the rotor of an induction generator used in a wind turbine, it is possible to manipulate the torque–speed curve and control the output power. To vary the effective value of the external resistance, a three-phase diode bridge and a dc chopper with a variable duty cycle as shown in Fig. 3.6 are commonly used. Thus, the wind turbine generator operates with a smooth variation of

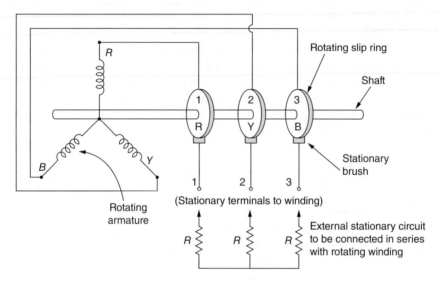

**Figure 3.5** Slip ring induction motor with rotor resistance control.

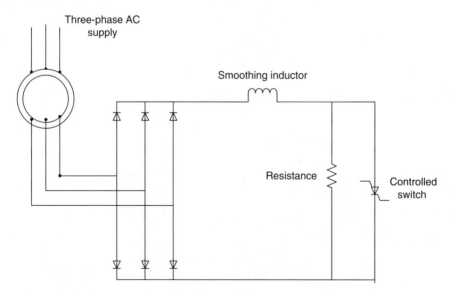

**Figure 3.6** Electronic control of rotor resistance.

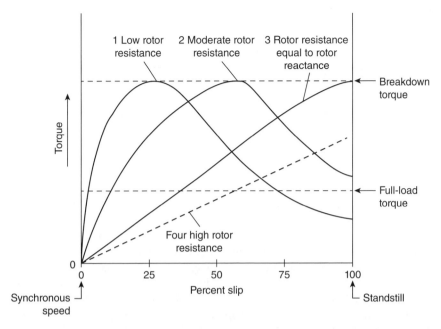

**Figure 3.7** Torque–slip characteristic with rotor resistance control for slip ring induction machine.

the slip above the synchronous speed by switching "on" and "off" the chopper circuit. The torque–slip characteristic is shown in Fig. 3.7.

### 3.3.3 Doubly Fed Induction Generator (DFIG)

DFIG systems are variable-speed power generation systems with partially rated power electronics or full-scale power electronics. They are cost-effective and provide simple pitch control. The controlling speed of the generator (frequency) allows the pitch control time constants of the turbine to become longer, reducing the pitch control complexity and peak power requirements. The DFIG is an induction machine with a wound rotor where both the rotor and stator are connected to electrical sources; hence, the term "doubly fed." The stator is directly connected to the ac mains, whereas the rotor is fed from the power electronics converter via slip rings to allow DIFG to operate at a variety of speeds in response to the changing wind speed.

The operation of DFIG is similar to that of a doubly fed induction motor. By connecting the supply to the rotor, it is possible to extract the slip energy from the rotor and supply it to the grid. Thus, power can be supplied to the grid through stator as well as rotor. The DFIG works as a synchronous generator whose synchronous speed can be varied by varying the frequency of the rotor circuit. Since the rotor is supplied with three-phase current, it also produces a rotating magnetic field. Therefore, the frequency of emf induced at the stator is given by

$$f_{\text{stator}} = \frac{P \times N}{120} \pm f_{\text{rotor}} \tag{3.5}$$

The + or − sign depends on the phase sequence of the rotor supply with respect to the phase sequence of the stator supply. In order to have constant frequency of the stator

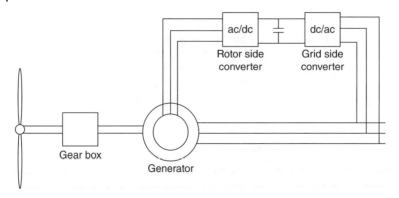

**Figure 3.8** Doubly fed induction generator.

equal to the grid frequency, the frequency of the rotor supply is to be adjusted continuously depending on the speed of the prime mover (wind turbine).

In the doubly fed induction generator shown in Fig. 3.8, a back-to-back voltage source converter feeds the three-phase rotor winding. The dc capacitor linking grid-side and rotor-side converters allows the storage of power from the induction generator for further generation. In this way, the mechanical and electrical rotor frequencies are decoupled and the electrical stator and rotor frequencies can be matched, independently of the mechanical rotor speed. In addition, the power can flow in both the directions, that is, to the rotor from the supply and from the supply to the rotor, and hence, the speed of the machine can be controlled from either rotor- or grid-side converter in both super- and sub-synchronous speed ranges.

The DFIG works as a synchronous generator if it is running at the synchronous speed. It can operate below as well as above the synchronous speed in the motoring mode or generating mode. Below the synchronous speed in the motoring mode and above the synchronous speed in the generating mode, rotor-side converter operates as a rectifier and grid-side converter as an inverter, where slip power is returned to the stator. Below the synchronous speed in the generating mode and above the synchronous speed in the motoring mode, rotor-side converter operates as an inverter and grid-side converter as rectifier and slip power is supplied to the rotor.

For super-synchronous speed operation, rotor power is transmitted to the dc bus capacitor and to the grid and tends to increase the dc voltage. Thus, under this condition, power flows to the grid from both stator side and rotor side. For sub-synchronous speed operation, rotor power is taken out of the dc bus capacitor (from grid) and tends to decrease the dc voltage. A converter of the grid side is used to generate or absorb the grid-side power in order to keep the dc voltage constant. In steady state for a lossless ac/dc/ac converter, the rotor power is equal to that supplied to or taken from the grid, and the speed of the wind turbine is determined by the power absorbed or generated by the rotor.

Modern high-power wind turbines are capable of adjustable the speed operation. Key advantages of adjustable speed generators (ASGs) compared to fixed-speed generators are [4, 5] as follows:

- A speed variation of ±30% around the synchronous speed can be obtained by the use of power converter of 30% of nominal generated power.

- Power electronic converters in DGIG can generate and absorb reactive power; thus, there is no need for capacitors.
- Torque pulsations can be reduced due to the elasticity of the wind turbine system. This eliminates electrical power variations, which results in less flicker.
- The turbine speed can be adjusted as a function of wind speed to maximize the output power. Thus, the operation at the maximum power point can be realized over a wide power range.
- They reduce acoustic noise, because low-speed operation is possible at low-power conditions.

### 3.3.3.1 Equivalent Circuit of DGIG

The dynamic behaviour of DFIG can be explained with an equivalent circuit obtained from voltage equations in $d-q$ axes as follows:

For stator

$$v_{ds} = r_s i_{ds} + \frac{d\lambda_{ds}}{dt} - \omega_e \lambda_{qs} \tag{3.6}$$

$$v_{qs} = r_s i_{qs} + \frac{d\lambda_{qs}}{dt} - \omega_e \lambda_{ds} \tag{3.7}$$

For rotor

$$v_{dr} = r_r i_{dr} + \frac{d\lambda_{dr}}{dt} - (\omega_e - \omega_r)\lambda_{qr} \tag{3.8}$$

$$v_{qr} = r_r i_{qr} + \frac{d\lambda_{qr}}{dt} - (\omega_e - \omega_r)\lambda_{dr} \tag{3.9}$$

where $\omega_e$ is the angular speed of the $d-q$ frame, $r_s$ and $r_r$ are the stator and rotor resistances. $\omega_r$ is the angular rotor speed. The stator values of the current and voltage are indicated by the subscript s and rotor quantities by the subscript r. $v$ is the voltage, $i$ the current and $\lambda$ the flux linkage.

The flux linkage equations in terms of currents are

$$\lambda_{ds} = L_s i_{ds} + M i_{dr} \tag{3.10}$$
$$\lambda_{qs} = L_s i_{qs} + M i_{qr} \tag{3.11}$$
$$\lambda_{dr} = L_r i_{dr} + M i_{ds} \tag{3.12}$$
$$\lambda_{qr} = L_r i_q + M i_{qs} \tag{3.13}$$

where $L_s$, $L_r$ and $M$ are stator leakage, rotor leakage and mutual inductance between the stator and the rotor. The equivalent circuit is shown in Fig. 3.9.

The torque $T_e$ is given by

$$T_e = \frac{3}{2}\frac{P}{2}(\lambda_{ds} i_{qs} - \lambda_{qs} i_{ds}) \tag{3.14}$$

**Figure 3.9** Equivalent circuit.

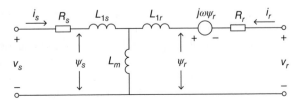

The stator and rotor active power are given by

$$P_s = \frac{3}{2}(v_{ds}i_{ds} + v_{qs}i_{qs}) \tag{3.15}$$

$$P_R = \frac{3}{2}(v_{dR}i_{dR} + v_{qR}i_{qR}) \tag{3.16}$$

The rotor will accelerate or decelerate depending on the following power equation.

$$P_a = P_s - P_m - P_r \tag{3.17}$$

In a wind turbine, when the wind speed is constant, the mechanical power is equal to the stator power after deducting the rotor power, and the machine runs at constant speed. When the wind speed changes, the machine will accelerate if there is gust of wind until the pitch angle control reduces the mechanical power.

### 3.3.3.2 Braking System

Braking systems for a wind turbine generating system must be able to reduce the speed of the aerodynamic rotor under abnormal conditions, such as over-speed and fault conditions. Normally, both aerodynamic and mechanical braking is used. There are usually combined with conventional mechanical shaft (disk) brakes and aerodynamic brakes (e.g. pitching mechanisms) for wind turbine brake systems. Aerodynamic brakes must be capable of reducing the wind turbine to a safe rotational speed at all anticipated wind speeds and under fault condition. The function of the mechanical brake in wind turbines having independent aerodynamic braking in each blade is simply to bring the rotor to rest. For a rapid response, electrodynamic braking can be used. Electrodynamic braking offers several advantages over conventional wind turbine brakes, such as controllable braking torque and high reliability, while requiring minimal maintenance. The technique has not been widely implemented in modern wind turbines, owing to a lack of appropriate design tools and limited operating experience.

## 3.4 Permanent Magnet Synchronous Generator

Direct-drive wind energy systems cannot employ a conventional high speed (and low torque) electrical generators. Permanent magnet synchronous generators with methods of speed control can be directly coupled to wind turbine, thus eliminating the gearbox. Permanent magnet machines use magnets instead of windings to produce the air-gap magnetic field. The replacement of rotor winding by magnets results in simplification in construction, reduction in losses and improvement in efficiency.

For wind turbine systems used for commercial power generation, the direct-driven wind turbine (WT) with a permanent magnet synchronous generator (PMSG) has the advantages of simple structure, low cost of maintenance, high conversion efficiency and high reliability. Moreover, the performance of vector control method of speed control is much less sensitive to the parameter variations of the generator. Therefore, a high-performance variable-speed generation including high efficiency and high controllability is possible by using a PMSG for a wind generation system. Use of permanent magnet (PM) generators for small wind turbines is very common. However, recently, PM generators of up to 3 MW capacity are being used in onshore wind power generation. Usually, an ac generator with many poles operates between 10 and 100 Hz. These generators are connected to the grid through an inverter to match the frequency of the grid. Because the generator is directly driven by the wind turbine, it is commonly known as a direct-drive generator [6–8].

The PM machines are basically synchronous machines and can be operated at unity power factor. The cost of a PM machine is higher than that of an induction machine due to the cost of magnets, but due to the high efficiency, the running cost is less. The PM machines are broadly classified on the basis of the direction of field flux as

1) radial field and
2) axial field.

In an axial-flux motor, the magnetic force (through the air gap) is along the same plane as the motor shaft, that is, along the length of the motor. A radial-flux motor is the more traditional design, in which the magnetic force is 90° (perpendicular) to the length of the motor/shaft or the flux is along the radius of the machine. These machines are more common than axial field machines. Radial-flux machines are conventional type of PMSGs. The manufacturing technology is well established which makes the production cost lower compared with the axial one. Furthermore, it is easy to design these machines for different power ratings, as the higher power rating can be achieved by increasing the length of the machine.

In axial-flux machines, the length of the machine is much smaller compared with radial-flux machines. Their main advantage is high torque-to-weight ratio, adjustable air-gap balanced rotor–stator attractive forces and better heat removal. These machines are therefore recommended for applications with space constraints especially in the axial direction. One of the disadvantages with the axial-flux machines is that they are not balanced in a single-rotor single-stator design. In addition, their rating cannot be changed by simply changing the length of the rotor as it is accompanied by change in the air gap. In order to achieve better performance, the rotor is normally sandwiched between two stators or vice versa.

Permanent magnet synchronous machine can be a surface permanent magnet type (SPM) or an interior or buried permanent magnet (IPM) type. In SPM, the magnets are glued and/or bandaged to the rotor surface in order to withstand the centrifugal force. This arrangement provides the highest air gap flux but lacks mechanical robustness. Usually, the magnets are oriented or magnetized in the radial direction and more seldom in circumferential direction. Construction of the rotor core in SPM is the easiest among different PM configurations due to the simple rotor geometry. Figure 3.10 shows a surface-mounted rotor for a PMSG. In an SPM machine, the $d$-axis and $q$-axis

**Figure 3.10** Surface-mounted PM machine.

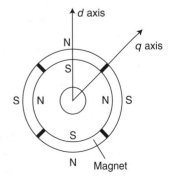

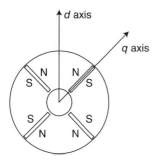

**Figure 3.11** Interior magnet PM.

reactances are almost equal as in cylindrical rotor synchronous machines. In an interior PM machine shown in Fig. 3.11, the magnets are placed in the middle of the rotor laminations, resulting in robust construction. Due to saliency, the *d*-axis and *q*-axis reactances are different in these machines.

The stator of the PMSM has conventional distributed three-phase windings, located in the stator slots. Magnets of alternately opposite magnetization produce radially directed flux density across the air gap. This flux density then interacts with the stator currents to produce torque. A PMSM used in wind energy system has to work with variable speed. A power electronic converter is therefore necessary to transform variable-frequency supply to fixed frequency. Matrix converters or ac-dc-ac two-stage conversion is commonly used for this purpose [9, 10].

### 3.4.1 Modelling of PMSG

The stator of PMSM and that of wound-rotor synchronous machine are similar, and therefore, the equivalent circuit shown in Fig. 3.12 is similar to the one for the synchronous machine. The equivalent circuit is based on the mathematical model based on the following assumptions:

1) Saturation is neglected (although it can be accounted by parameter changes).
2) The back EMF is sinusoidal.

The *d*- and *q*-axis voltage equations in rotor reference are given by

$$v_{sd} = R_s i_{sd} + L_d \frac{di_{sd}}{dt} - \omega L_q i_{sq} \tag{3.18}$$

$$v_{sq} = R_s i_{sq} + L_q \frac{di_{sq}}{dt} + \omega L_d i_d + \omega \psi_{pm} \tag{3.19}$$

where $R_s = R_d = R_q$; and $L_d$ and $L_q$ are the *d*-axis and *q*-axis winding inductances, respectively; and $L_m$ is the mutual inductance between stator winding and rotor magnets. For surface-mounted permanent magnet motor, the inductance $L_d = L_q = L_s$, where $L_s$ is stator inductance.

The torque is given by

$$T_e = \frac{3}{2} P \psi_{pm} i_{sq} \tag{3.20}$$

Similarly as separately excited synchronous machines, the PM synchronous machines are usually treated in a *d*–*q* reference frame fixed to the rotor. The equivalent circuit

**Figure 3.12** Equivalent circuit of PMSG.

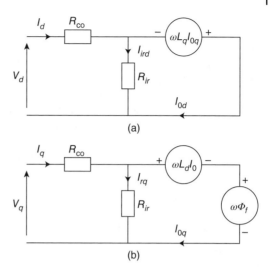

(a)

(b)

of the machine shown in Fig. 3.12 is almost the same as for a separately excited synchronous machine. The PMSG is connected to the grid through power electronic circuit which is controlled to obtain stable operation and discussed later.

## 3.5 Wave Energy Conversion (WEC) Technologies

Wave energy can be considered as a concentrated form of solar energy. Due to differential heating of the Earth, winds are created which pass over the oceans, transferring some of the energies to water to form waves. The amount of energy transferred by the wind results in a different pattern of sea waves. The power takeoff system for wave energy depends on the operating principle of wave energy conversion systems. The generators used in most of the wave energy systems use rotating generators; the asynchronous generator or induction machine is a preferred machine.

Linear generators are also being considered for extraction of energy from the waves because they do not require any other interface or power conversion (e.g. hydraulic) which reduces the transferred power in the transmission process. Linear generator is connected to the WEC via translator and displaces with it. Three topologies of linear generators are considered in the wave energy industry [11–14].

- Linear permanent magnet synchronous machine
- Air-cored PM tubular linear machines
- Transverse-flux permanent magnet machine
- Linear induction machines

Linear motors or actuators are used in transportation systems (including maglev trains), robotic systems, position control system and so on. Mostly, these systems have a low power level. However, there are a few applications with power levels comparable to the power levels of wave energy converters, such as maglev trains, aircraft launching systems for future aircraft, carriers and roller coasters.

### 3.5.1 Linear Permanent Magnet Synchronous Machine

In permanent magnet linear generator, the stator has a winding, whereas the translator (moving part) has permanent magnets mounted with alternate polarity and is directly coupled to the heaving buoy. It cancels intermediate-drive device of traditional wave power generator, which simplifies the system, and reduces the losses as the heaving boy oscillates the voltage induced in the stator winding which is then connected to the grid through a PE converter. In order for a linear generator to be suitable for a WEC, it should be able to provide high performances at low speeds with simple structure and high reliability. A linear PM generator has the following advantages [12]:

- High peak force
- Good efficiency at low speed
- Irregular motion
- Low cost as magnet cost has reduced now

A two-sided linear permanent magnet generator with surface-mounted magnets is shown in Fig. 3.13. As shown in the figure, the double-sided stator is on the outside and the translator with permanent magnets is inside. The stator has a three-phase full-pitch winding with one slot per pole per phase. The double-sided stator is used to balance the attractive forces and balance the bearing loads. When a wave crest is above the AWS, the floater is pushed down by the added pressure of water, and when a wave trough is above it, the floater is moved up by the reduced pressure of water. The motion of floater is resisted by a two-sided permanent magnet linear generator, and the sketch of this two-sided permanent magnet linear generator is shown in Fig. 3.13. There are several advantages of a two-sided permanent linear generator.

- High force density.
- Reasonable work efficiency.
- Permanent magnet material is cheap.
- The electricity is only restricted in the stator.

A study concerning the shear stress developed by various linear generator topologies pointed out that the transverse-flux machines provide the highest force density in the air gap. These machines have 3 to 5 times higher specific torque compared to conventional synchronous machines. However, due to their complicated structures resulting in high

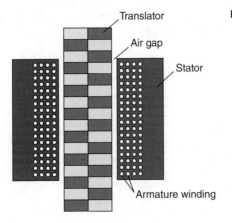

**Figure 3.13** Double-sided linear PM generator.

manufacturing cost, these are not common. At the same time, because of their structure, only one- or two-sided structures have been reported, so high magnetic forces appear between the stator and the translator.

In a transverse-flux permanent magnet (TFPM), the main flux flow is in the direction perpendicular to the direction of travel as shown in Fig. 3.14. A transverse-flux permanent magnet machine (TFPM) is ideally suited to high-torque and lower-speed applications such as the generation of electricity by wind energy and the electric propulsion of naval ships. The TFPM topology seems to have great potential for ocean wave energy conversion, showing better use of the machine size and permanent magnet materials when compared to conventional and other non-conventional topologies. Besides the possibility of reaching a very high torque density, by increasing the number of poles, it allows to set the electric current density regardless of the magnetic flux, unlike conventional radial topology machines, where the cross-sectional area of air gap competes directly with the windings for the available space. In TFPM, the air gap area defines the current density, while the axial length defines the magnetic flux density.

Another solution is the vernier hybrid generator, with a simplified construction at the cost of lower shear stress forces. It has a passive translator with teeth, while the stator is made of "C-"shaped parts, which have both the coils and the permanent magnets. The major drawback of such a structure is the large mass of moving parts and large inertia.

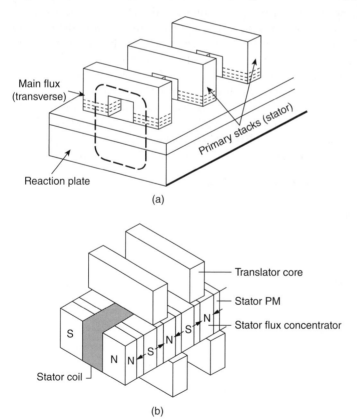

**Figure 3.14** Transverse-flux PM generator: (a) single-sided, (b) double-sided.

Other characteristics of linear generator systems in wave energy conversion systems are as follows:

1) There is a high attractive force between the translator and the stator. This complicates the mechanical design and the bearing design.
2) The air gap between the stator and the rotor is relatively large as it is difficult to build a generator with a small air gap because of manufacturing tolerances, the limited stiffness of the complete construction, the large attractive forces between the stator and the translator, thermal expansion and so on.
3) Because of the irregular motion of continuously varying speed, the grid connection of the wave energy converter always has to be done using a power electronic converter that connects the voltage of the wave energy converter with a varying frequency and amplitude to the grid with a fixed frequency and amplitude.

### 3.5.2 Tubular Permanent Magnet Linear Wave Generator (TPMLWG)

Since the displacement in wave energy generator is short, the best possible solution is to use PM tubular linear generator. The generator shown in Fig. 3.15 is one of the few types of tubular PM linear generators. As the name suggests, the shape of the generator is cylindrical. TPMLWG consists of a primary part and a secondary part. The primary part is made up of primary iron core, armature windings, and central axis. The armature windings which are fixed with epoxy resin or slot wedge are embedded in the primary iron core slot in a certain way, and the primary iron core is located in the middle of the central axis which is divided into two parts. The secondary part or translator consists of a series of alternate steel pole pieces and Nd-Fe-B magnets. The relative movement of the translator due to waves generates magnetic field lines and induces electromotive force in the armature windings [13–15]. The magnets have axial magnetization and are mounted in such a manner so that the steel pieces form alternate north and south poles. Tubular machines encapsulate the flux more completely. Unlike flat designs, there exist no transverse edges from which flux may leak. Thus, the flux utilization of tubular machines is

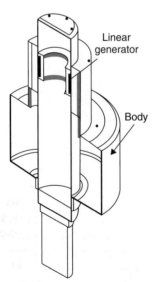

**Figure 3.15** Tubular linear PM generator.

Linear generator

Body

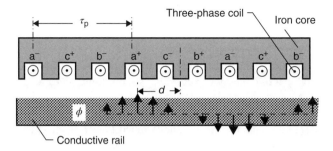

**Figure 3.16** Linear induction generator.

better than that of flat machines. This results in a better utilization of the permanent magnet material. The superior flux utilization in a tubular machine means that less permanent magnet material is required per unit of force capability. The double-sided layout reduces the size of the machine through sharing the magnetic material between the two stators.

Significant structural savings can be made if the magnetic forces can be reduced or eliminated. This can be achieved by constructing a stator which contains no iron. The translator consists of north and south magnetized PMs separated by steel spacers, both mounted on a non-magnetic shaft.

### 3.5.3 Linear Induction Machines

Linear induction generators can also be used for conversion of wave energy into electrical energy. The single-sided stator configuration consists of a single coil assembly that is used in conjunction with aluminium or copper plate which may be backed with either steel or iron plate. A three-phase ac is used to excite the stator windings which produces travelling magnetic field in the air gap. The translator consists of aluminium plates which act as a short-circuit path for the induced voltage due to the field of the stator. The principle of operation of linear induction generator is the same as for rotary generator. In linear induction machines, the synchronous speed does not depend on the number of poles. It only depends on the frequency of supply and the pole pitch. If the translator moves at a speed greater than the speed of the electromagnetic field produced by the stator, it works as a generator. By controlling the frequency of the stator supply through electronic circuits, power can be generated at any speed of the translator. Normally, a short-stator-type linear machine as shown in Fig. 3.16 will perform better in the extraction of wave energy, because a larger size translator is easy to construct.

## 3.6 Summary

Renewable energy systems use either static systems or rotating generators to convert the renewable energy in various forms to electrical energy. The different types of generators used are induction generators, singly fed and doubly fed permanent magnet generators and so on. These generators are described for applications in wind energy.

For wave energy conversion, linear generators are more suitable and have been discussed.

## References

1 Ahmad, M. (1996) *Industrial Drives*, Macmillan India.
2 Fitzgerald, A.E. *et al.* (1971) *Electric Machinery*, McGraw-Hill.
3 Lara, O.A. and Jenkins, N. (2009) *Wind energy Generation*, 1st edn, John Wiley & sons.
4 Yang, S.Y. *et al.* (2007) Development of a variable-speed wind energy conversion system based on doubly-fed induction generator, IEEE 22nd Applied Power electronics Conference, APEC 2007, pp. 1334–1338.
5 Eriksen, P.B. *et al.* (2005) System operation with high wind penetration. *IEEE Power and Energy Magazine*, **3** (6), 65–74.
6 Petersson, A. (2005) *Analysis, Modeling and Control of Doubly-Fed Induction Generators for Wind Turbines*, Chalmers University of Technology, Goteborg, Sweden.
7 Ahmad, M. (2010) *High Performance ac Drives: Modelling Analysis, and Control*, Springer.
8 Wing, M. and Gieras, J.F. (2002) *Permanent Magnet Motor Technology: Design and Applications*, 2nd edn, Marcel Dekker Inc., Basel, Switzerland.
9 Chinchilla, M. *et al.* (2006) Control of permanent magnet generators applied to variable speed wind energy systems connected to the grid. *IEEE Transaction on Emersion*, **21** (1), 130–135.
10 Islam, S. and Tan, K. (2004) Optimum control strategies in energy conversion of PMSG wind turbine system without mechanical sensors. *IEEE Transactions on Energy Conversion*, **19**, 392–399.
11 Kim, H.W. *et al.* (2010) Modeling and control of PMSG-based variable-speed wind turbine. *Electric Power System Research*, **80** (1), 46–52.
12 Polinder, H. *et al.* (2005) Linear generators for direct drive wave energy conversion. *IEEE Transactions on Energy Conversion*, **20** (2), 260–267.
13 Polinder, H. *et al.* (2005) Conventional and TFPM linear generators for direct-drive wave energy conversion. *IEEE Transactions on Energy Conversion*, **20** (2), 260–267.
14 Oprea, C. A. *et al.* (2011) Permanent magnet linear generator for renewable energy applications: tubular vs. four-sided structures. 3rd International Conference on Clean Electrical Power: Renewable Energy Resources Impact, ICCEP 2011.
15 Lu, Q. and Yunyue, Y. (2009) Design and analysis of tubular linear PM generator. *IEEE Transaction Magnetics*, **45** (10), 4716–4719.

# 4

# Grid-Scale Energy Storage

## 4.1 Requirement of Energy Storage

The necessity of energy storage in renewable systems is mainly due to the fact that most of the renewable sources of energy such as solar, wind are intermittent in nature. When the renewable energy exceeds 10–30% of the power supply, additional resources are needed to match the fluctuating supply to the already fluctuating load. The main purpose of energy storage device is to store the energy when production is more than the load and return it back to the grid during shortage in production.

Battery or other form of energy storage systems are used to regulate frequency, act as reserve capacity, improve power quality, support voltage and help to reduce peak load. Energy storage system can also be used as emergency power source in long-duration outages [1]. If the cost of energy storage can be reduced substantially, it can bring revolutionary change in the operation and control of energy systems. Storage can be applied at the power plant, in support of the transmission system, in the distribution system and on the customer side.

## 4.2 Types of Energy Storage Technologies

Electrical energy can be stored in various forms which is suitable for grid-scale energy use. It can be broadly classified as [1–4] the follows:

A) Electromechanical Storage
  - Pumped hydro storage
  - Compressed air energy storage
  - Flywheel storage
B) Superconducting Magnetic Energy Storage
C) Chemical Storage
D) Supercapacitors
  - Batteries
E) Thermal Storage
  - Sensible heat storage
  - Latent heat storage
F) Hydrogen Energy Storage

*Operation and Control of Renewable Energy Systems*, First Edition. Mukhtar Ahmad.
© 2018 John Wiley & Sons Ltd. Published 2018 by John Wiley & Sons Ltd.

## 4.3 Electromechanical Storage

The most common electromechanical storage systems are pumped hydroelectric power plants (pumped hydro storage, PHS), compressed air energy storage (CAES) and flywheel energy storage systems. For the past few years, superconducting magnetic energy storage and supercapacitors are also being developed for energy storage.

### 4.3.1 Pumped Hydro Storage (PHS) System

PHS plants are the most commonly used energy storage systems because of their large capacity, high efficiency and economy. The first hydro storage plants were built in Switzerland and Italy in the 1890s. Pumped storage hydropower is a modified use of conventional hydropower technology to store the energy during light load period and use it when more power is required.

Pumped storage is the largest-capacity form of grid energy storage available. As reported by the Electric Power Research Institute (EPRI), PHS accounts for more than 98% of bulk storage capacity worldwide, with total capacity of around 127 GW [4, 5]. Japan currently has the largest PHS capacity in the world followed by the United States. Table 4.1 lists the installed PHS capacities in major countries of the world.

As the net electricity output of PHS operation is negative, a PHS facility usually cannot qualify as a source of power. The energy for pumping the water is used only when it is available in surplus or for the purpose of load levelling (peak shaving). However, now it is also being considered for storing the energy generated by wind and solar power plants so that the supply may be used when the power is not available from these sources. The wind power and PHS integrated power systems are the most economically and technically competitive technologies particularly where large wind energy is available. Similarly, the energy from the Sun is intermittent in nature and also available only during daytime. Hence, to make its best and continuous use, an energy storage system such as PHS is most suitable. During daytime, it can store the energy when excess energy is available and then use the stored energy during night.

**Table 4.1** Installed PHS capacity in major countries of the world till 2014.

| Country | Installed PHS capacity (MW) |
| --- | --- |
| Japan | 27, 438 |
| United States | 22, 365 |
| China | 21, 600 |
| Italy | 7, 365 |
| Spain | 6, 889 |
| Germany | 6, 388 |
| France | 5, 894 |
| India | 5, 072 |
| Austria | 4, 808 |
| South Korea | 4, 700 |

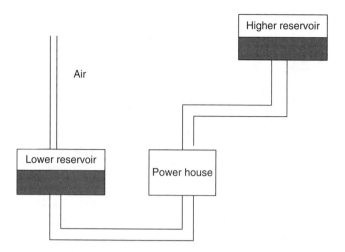

**Figure 4.1** Pumped hydro storage system.

The principle of pumped storage system is fairly simple. It is made up of two reservoirs as shown in Fig. 4.1. One reservoir is located at a higher elevation than the other. There is a system of water tunnel and piping connecting the two reservoirs. When demand is low and electricity is cheap, it uses energy to pump water from the lower reservoir to the upper reservoir. When water of a certain mass is lifted to a height, force $F = m \times a$, *where m* is mass, and *a* is acceleration is required, In this case, *a* is replaced by *g* for the gravitational acceleration (9.81 m/s²). This energy supplied is stored in the water as potential energy and is equal to $W = F \times h = m \times g \times h$, where *h* is the height to which water is lifted. When demand is high and electricity is more expensive or when power supply is not available, water from the upper reservoir is released back into the lower reservoir, thus releasing energy to produce electricity. Francis turbine is normally used for both pumping and generating. However, the types of turbines used depend on the height of the reservoir, and Pelton wheel or Kaplan turbines are also used.

The efficiency of PHS system is defined as the electricity generated divided by the electricity used to pump water. The efficiency depends on the technology and the life of a facility which may be as low as 60%, for older designs, while a state-of-the-art PHS system may achieve over 80% efficiency.

In the past, almost all the pumped storage projects required the construction of at least one dam along the main-stem rivers. These systems were not considered eco-friendly. Therefore, modern PHS projects have reservoirs located in areas that are physically separated from the existing river systems. These projects are called "closed-loop" pumped storage system. These systems present minimal tor almost no impact to the existing river systems. New systems have improved efficiencies with modern reversible pump turbines, adjustable-speed pump turbines, new equipment controls such as static frequency converter systems, as well as improved underground tunnelling construction methods and design capabilities [6].

### 4.3.2 Underground Pumped Hydro Energy Storage

Underground pumped hydroelectric energy storage is an adaptation of the conventional surface pumped hydroelectric that uses a underground cavern as the lower reservoir.

This concept is known as "aquifer underground pumped hydroelectric energy storage (UPHS)". It stores electricity as gravitational potential energy between a surface reservoir and an underlying subterranean aquifer water table.

This concept has not been implemented anywhere so far. Pumped hydroelectric energy storage using old unused coal or salt mines as the lower reservoirs has been suggested by German scientists and engineers. However, realizing such a project underground is more difficult and expensive compared to conventional pumped storage plants. More research and at least one field test will need to be conducted before larger projects can be considered. Presently, energy companies still prefer conventional plants as they are cheaper and easier to realize.

### 4.3.3 Compressed Air Energy Storage

Compressed Air Energy Storage (CAES) is a modification of the basic gas turbine (GT) technology, in which low-cost electricity is used for storing compressed air in an underground cavern. The air is then heated and expanded in a gas turbine in order to produce electricity during peak demand hours. A compressed air energy storage plant is shown in Fig. 4.2. The basic idea is to use excess energy in an electric compressor to compress air to a pressure of about 60 bars and store it in giant underground spaces such as old salt caverns, aquifers or pore storage sites when electricity demand is less. When demand is high, energy stored in compressed air is used to power a gas turbine to generate electricity.

Only two plants of this type exist worldwide: the first one built in 1978 in Huntorf, Germany, with a power output of 320 MW and a storage capacity of 580 MWh [7]. The plant, located in Bremen, Germany, is used to provide peak shaving, spinning reserves and VAR support. A total volume of 11 million cubic feet is stored at pressures up to 1000 psi in two underground salt caverns, situated 2100–2600 feet below the surface. It requires 12 h of off-peak power to fully recharge and then is capable of delivering full output (290 MW) for up to 4 h.

The second one is located in McIntosh, Alabama, United States, and began operation in 1991 with a 110 MW output and 2860 MWh of storage capacity. 19 million cubic feet air is stored at pressures up to 1080 psi in a salt cavern up to 2500 feet deep and can provide full power output for 26 h. Both plants are still in operation. The main advantage of this technology is that it is relatively low cost and can store many kilowatt-hours of energy. The main problem with compressing the air is the amount of heat generated during compression. The heat of compression is therefore extracted during

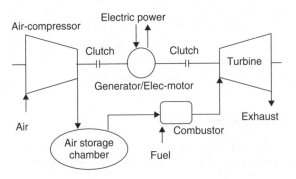

**Figure 4.2** Compressed air energy storage.

compression process or removed by an intermediate cooler. This loss of heat energy has to be compensated during the expansion stage of gas turbine power generation.

### 4.3.4  Flywheel Storage

In a flywheel, the inertia of a rotating mass is used to store energy. Basically, a flywheel is a disk with a certain amount of mass that spins, holding the kinetic energy. The potter's wheel is a low-tech example of this. In a potter's wheel, the potter spins the wheel to a desirable speed so that it continues to rotate because of the energy stored in its mass. The purpose of the wheel is to keep the clay moving in a circular path so that it can be shaped into a vessel of cylindrical symmetry. The heavier the wheel, the more energy is stored and the velocity of rotation is also more uniform.

A mechanical flywheel is coupled to an electrical machine which can work as a motor or generator. When working as a motor, it drives the flywheel to a certain speed, converting electrical energy into kinetic energy. It is equivalent to charging of a battery from the grid. The maximum spinning speed of the flywheel is determined by the tensile strength of the material to withstand the centrifugal forces. When energy is required, the flywheel slows down to release kinetic energy which drives the machine that now acts as a generator producing electricity. In high-speed flywheel-based energy systems, the electrical machine and the flywheel are fully integrated, forming a single compact element. In low-speed system, the flywheel and machine are separate parts or are just partially integrated in a common enclosure.

Modern high-tech flywheels are built with the disk attached to a rotor in upright position to prevent gravity influence. Typical electromechanical energy storage system consists of a high-speed inertial composite rotor, integral drive motor/generator, power converters, vacuum support compartment, heat exchangers, and monitoring and control. Figure 4.3 shows a typical electromechanical energy storage system. The flywheel is connected to a motor/generator with magnetic bearings in a vacuum housing. The motor/generator is used to convert electrical energy into stored kinetic energy by spinning a composite flywheel up to the rated speed. The flywheel is able to spin for long periods efficiently because there is no friction or drag by employing magnetic bearings and a vacuum system. When the dc voltage is above a certain

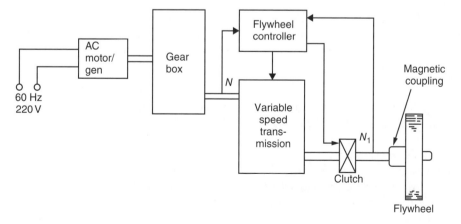

**Figure 4.3** Flywheel energy storage system.

level, the machine operates as a motor (charging), and the speed increases up to the maximum level, determined by the construction of the flywheel. Charging of the flywheel is at a constant power, which can be adjusted by motor control, and it can be up to the full rating of the electrical machine. As friction is very small, the fly wheel can run at a constant speed without taking much power during running. When electric power is needed by the system, the spinning flywheel drives the motor/generator, which instantaneously converts the kinetic energy into electrical energy.

As the kinetic energy is linearly proportional to the flywheel mass and is proportional to the square of the rotational speed, it is more effective to have higher speed instead of having lager mass to achieve higher energy content. However, the maximum speed is limited by the tensile strength of the flywheel material. For a given geometry, the limiting energy density (energy per unit mass) of a flywheel is proportional to the ratio of the material strength to the weight density, also known as the specific strength. Now with advances in composite materials technology, the specific strength of composite materials is quite large high which may be nearly an order of magnitude more compared to even the best engineering metals.

The generator rotor can be made of non-magnetic material in order to reduce the hysteresis losses. The generator of these systems is usually permanent magnet generator. The charge and discharge power of the electromechanical energy storage system is only determined by the type and size of the electrical machine and not by the flywheel. The charging time is closely related to the acceleration torque of the electrical machine, and the discharge power is closely related to the current rating of the machine.

### 4.3.4.1 Energy Stored in Flywheel

The expression for the energy of a rotational system is [8]

$$E = \frac{1}{2}I\,\omega^2 = \frac{1}{2}Mr^2\omega^2 \tag{4.1}$$

where $Mr^2 = I$ is the moment of inertia, $E$ energy, $M$ mass, $r$ radius, $\omega$ angular velocity.

In order to determine the limits to the angular velocity due to the material used: $\rho$ = density, $r$ = radius, $\omega$ = angular velocity, $\sigma$ = tensile stress (maximum before breaking).

$$\sigma = \rho r^2 \omega^2 \tag{4.2}$$

Substituting the maximum angular velocity into energy equation, the maximum energy stored in the flywheel is

$$E_{max} = \frac{1}{2}\frac{M\sigma}{\rho} \tag{4.3}$$

Table 4.2 provides the information about the maximum energy stored in the flywheel for the materials used for 450 kg of mass.

### 4.3.4.2 Motors for Flywheels

The motor required to drive a flywheel must have high efficiency, robust structure and low rotor losses. Permanent magnet (PM) motors are currently the most commonly used motors for flywheel systems. They have the advantages of high efficiency and low rotor loss, but their cost is higher for high-temperature working. Synchronous reluctance motors and homopolar induction motors are also being considered for

**Table 4.2** Energy storage in flywheel of different materials.

| Material | $M$ (kg) | $\sigma$ (Pa) | $\rho$ (kg/m$^3$) | $E_{max}$ (J) | $E_{max}$ (kWh) |
|---|---|---|---|---|---|
| Titanium | 450 | $8.8 \times 10^8$ | 4506 | $4.4 \times 10^7$ | 12 |
| Carbon fibre | 450 | $4.0 \times 10^9$ | 1799 | $5.0 \times 10^8$ | 139 |
| Steel | 450 | $6.9 \times 10^8$ | 8050 | $1.9 \times 10^7$ | 5 |
| Aluminium | 450 | $5.0 \times 10^8$ | 2700 | $4.2 \times 10^7$ | 12 |

driving flywheels. The flywheel energy subsystem consists of a back-to-back converter, a synchronous motor with its exciter or a PM motor and a flywheel. A back-to-back converter is required to provide bidirectional power flow, which acts as a bridge between the motor and the pulsed load.

## 4.4 Superconducting Magnetic Energy Storage

Superconducting magnetic energy storage (SMES) is an energy storage technology that stores energy in the form of dc electricity that is the source of a dc magnetic field. SMES is a method of energy storage based on the fact that current will continue to flow in a superconductor even after the voltage across it has been removed. When the superconductor coil is cooled below its superconducting critical temperature, it has negligible resistance; hence, current will continue to flow (even after a voltage source is disconnected). The system consists of three major components: the superconducting coil, the power conditioning system (PCS) and a cooling system [8, 9].

Originally, it was considered as a large-scale load-levelling device, but nowadays, it has many other applications that include transmission-line voltage stabilization, spinning reserve and to improve power quality. The components of an SMES system are shown schematically in Fig. 4.4. The heart of the system is the superconducting coil. The other components are the cryogenic system and the power conversion/conditioning system

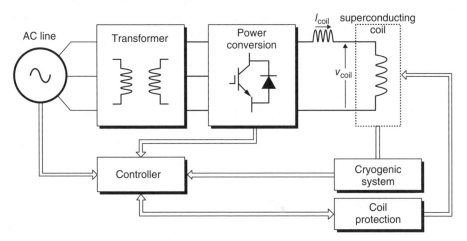

**Figure 4.4** Superconducting magnetic energy storage system.

along with control. The dimensions of the coil are determined by the desired energy storage capacity and the selection of coil design. The inductively stored energy $E$ (in J) and the rated power $W$ (in W) are commonly given specifications for SMES devices and can be expressed as

$$E = \frac{1}{2}LI^2 \quad \text{and} \quad P = LI\frac{dI}{dt} = VI \tag{4.4}$$

where $L$ is the inductance of the coil, $I$ is the dc current and $V$ is the dc voltage.

An SMES is economical only if it is designed for large power. In order to design a superconducting magnetic storage of capacity of 5000 MWh, the current in the superconducting coil may be of the order of 100 kA. The superconductor material should preferably be a Nb-Ti/copper composite stabilized by high-purity aluminium which has a critical temperature of around 9 K. Superconductor wires have thousands of strands of superconductors. These strands are supported by copper or aluminium. These materials are called stabilizers since they help maintain superconductivity when the conductors experience large heat pulses. In addition, because of their high resistivity compared to the superconductor, they electrically isolate the Nb-Ti filaments.

The current level required in SMES is much higher than the normal power system current. A transformer is therefore needed to convert the high voltage and low current of the ac system to the low voltage and high current required by the coil. The dc magnetic coil is connected to the ac grid by the power conditioning system (PCS). The PCS has two power converters, which are used to convert between alternating and direct current sides. When the direct current passes through the superconducting coil, magnetic energy is stored around it. When the load is suddenly increased, the SMES in its discharging mode releases immediately its stored energy through the power conditioning system as an alternating current to the grid, whereas if a sudden decrease in the load occurs, the superconducting coil rapidly changes to the charging mode to absorb the excess energy from the power system. The overall efficiency of SMES system is very high (90–95%) as it does not require any conversion of energy into mechanical or chemical form.

A low-temperature SMES requires liquid helium for its operation, which makes it expensive to operate, particularly because of the cryogenic system. With the availability of a high-temperature superconducting coil, only liquid nitrogen is required, which is readily available and much cheaper than liquid helium. Because of the difficult and expensive procedure to process high-temperature superconductors, it is expected that low-temperature materials will continue to be used in short and medium terms. Right now, the development focus lies on micro-SMES systems with capacities up to 10 kWh, applied mainly for power quality and uninterrupted power sources (UPS).

In 1982–83, an SMES system of rated 30 MJ was assembled in Bonneville Power Administration, Tacoma. The installed configuration functioned for 1200 h, and from the various results obtained, it can be concluded that the SMES configuration had successfully met the design requirements [9, 10].

## 4.5  Supercapacitors

Supercapacitors, also known as ultracapacitors or electrochemical capacitors and electric double-layer capacitors (EDLC) utilize high-surface-area electrode materials and thin electrolytic dielectrics to achieve capacitance values greater than that of any other capacitor type available today. Capacitance values up to 400 F in a single standard case size are now available. Energy storage in capacitors is by means of static charge and does not involve an electrochemical process inherent to the battery. The basic principle of a supercapacitor is the same as that of conventional capacitors. A conventional capacitor consists of two conducting electrodes separated by an insulating dielectric material. When a voltage is applied to the electrodes, opposite charges accumulate on the surfaces of each electrode. The charges are separated by the dielectric, thus producing an electric field that allows the capacitor to store energy. The energy stored is given by

$$E = \frac{1}{2}CV^2 \tag{4.5}$$

where $C$ is the capacitance and $V$ is the voltage across the capacitor.

The supercapacitors incorporate very high surface area and a very thin layer of electrolyte. Thus, the distance between the electrodes is very small. Supercapacitors have separation of charge at an electric field which is only fractions of a nanometre, compared with micrometres for most polymer film capacitors. In doing so, supercapacitors are able to attain greater energy densities while still maintaining the characteristic of high-power density of conventional capacitors. While the energy density of a supercapacitor is much higher compared to a conventional dielectric capacitor, it is still lower than that of batteries and fuel cells.

Energy storage with high storage capacity and a quick delivery over a short span of time is required to support the power system during short-time fluctuations in energy supply. The energy stored in supercapacitors can perform this function very well. Depending on the capacity, they can even support the power system over long-duration fluctuations.

Ultracapacitor consist of a porous electrode, electrolyte and a current collector (metal plates). A paste of activated and porous carbon is applied on both sides of an aluminium film, which acts as a collector. The carbon acts as an electrode. Strips of conductors are attached to the aluminium for taking out the current. There is a porous separating plate between two such pasted aluminium collectors. These are then wound in a coil. The wound coil is impregnated with an electrolyte of high electronic activity. Again, the thickness over which this charge builds up in the electrolyte has molecular dimensions as low as $10^{-10}$ m. This gives a very large capacitance for a small volume and small weight [11].

When the voltage is applied to the positive plate, it attracts negative ions from the electrolyte. When the voltage is applied to the negative plate, it attracts positive ions from the electrolyte. When two aluminium plates are connected to a dc source, the capacitance obtained is a result of two capacitances in series across the electrolyte. Hence, these capacitors are also called double-layered capacitors. Because of the very small distance, the applied voltage is limited to not more than about 3 V, for a given electrolyte. This gives a high-capacitance, low-voltage capacitor cell. The number of cells connected in parallel to form a group and the groups are balanced and connected in series to obtain the desired capacitance and voltage.

The equivalent capacitance $C_{eq}$ of a supercapacitor in Farad is given by

$$C_{eq} = \frac{(n_p \times C_{cell})}{c_s} \tag{4.6}$$

where $n_p$ = number of cells connected in parallel, $C_{cell}$ = the capacity of each cell in Farad and $C_s$ = number of cells connected in series.

Typical values of such supercapacitors are as follows:

1) $10\,F \times 2.9\,V$ for each element. Equivalent series resistance (ESR) also known as leakage resistance $60\,m\Omega$. 12 elements in series form a unit. Array voltage is $35.2\,V$, capacity = $0.833\,F$ and ESR = $720\,m\Omega$.
2) $1800\,F \times 2.5\,V$ unit. Three units in parallel per group $\times 40$ groups in series form an array. Array design voltage of $40 \times 2.9\,V$ is scaled down to $100\,V$, operating voltage. Capacitance = $1800 \times 3/40 = 135\,F$. Stored energy = $\frac{1}{2}CV^2 = \frac{1}{2} \times 135 \times 100^2 \times 10^{-3}\,kJ = 675\,kJ$.

The most significant advantage of supercapacitors over batteries is their ability to be charged and discharged continuously without degrading unlike batteries. The batteries and supercapacitors are generally used in conjunction with each other. The

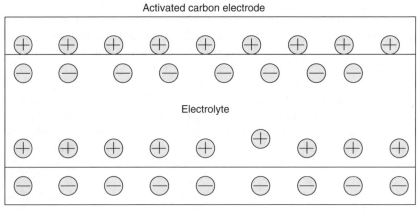

Activated carbon electrode
(a)

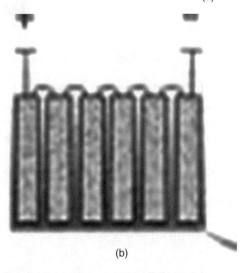

(b)

**Figure 4.5** (a) Supercapacitor cell. (b) Module.

**Figure 4.6** Simplified equivalent circuit of ultracapacitor.

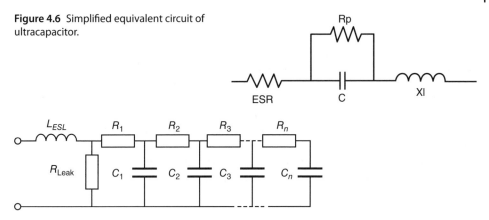

**Figure 4.7** Equivalent circuit of ultracapacitor (transmission line model).

supercapacitors will supply power to the system when there are surges or energy bursts since supercapacitors can be charged and discharged quickly while the batteries can supply the bulk energy.

### 4.5.1 Equivalent Circuit

An individual ultracapacitor cell is shown in Fig. 4.5(a), and a module schematic is shown in Fig. 4.5(b). The simplified equivalent circuit of a supercapacitor is shown in Fig. 4.6. It is only a simplified model of a supercapacitor. Practical supercapacitors exhibit a non-ideal behaviour due to the porous materials used to make the electrodes. It therefore resembles transmission-line model of ladder circuits as shown in Fig. 4.7.

The equivalent series resistance (ESR) prevents supercapacitors from achieving power densities closer to the theoretical limits. Several methods for reducing the ESR have already been developed, including polishing the surface of the current collector, chemically bonding the electrode to the current collector and using colloidal thin-film suspensions [11].

One of the disadvantages with supercapacitors is that there is always some discharging even on an open circuit. Improving the material purity has been found to reduce the discharging of the capacitors.

## 4.6 Chemical Storage (Batteries)

Due to their high cost and/or low conversion efficiencies, alternative energy storage technologies, such as flywheels, capacitors, hydrogen and superconductors, are not yet fully developed for utility-scale electricity storage. Two main utility-scale energy storage systems apart from pumped hydro storage are (i) compressed air energy storage (CAES) and (ii) advanced battery energy storage (BES). Pumped hydro and CAES have been utilized at large-sized wind energy systems or for geothermal or nuclear power generation. Advanced battery energy storage is primarily used for load levelling and regulation and grid stabilization. These can also be used for frequency regulation and integration of renewable energy sources. In recent years, much of the focus in the development of electric energy storage technology has been cantered on battery storage devices. There are currently a wide variety of batteries available commercially, and many more are in the development stage. In a chemical battery, charging from a power source causes reactions in the electrochemical compounds to store energy in chemical form. Upon demand, reverse chemical reactions cause electricity to flow out of the battery and back to the load or grid called discharging. In a battery, the chemical nature of positive and negative electrodes decides the energy output, while the electrolyte defines how fast the energy could be released by controlling the rate of mass flow within the battery. Thus, the type of electrodes and the electrolyte used in different batteries produce different characteristics in batteries.

There are three main types of conventional storage batteries that are used extensively: the lead–acid batteries, the nickel-based batteries and the lithium-based batteries [12, 13].

### 4.6.1 Lead–acid Battery

The lead–acid battery is the oldest known type of rechargeable battery and was invented in 1859 by the French physicist Gaston Planté. Even though developed over 150 years ago, the lead–acid battery is used even now because of its cost-effectiveness. The most common use of these batteries is in automobiles and trucks for starting of engines. These are also used in golf cars, lawn mowers and so on.

The main components of a lead–acid battery are lead dioxide $PbO_2$ (which forms the cathode electrode), sponge lead Pb (which forms the anode electrode) and sulfuric acid $H_2SO_4$ which acts as the electrolyte. The battery is formed by dipping lead peroxide plate and sponge lead plate in dilute sulfuric acid. The chemical reactions at the electrode surfaces introduce electrons into the Pb electrode and create a deficit of electrons in the $PbO_2$ electrode. Thus, a voltage is created between the two electrodes. The rated voltage of a lead–acid cell is 2 V, and typical energy density is around 30 Wh/kg with power density around 180 W/kg. The voltage rating is based on the number of cells connected in series. The ampere-hour (Ah) capacity available from a fully charged battery depends on its temperature, rate of discharge and age.

Lead–acid batteries have high energy efficiencies (between 85 and 90%), are easy to install and require relatively low level of maintenance and low investment cost. In addition, the self-discharge rates for this type of batteries are very low, around 2% of rated capacity per month (at 25°) which makes them ideal for long-term storage applications [14]. Lead–acid batteries generally are rated at 25 °C (77 °F) and operate best around this temperature. Exposure to low ambient temperatures results in performance decline, whereas exposure to high ambient temperatures results in shortened life.

However, the limiting factors for these batteries are low energy density and short service life. Typical lifetimes of lead–acid batteries are between 1200 and 1800 charge/discharge cycles or 5–15 years of operation. The cycle life is negatively affected by the depth of discharge and temperature. They are not suitable for discharges over 20% of their rated capacity.

A lead–acid battery is shown in Fig. 4.8. The chemical reactions that occur in charging and discharging of a lead–acid battery are represented by the following equations:

At the anode:

$$PbO_2 + H_2SO_4 + 2H^+ + 2e^- \xrightleftharpoons[\text{charge}]{\text{discharge}} PbSO_4 + 2H_2O \tag{4.7}$$

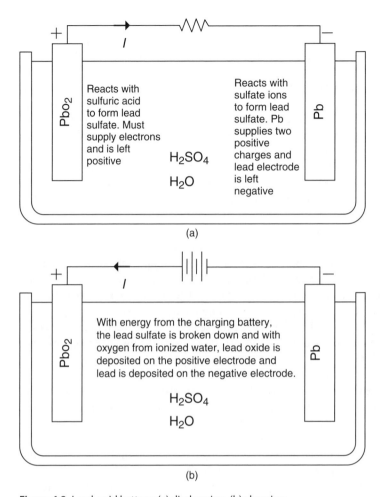

**Figure 4.8** Lead–acid battery; (a) discharging, (b) charging.

And at the cathode:

$$Pb + H_2SO_4 \xrightleftharpoons[\text{charge}]{\text{discharge}} PbSO_4 + 2H^+ + 2e^- \tag{4.8}$$

Overall cell reaction:

$$PbO_2 + Pb + 2H_2SO_4 \xrightleftharpoons[\text{charge}]{\text{discharge}} 2PbSO_4 + 2H_2O \tag{4.9}$$

As the cell is charged, the sulfuric acid ($H_2SO_4$) concentration increases and becomes the highest when the cell is fully charged. It has rated voltage at the fully charged condition. Similarly, when the cell is discharged, the acid concentration decreases and becomes most dilute when the cell is fully discharged. In addition, the voltage of the cell decreases as the battery is discharged. The acid concentration generally is expressed in terms of specific gravity, which is the weight of the electrolyte compared to the weight of an equal volume of pure water.

The problem with lead–acid batteries has been accumulation of lead sulfate which has prevented them from achieving a sustained level of operation required in hybrid electric vehicles and other energy storage applications. This problem was solved in the 1960s and 1970s when a lead–acid battery was modified in construction to support significant oxygen recombination in a similar manner to recombination in sealed nickel–cadmium cells. Thus, a valve-regulated lead–acid battery (VRLA) was developed. A special feature of VRLA batteries consists in the recombination of oxygen during charge. If the positive plate is fully charged, water is decomposed into oxygen, hydrogen ions and electrons. In flooded cells, the oxygen gas is released from the cell, and the hydrogen ion moves in the electrolyte to the negative plate and is reduced there to hydrogen gas, which also leaves the cell. This results in water loss. Only a very small amount of the oxygen gas finds its way to the negative plate for recombination with water, because the solubility of oxygen in water is very low. These designs are commonly referred to as gel batteries (GEL VRLA) and *Absorbed Glass Mat* (AGM VRLA), respectively [15].

AGM is an improved lead–acid battery with higher performance compared to the regular flooded type. Instead of submerging the plates into a liquid electrolyte, the electrolyte is absorbed in a mat of fine glass fibres. This makes the battery spill-proof, allowing shipment without hazardous material restrictions. The plates can be made flat as the standard flooded lead–acid and placed in a rectangular case or wound into a conventional cylindrical cell. With the development of the sealed, low-maintenance affordable, efficient and environmentally safe lead–acid batteries, these are now available for use in hybrid electric vehicles and other energy storage systems.

Another development in lead–acid batteries occurred when it was found that adding carbon to the battery dramatically reduces the formation of deposits, thereby increasing the performance and lifetime. This led to the development of the UltraBattery, which combines a lead–acid battery and a capacitor into single cell, with the capacitor acting as a buffer of high rates of charge/discharge.

Advanced carbon lead–acid battery is different from a conventional VRLA cell in the sense that the negative electrode is purely a carbon electrode, which is more in similarity with the design of an asymmetric supercapacitor. The integration of carbon and supercapacitor features allows UltraBattery to handle high power peaks and operate for extended periods in the partial stage of charge operation.

### 4.6.2 UltraBattery

UltraBattery combines an asymmetric ultracapacitor and a lead–acid battery in each cell. The capacitor enhances the specific power by acting as a buffer during charge and discharge. This produces 50% more power compared to a regular lead–acid battery. Furthermore, the ultracapacitor/lead–acid combination is said to prolong the battery service life by a factor of 4. In the UltraBattery, a supercapacitor electrode composed of carbon is combined with the lead–acid battery negative plate in a single cell. It has the capability of cycling for extended periods at a partial state of charge. This ability is essential for an energy storage system that is ready to either accept or release energy. This hybrid device combines the advantages of high energy density of a battery of advanced lead–acid technology with the advantages of the high specific power of a supercapacitor in a single low-cost device.

The primary goals of lead–carbon research have been to extend the cycle lives of lead–acid batteries and increase their power. The following innovations are being tried:

- Blending carbon additives into the lead sulfate paste that is used for negative electrodes
- Developing split electrodes where half of the negative electrode is lead and the other half is carbon
- Completely replacing the lead-based negative electrode with a carbon electrode assembly

The UltraBattery, constructed from VRLA cell, is illustrated in Fig. 4.9. It is a 12 V, 6.7 A h consisting of a hybrid lead–acid and supercapacitor. The UltraBattery has a smaller dc voltage compared to a supercapacitor and, therefore, can use lower cost dc/ac conversion systems. It results in strings of cells being better balanced than similarly sized strings of conventional VRLA cells. Figure 4.10 shows the comparison of UltraBattery power handling with VRLA.

UltraBatteries achieved a round-trip efficiency of around 90%, compared to 70% for conventional lead–acid batteries under both low-current and high-current charge–discharge cycles. In addition to its superior electrical characteristics, the UltraBattery is non-flammable and made from materials that are easily available, fully recyclable and non-hazardous [16, 17].

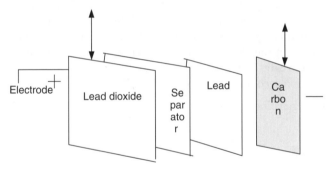

**Figure 4.9** UltraBattery.

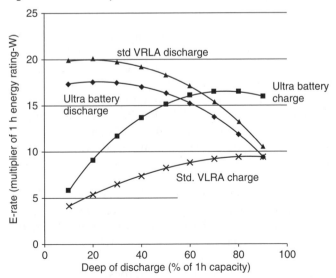

**Figure 4.10** UltraBattery power handling compared to VRLA.

### 4.6.3 Lithium-ion Battery

Lithium-ion battery currently represents the state-of-the-art technology in small rechargeable batteries because of its many merits (e.g., higher voltage, higher energy density, and longer cycle life) compared with traditional rechargeable batteries such as lead–acid and Ni-Cd batteries. These are very common in the portable/consumer electronics markets and are now being used in hybrid and electric vehicle applications [18]. Now these are also employed in energy storage applications ranging from few kWh to multi MWH. Li ion has made significant inroads into grid energy storage due to its strong performance in high-power grid applications such as frequency regulation, black start and spinning reserves. Lithium technology batteries consist of two main types: lithium-ion and lithium–polymer cells.

Over 200 MWs of Li ion is deployed for grid applications globally. Li ion has made significant inroads into grid energy storage given its strong performance in high power grid applications such as frequency regulation that require less than 30 min of power at any given time, black start and spinning reserves.

However, in order to have application in grid-scale energy storage, the long-duration technologies that can address over 4 h of energy storage are required. Nanomaterials can play a large role in improving the performance of lithium-ion batteries, because in nanoparticle systems, the distances over which $Li^+$ must diffuse are dramatically decreased. The nanoparticles can quickly absorb and store vast numbers of lithium ions without causing any deterioration in the electrode and have large surface areas, short diffusion lengths and fast diffusion rates along their many grain boundaries.

Generally speaking, a lithium-ion secondary battery has the following features:

- High voltage (nominal discharge voltage: 3.6–3.8 V)
- High energy density
- No memory effect
- Wide usable temperature range
- Maintenance-free operation
- High efficiency and rapid rate of charge and discharge,
- Excellent storage characteristic

Lithium-ion batteries are available in variety of types with slightly different chemical compounds. The construction of these batteries is similar to that of a capacitor, using three different layers curled up in order to minimize space. The first layer acts as the anode and is made of a lithium compound; the second one is the cathode and is usually made of graphite. Between the anode and the cathode is the third layer – the separator that, as suggested by the name, separates them while allowing lithium ions to pass through. The separator can be made of various compounds allowing different characteristics and, with that, different benefits and problems. In addition, the three layers are submerged in an organic solvent – the electrolyte, allowing the ions to move between the anode and the cathode.

Figure 4.11 shows a lithium battery with charging and discharging processes. In the charging process, the lithium ions pass through the porous separator into the spaces between the graphite, receiving an electron from the external power source. During the discharge process, the lithium atoms located between the graphite release its electrons again which travel over the external circuit to the anode providing a current. The lithium ions also move back to the anode, parallel to their released electrons.

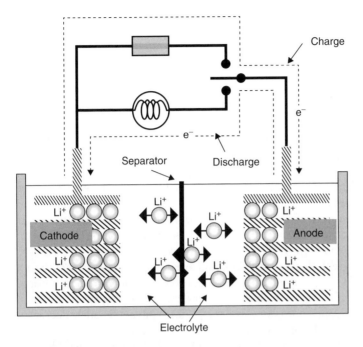

**Figure 4.11** Lithium-ion battery.

**Figure 4.12** Lithium-ion battery storage in California.

A new report from Lux Research has found that lithium-ion (Li-ion) battery technology is set to dominate the emerging grid-tied storage market. The report finds that 90% of the proposed grid storage projects in 2014 are set to be supplied by Li-ion technology.

To maintain its stability and lifespan, the lithium-ion battery must be controlled with a monitoring unit. In particular, to avoid over-charging and over-discharging, a voltage balance circuit is often installed to monitor the voltage of each individual cell and prevent voltage deviations between cells. A large storage using number of lithium-ion batteries is shown in Fig. 4.12.

### 4.6.4 Liquid metal battery

A lithium–antimony–lead (Li-Sb-Pb) battery comprises a liquid lithium negative electrode, a molten salt electrolyte and a liquid antimony–lead alloy positive electrode. These are automatically segregated by density into three distinct layers owing to the non-mixing nature of the contiguous salt and metal phases. The liquid metal batteries differ from conventional batteries because the electrolyte and both electrodes operate in the liquid state. Thus, to understand their design and working, study of flow physics is required. In order to keep the batteries in liquid form, maintaining elevated temperatures is necessary which unavoidably introduces thermal gradients and therefore thermal convection. In addition, these batteries carry large electrical currents which in turn cause electromagnetic (Lorentz) forces. Additional thermal gradients can arise locally from local and entropic heating.

The all-liquid construction confers the advantages of higher current density, longer cycle life and simpler manufacturing of large-scale storage systems (because no membranes or separators are involved) relative to those of conventional batteries. At charge–discharge current densities of 275 milliamperes per square centimetre, the cells cycled at 450 °C were tested and obtained 98% Coulombic efficiency and 73% round-trip energy efficiency. Alloying a high-melting-point, high-voltage metal (antimony) with a low-melting-point, low-cost metal (lead) advantageously decreases the operating temperature while maintaining a high cell voltage.

A high-temperature (700 °C) magnesium – antimony (Mg||Sb) liquid metal battery comprising a negative electrode of Mg, a molten salt electrolyte ($MgCl_2$–KCl–NaCl) and a positive electrode of Sb is proposed and characterized. Because of the immiscibility of the contiguous salt and metal phases, they stratify by density into three distinct layers. Cells were cycled at rates ranging from 50 to 200 $mA/cm^2$ and demonstrated up to 69% dc–dc energy efficiency. The self-segregating nature of the battery components and the use of low-cost materials result in a promising technology for stationary energy storage applications. Extensive testing has shown that even after 10 years of daily charging and discharging, the system should retain about 85% of its initial efficiency: a key factor in making such a technology an attractive investment for electric utilities.

### 4.6.5 Flow Battery

Flow batteries are relatively new battery technology designed for high-power applications as well as for high-capacity electricity storage. A flow battery as shown in Fig. 4.13 is a type of battery that can be designed very flexibly. These batteries technically are similar to conventional batteries, except that the electrolytes use one or two different fluids. The redox flow battery (RFB) is a highly efficient energy storage technology that uses the redox states of various soluble species for charge/discharge purposes. These electrolytes are not stored in the power cell of the battery as in a conventional battery, but in separated storage tanks. During operation, these electrolytes are pumped through the electrochemical reactor, These batteries therefore can be recharged almost instantly by replacing the electrolyte liquid. This concept could, in theory, become very handy for electric cars as they can be charged simply by refuelling just as it is done now.

One distinguished advantage of flow battery technologies is that the power and energy ratings can be sized independently [18]. The power rating is determined by the design of the electrode cells, and the energy capacity depends on the volume of the electrolytes.

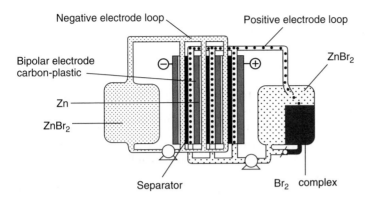

**Figure 4.13** Schematic of a zinc bromine flow battery.

Therefore, flow battery can be easily designed to meet specific energy capacity or power rating requirements. These characteristics make them suitable for a wider range of applications compared to conventional batteries. Other advantages are the long service life of about 10,000 cycles at 75% depth of discharge, high safety, negligible degradation of deep discharge and negligible self-discharge. The major disadvantage is that the flow battery system involves pump systems. In addition, the toxicity of some of the electrolytes is a matter of concern.

The battery is charged and discharged by a (reversible) chemical reaction between the two liquid electrolytes of the battery. One of the problems with flow batteries is that their overall power density is low compared to lead–acid and lithium ion batteries. They can only store about a quarter of the energy compared to lithium-ion batteries and require frequent recharging. This makes them impractical.

### 4.6.6 Nickle-Based Battery

The nickel-based batteries are mainly the nickel–cadmium (NiCd), the nickel–metal hydride (NiMH) and the nickel–zinc (NiZn) batteries. All three types use the same material for the positive electrode and the electrolyte which is nickel hydroxide and an aqueous solution of potassium hydroxide with some lithium hydroxide, respectively. The negative electrode of the NiCd battery uses cadmium hydroxide, the NiMH uses a metal alloy and the NiZn uses zinc hydroxide. Nickel–cadmium batteries have been in development for almost as long as lead–acid batteries. This battery is the most popular alkaline rechargeable battery and is available in several cell designs and in a wide range of sizes. They have very long lives and require little maintenance beyond occasional topping with water. This type of battery is used in heavy-duty industrial applications. However, the NiCd battery may cost up to 10 times more than the lead–acid battery. Apart from that, the energy efficiencies for the nickel batteries are lower than for the lead–acid batteries.

Another problem with some NiCad batteries is a phenomenon known as the "memory effect". It can be described as an apparent reduction in cell capacity to a predetermined cutoff voltage resulting from highly repetitive patterns of use. If these batteries are recharged repeatedly after being only partially discharged, they gradually lose usable capacity owing to a reduced working voltage.

Crystalline formation occurs if a nickel-based battery is left in the charger for days or repeatedly recharged without a periodic full discharge. Since most applications fall into this user pattern, NiCd requires a periodic discharge to 1 volt per cell to prolong service life. A discharge/charge cycle as part of maintenance, known as *exercise*, should be done every 1–3 months.

The rechargeable sealed **nickel–metal hydride** battery is also known as alkaline battery as it uses potassium hydroxide as electrolyte. It uses hydrogen, absorbed in a metal alloy, for the active negative material in place of the cadmium used in the nickel–cadmium battery. The replacement of cadmium not only increases the energy density but also produces a more environmentally friendly power source with less severe disposal problems. However, the problem with these batteries is storage of hydrogen which is a gas and must be stored at a high pressure. The voltage of a nickel–metal–hydride cell is 1.2 V which is somewhat less than the 1.5 V of an alkaline battery. But almost all devices powered by batteries can cope with this voltage difference. They have the disadvantage of high self-discharge (which means that they lose energy even without being used). No battery is free of self-discharge, but with nickel–metal–hydride batteries, the energy loss is higher than with other battery types, about 1% per day. The nickel–metal hydride battery, owing to its higher energy density and other comparable characteristics, has been able to replace NiCd battery in many applications.

**Ni-Zn batteries** do not use mercury, lead or cadmium or metal hydrides that are difficult to recycle. Both nickel and zinc are commonly occurring elements in nature and can be fully recycled. The nickel/zinc battery is very attractive for electric vehicle applications because it is up to 25% smaller, 30% lighter, 25% more powerful and less expensive than other nickel-based technologies.

The nickel–zinc cells have an open circuit voltage of 1.85 V when fully charged, and a nominal voltage of 1.65 V.

The NiMH batteries have energy efficiency between 65% and 70% while the NiZn batteries have 80% efficiency. The energy efficiency of the NiCd batteries varies depending on the type of technology used during manufacture. For the vented type, the pocket plate has 60%, the sinter/PBE plate has 73%, the fibre plate has 83% and the sinter plate has 73% energy efficiency.

## 4.7 Thermal Storage

The thermal energy storage (TES) can be defined as the temporary storage of thermal energy at high or low temperatures. The TES is not a new concept and has been used for centuries [19]. As shown in Fig. 4.14, there are three basic methods for storing thermal energy.

1) Sensible heat storage that is based on storing thermal energy by heating or cooling a liquid or solid storage medium (e.g. water, sand, molten salts, rocks), with water being the cheapest option
2) Latent heat storage using phase-change materials or PCMs (e.g. from a solid state into a liquid state)
3) Thermochemical storage (TCS) using chemical reactions to store and release thermal energy

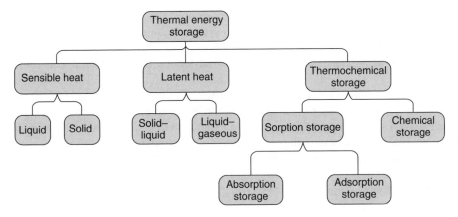

**Figure 4.14** Different types of thermal storage.

Thermal energy storage for solar thermal power plants involves heating synthetic oil or molten salt using solar energy. Once the substance heats, it can support electricity generation during cloudy periods and up to 10 h after sunset. One of the advantages of the solar thermal approach, versus conventional photovoltaics that convert sunlight directly into electricity, is that heat can be stored cheaply and used when needed to generate electricity. By adding storage systems, the power can be generated over longer periods of cloud cover and, or even throughout, at night. Such long-term storage could be needed if solar is to provide a large share of the total power supply.

### 4.7.1 Sensible Heat Storage

In the case of sensible heat storage systems, energy is stored or extracted by heating or cooling a liquid or a solid, which does not change its phase during this process. A variety of substances have been used in such systems. These include liquids such as water, heat transfer oils and certain inorganic molten salts and solids such as rocks, pebbles and refractory. In sensible heat storage (SHS), thermal energy is stored by raising the temperature of a solid or liquid called charging. The energy is released when the temperature of the material drops, called discharging. SHS system utilizes the heat capacity and the change in temperature of the material during the process of charging and discharging. The amount of heat stored depends on the specific heat of the medium, the temperature change and the amount of storage material. The charging and discharging operations in a sensible heat storage system are completely reversible for unlimited cycles.

The choice of the substance used depends largely on the temperature level of the application, the water being used for temperature below 100 °C and the refractory bricks being used for temperatures around 1000 °C. Sensible heat storage systems are simpler in design compared to latent heat or thermochemical storage systems. However, they suffer from the disadvantage of being bigger in size.

In the case of solids, the material is invariably in porous form and heat is stored or extracted by the flow of a gas or a liquid through the pores. With its highest specific heat, water is the most commonly used medium in a sensible heat storage system. Water has the following advantages:

It is abundant and inexpensive, non-toxic and non-combustible. It can be used both as storage medium and as a working medium, thus eliminating the need for a heat exchanger. However, it has a limited temperature range ≤100 °C. Heat transfer oils are used in sensible heat storage systems for intermediate temperatures ranging from 100 to 300 °C. Some of the heat transfer oils used for this purpose are Dowtherm and Therminol. The problem associated with the use of heat transfer oils is that they tend to degrade with time.

The difficulties and limitations relative to liquids can be avoided by using solid materials for storing thermal energy as sensible heat. But larger amounts of solids are needed than using water, due to the fact that solids, in general, exhibit a lower storing capacity than water. Solid media storage using concrete as solid storage material is most suitable, as it is easy to handle, the major aggregates are available all over the world and there are no environmentally critical components. Long-term stability of concrete has been proven in oven experiments and through strength measurements up to 500 °C.

### 4.7.2 Latent Heat Storage

Latent heat storage (LHS) is based on the heat absorption or release when a storage material undergoes a phase change from solid to liquid or from liquid to gas or vice versa. The thermal energy transfer occurs when the material changes from solid to liquid or from liquid to solid. Initially, the PCM absorbs heat as any other heat storage material, and as its temperature increases, it changes its phase from solid to liquid by storing the heat as latent heat of fusion or from liquid to vapour as latent heat of vaporization. Upon storing heat in the storage material, the material begins to melt when the phase-change temperature is reached. The temperature then stays constant until the melting process is finished as shown in Fig. 4.15.

The heat stored during the phase-change process (melting process) of the material is called latent heat. The effect of latent heat storage has two main advantages:

1) It is possible to store large amounts of heat with only small temperature changes and therefore to have a high storage density.
2) Because the change of phase at a constant temperature takes some time to complete, it becomes possible to smooth temperature variations.

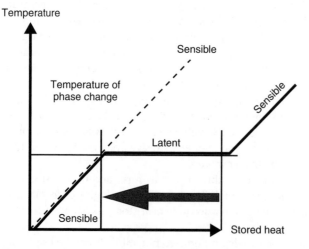

**Figure 4.15** Phase-changing material storing latent heat.

They store 5–14 times more heat per unit volume than sensible storage materials such as water, masonry or rock. A large number of PCMs are known to melt with a heat of fusion in any required range. Many substances have been studied as potential PCMs, but only a few of them are commercially tested. For their employment as latent heat storage materials, these materials must exhibit certain desirable thermodynamic, kinetic and chemical properties. Moreover, economic considerations and easy availability of these materials have to be kept in mind. While selecting a PCM for a particular application, the operating temperature of the heating or cooling should be matched to the transition temperature of the PCM. The latent heat should be as high as possible, especially on a volumetric basis, to minimize the physical size of the heat store. High thermal conductivity would assist the charging and discharging of the energy storage.

The phase-changing materials are available in large numbers and are classified as organic, inorganic and eutectic. Organic materials are further classified as paraffins and non-paraffin inorganic materials are further classified as hydrates and metallics. For details, Ref. [20] may be consulted.

### 4.7.3 Thermochemical Energy Storage (TES)

Thermal energy can be stored in chemical bonds by means of reversible thermochemical reactions. Energy may be stored in systems composed of one or more chemical compounds that absorb or release energy through bond reactions. There are many forms in which energy can be stored through bond reactions. The thermochemical storage involves an endothermic reversible reaction, which can be reversed when required to release heat.

These materials absorb or release energy by breaking or reforming molecular bonds in a reversible chemical reaction. The reaction can be described as

$$C + Heat \leftrightarrow A + B$$

In this reaction, a thermochemical material (C) absorbs energy and is converted chemically into two components (A and B), which can be stored separately. The reverse reaction occurs when materials A and B are combined together, and C is formed. Energy is released during this reaction and constitutes the recovered thermal energy from the TES. The storage capacity of this system is the heat of reaction when material C is formed. Here is no restriction on phases, but usually C is a solid or a liquid, and A and B can be any phase.

Thermochemical materials have higher energy densities relative to PCMs and sensible storage media. Because of higher energy density, thermochemical TES systems can provide more compact energy storage relative to latent and sensible TES. This attribute is particularly beneficial where space for the TES is limited.

## 4.8 Hydrogen Energy Storage Technology

In comparison to commonly used battery storage, hydrogen energy storage is well suited for seasonal storage applications because of its inherent high mass energy density, insignificant leakage from the storage tank and easy-to-install properties. Hydrogen gas with the ability to hold 120 MJ/kg has the largest energy content of any fuel, making it a good choice for holding and distributing energy. The hydrogen serves as an energy carrier that can be stored as a liquid or a gas to be later utilized by a fuel cell to produce electricity. In combination with power electronics, hydrogen electrolyzer/fuel cell systems could provide wind with peak shaving, ancillary services and greater despatch ability. However, hydrogen electrolyzer/fuel cell technology also can provide important services to wind in the damping of power fluctuations.

It is possible to use the excess electricity of renewable energies to make hydrogen (H2) through electrolysis of water and, in further steps, methane. Both methane and hydrogen could be stored in existing natural gas grids, Currently, there are four main technologies for hydrogen storage, out of which two are more mature and developed. These are the hydrogen pressurization and the hydrogen adsorption in metal hydrides. The remaining two technologies that are still in research and technological development phase are the adsorption of hydrogen on carbon nanofibres and the liquefaction of hydrogen. Pressurized hydrogen technology relies on high material permeability to hydrogen and to their mechanical stability under pressure. Currently, steel tanks can store hydrogen at 200–250 bar but present very low ratio of stored hydrogen per unit weight. Storage capability increases with higher pressures, but stronger materials are then required. Storage tanks with aluminium liners and composite carbon fibre/polymer containers are being used to store hydrogen at 350 bar, providing higher ratio of stored hydrogen per unit weight (up to 5%). In order to reach higher storage capability, higher pressures are required in the range of 700 bar with the unavoidable auxiliary energy requirements for the compression. Research is currently under way to materials that are adequate for use in such high pressures. The use of metal hydrides as storage mediums is based on the excellent hydrogen absorption properties of these compounds. These compounds, obtained through the direct reaction of certain metals or metal alloys to hydrogen, are capable of absorbing the hydrogen and restoring it when required. These compounds have a low equilibrium pressure at room temperature (lower than atmospheric pressure) in order to prevent leaks and guarantee containment integrity and a low degree of sensitivity to impurities in the hydrogen stored. The use of such a storage technology is safe. The hydrogen is released by heating the hydride. The reversible reaction is as follows:

$$Mg_2Ni + 2H_2 \rightleftharpoons (Mg_2Ni)H_4 + Heat \tag{4.10}$$

The hydride store can be replenished with hydrogen. A portable hydride store can be used for distribution of energy as convention fuel tank.

## 4.9 Summary

The increasing interest in energy storage for the grid can be attributed to multiple factors, including the capital costs of managing peak demands, the investments needed for grid reliability and the integration of renewable energy sources. Although the existing energy storage is dominated by pumped hydroelectric, there is the recognition that battery systems can offer a number of high-value opportunities, provided that lower costs can be obtained. In this chapter, energy storage techniques using flywheel, supercapacitor, superconducting magnetic storage apart from batteries are presented. The battery systems discussed in this chapter here are include lead–acid, ultra-power, flow battery and other types of batteries. In the end, energy storage using hydrogen technology is also described.

# References

1 Robert, H. (2010) *A Energy Storage*, Springer.
2 Andreas, O. (2012) *Energy Storage Technologies & Their Role in Renewable Integration*, Global Energy Network Institute.
3 Barton, J.P. and Infield, D.G. (2004) Energy storage and its use with intermittent renewable energy. *IEEE Transactions on Energy Conversion*, **19** (2), 441–448.
4 Barnes, F. (2011) *Levine Jonah: Large Energy Storage Systems Handbook*, CRC Press.
5 Hadjipaschalis, I. *et al.* (2009) *Overview of Current and Future Energy Storage Technologies for Electric Power Applications Renewable and Sustainable Energy Reviews*, Elsevier.
6 Figueirdo, F.C. and Flynn, P.C. (2006) Using diurnal power to configure pumped storage. *IEEE Transactions on Energy Conversion*, **21**, 804–809.
7 LaMonica, M. (2013) *Compressed Air Energy Storage Makes a Comeback*, IEEE Spectrum.
8 Pena-Alzola, R. *et al.* (2011) Review of flywheel based energy storage systems, International conference on Power engineering Energy and Electric Drives, Spain, pp. 1–6.
9 Rogers, D. and Boenig, H.J. (1985) Operation of 30MJ SMES in BPA electrical grid. *IEEE Transactions on Magnetics*, **2**, 752–755.
10 Tixador, P. *et al.* (2008) First tests of a 800 kJ HTS SMES. *IEEE Transactions on Applied Superconductivity*, **18** (2), 774–778.
11 Halper, M.S. and Ellenbo, J.C. (2006) *Super Capacitors a Brief Overview*, MITRE Nano Systems Group.
12 Buller, S. *et al.* (2002) Modeling the dynamic behaviour of super-capacitors using impedance spectroscopy. *IEEE Transactions on Industry Applications*, **38** (6), 1622–1626.
13 Menictas, C. *et al.* (eds) (2014) *Advances in Batteries for Medium and Large Scale Energy Storage*, Woodhead Publishing Elsevier.
14 Yang, Z. *et al.* (2011) Electrochemical energy storage for green grid. *Chemical Reviews*, **111**, 3577–3613.
15 Mckon, B.B. and Fukukawa, J. (2014) Advanced Lead Acid batteries and the development of Grid scale energy storage. *Proceedings of IEEE*, **102** (6), 951–963.
16 Onoda, Y. *et al.* (2001) Development of reliable and long life VRLA batteries, Telecommunications energy conference, INTELEC 2001, pp. 12–16.
17 Hund, T. (2008) Ultrabattery test results for utility cycling applications, The 18th International Seminar on Double Layer Capacitors and Hybrid Energy Storage Devices.
18 McDowell, J (2008) Lithium-ion technologies–gaining ground in electricity storage, 2008 Energy Storage Association Annual Meeting, Orange, CA, May 20–21, 2008.
19 Hasnain, S.M. (1998) Review on sustainable thermal energy storage technologies, Part I: heat storage materials and techniques. *Energy Conversion and Management*, **39** (11), 1127–1138.
20 Sharma, A. *et al.* (2009) Review on thermal energy storage with phase change materials and applications. *Renewable and Sustainable Energy Reviews*, **13**, 318–345.

# 5

# Solar Energy Systems

## 5.1 Sun as Source of Energy

Solar energy is available directly from the energy produced by the Sun and is used to produce electricity, heat and light in solar energy systems. The Sun gets its energy from the nuclear fusion occurring in its core by converting millions of tons of hydrogen into helium and converting the difference in mass into energy. The Sun is a sphere of diameter 1,400,000 km, and because of nuclear fusion in its core, its outer region comprises hot gases mostly hydrogen and helium and has a surface temperature of 600 K. These gases are confined to the Sun due to the gravitational forces acting on the Sun. At this temperature, the Sun radiates the heat and light which travels through space and reaches the Earth. The rate of energy radiated from the Sun is $3.8 \times 10^{23}$ kW. The energy received by the Earth's atmosphere is at an annual average rate of about 1.3–1.4 kW/m$^2$. Out of this energy received, a fraction is reflected by the atmosphere to the space. Still it is estimated that the maximum influx at the Earth's surface is about 1 kW/m$^2$. This amount is so large that at present rate of consumption, the total energy received in 1 h is sufficient to provide energy to the whole world for 1 year. But the problem is that not all the energy received from the Sun can be effectively used because it is thinly spread. In addition, the amount of solar energy which can be used is limited by diurnal and geographic variations and weather conditions. The yearly average values of effective solar irradiation are different at different locations on the Earth. To convert solar energy into useful energy using solar equipment, the precise values of the quality and quantity of solar radiation at a specific location are required.

## 5.2 Solar Radiations on Earth

The solar radiation reaching the Earth's surface varies because of weather conditions and changing position of the Sun. In 1 year, that is, in 365.25 days, the Earth rotates around the Sun in slightly elliptical orbit. This elliptical orbit has major and minor axes equal to $1.54 \times 10^{11}$ and $1.45 \times 10^{11}$ m, respectively. The Earth also rotates along its polar axis from the North Pole to the South Pole in 24 h. The polar axis is inclined at 23.45° to the plane of elliptical orbit. Because of Earth's inclination, there is difference in energy absorption by the two hemispheres of the Earth and is responsible for seasonal variations in solar radiation. It is also responsible for variations in local temperatures, local weather and wind pattern. Due to different positions of the Earth around the Sun in a year, four

*Operation and Control of Renewable Energy Systems*, First Edition. Mukhtar Ahmad.
© 2018 John Wiley & Sons Ltd. Published 2018 by John Wiley & Sons Ltd.

different seasons are observed, namely winter, summer, spring and autumn. When the northern hemisphere is having winter, the southern hemisphere will be having summer. Thus, the solar radiation reaching different parts of the globe is different. The annual mean of the solar radiation reaching the Earth's surface is $170\,W/m^2$ for the oceans and $180\,W/m^2$ for the continents. The extraterrestrial solar radiation is separated into different components when it passes through the atmosphere. Solar radiation propagating in straight line and received at the Earth's surface without change of direction is called direct radiation or beam radiation. Diffused solar radiation is solar radiation received by the Earth after its direction is changed because of scattering by aerosols, dust and molecules. The sum of direct radiation and diffused radiation is the global (total) radiation. A part of the solar radiation reflected by the ground may also be present in total radiation.

For the utilization of solar energy, it is desirable to measure the diffused solar radiation and the global solar radiation [1–3]. Solar irradiance $G$ in $w/m^2$ is the rate at which the radiant energy is incident on a unit area of a surface. It is also known as *insolation*.

### 5.2.1 Spectral Distribution of Solar Energy

The Sun emits a continuous spectrum of electromagnetic radiation in a wide frequency range: from very short waves of high energy to long waves with low energy. The spectral distribution of extraterrestrial radiation has a wide spectrum; about half of it lies in the visible part of the electromagnetic spectrum. It produces daylight and can be seen by the human vision system. But nearly half of it also contains the near-infrared and ultraviolet radiations. This spectral distribution is modified in intensity as the radiation crosses the atmosphere downwards, due to scattering by gases and aerosols and by absorption due to atmospheric gases. The energy flow within the Sun results in a surface temperature of 6000 K; hence, the radiation of the Sun resembles the 5800 K blackbody as shown in Fig. 5.1. Electromagnetic radiation as well as solar radiation is commonly classified on

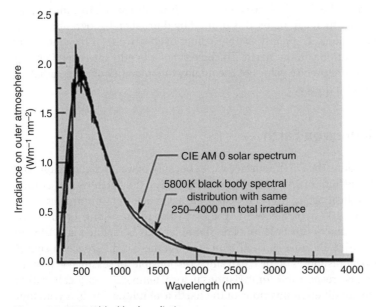

**Figure 5.1** 5800 K blackbody radiation.

**Table 5.1** Wavelength band of solar radiation.

| Band | Wavelength (nm) | Atmospheric effects |
| --- | --- | --- |
| Gamma ray | <0.03 | Completely absorbed by the upper atmosphere |
| X-Ray | 0.03–3 | Completely absorbed by the upper atmosphere |
| Ultraviolet (UV) | | |
| UV (B) | 3–300 | Completely absorbed by oxygen, nitrogen and ozone in the upper atmosphere |
| UV (A) | 300–400 | Transmitted through the atmosphere, but atmospheric scattering is severe |
| Visible | 400–700 | Transmitted through the atmosphere, with moderate scattering of the shorter waves |
| Infrared (IR) | | |
| Reflected (IR) | 700–3000 | Mostly reflected radiation |
| Thermal IR | 3000–14,000 | Absorption at specific wavelengths by carbon dioxide, ozone, and water vapour, with two major *atmospheric windows* |

the basis of radiation wavelength into several regions or *bands*. The wavelength bands of solar radiation, both visible and invisible, are mentioned in Table 5.1.

As can be seen from Table 5.1, about 99% of extraterrestrial radiation has wavelength in the range of 200–4000 nm with a maximum spectral intensity at 480 nm, which falls within the band of green light. There is almost complete absorption of short-wave radiation of less than 290 nm and infrared radiation above 2300 nm. Thus, the radiation of solar energy in the wavelength range of 290–2300 nm is most significant which can be utilized for different applications.

## 5.3 Measurement of Solar Radiation

In order to convert the solar energy into useful energy, it has to be properly focussed on a collecting device. All the irradiated energy of the incident solar radiation does not reach the solar collectors when it passes through the earth atmosphere. In order to design a solar thermal system or photovoltaic system, it is important to know the amount of total energy that is received at the solar collector/PV panel. Although solar radiation data is now available for most of the localities, this data is not sufficient because it is mainly about total radiation (direct plus diffuse). But in most cases, separate data of diffuse radiation is required, the computation of which is extremely difficult. However, instruments are available for the measurement of both types of solar radiation [4, 5]. The radiometers used for ordinary observation are pyrheliometers and pyranometers that measure direct solar radiation and global solar radiation, respectively. These are described here.

### 5.3.1 Pyrheliometer

A pyrheliometer is used to measure direct beam solar radiation from the Sun. In order to measure the direct solar radiation correctly, the instrument is required to

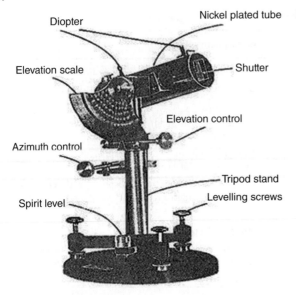

Diopter

Nickel plated tube

**Figure 5.2** Angstrom pyrheliometer.

Elevation scale

Shutter

Elevation control

Azimuth control

Tripod stand

Levelling screws

Spirit level

be permanently pointed towards the Sun. A two-axis Sun-tracking mechanism that continuously follows the Sun is most often used for this purpose. Different companies manufacture pyrheliometers, and International Pyrheliometer Comparisons take place every 5 years at the WRC in Davos to ensure the worldwide transfer of the world radiometric reference.

Here, Angstrom electrical compensation pyrheliometer which is considered quite simple and shown in Fig. 5.2 is described. It is used for instantaneous measurement of direct solar irradiance from 300 to 4000 nm and is capable of very high accuracy and has very high stability. This pyrheliometer has a rectangular aperture, two Manganin-strip sensors ($20.0 \, mm \times 2.0 \, mm \times 0.02 \, mm$) and several diaphragms that allow entry of only direct sunlight. The sensor is mounted in a long metallic tube to collimate the beam and minimize the effect of scattered irradiance. The collimating tube is blackened from the inside and has a about 5.44° field of view. The tube is filled with dry air at atmospheric pressure and its viewing end is sealed by a removable 1-mm-thick crystal quartz window. When solar irradiance is measured with this type of pyrheliometer, the small shutter on the front face of the cylinder shields one sensor strip from sunlight. Thus, sunlight can reach only the other sensor. A temperature difference is therefore produced between the two sensor strips because one gets heated due to solar radiation whereas the other does not. A thermo-electromotive force proportional to this difference in temperature induces current flow through the galvanometer. The heating by direct irradiation received by the exposed strip is compensated by electrically heating the shielded strip till the galvanometer current becomes zero. Electrical power required for heating the shielded strip is proportional to the incident irradiance. If $S$ is the intensity of direct solar irradiance and $i$ is the current in the galvanometer, then $S = Ki^2$, where $K$ is a constant whose value is determined by comparing with a standard pyrheliometer.

## 5.3.2  Pyranometer

The pyranometer is used to measure the total or global solar radiation incident on a horizontal surface. It is capable of measuring solar radiation in the wavelength range of about 130–3000 nm. The working principle of a pyranometer is similar to that of a pyrheliometer, except that the sensitive surface is exposed to total solar radiation. A thermoelectric pyranometer is shown in Fig. 5.3. Several pairs of thermocouples are connected in series to make a thermopile that detects the temperature difference between the black hot junction exposed to the Sun and the cold junctions which are completely shaded. The temperature difference between the hot and cold junctions is the function of radiation incident on the sensitive surface. Two concentric hemispherical glass domes cover the sensitive surface to protect it from rain and wind and also reduce the convection currents. When sunlight falls on a pyranometer, the thermopile sensor produces a proportional response typically in 30 s or less. If the sunlight is more, the sensor is hotter, and it generates a larger voltage. The thermopile is designed to be precisely linear (so a doubling of solar radiation produces twice as much current) and also has a directional response: it produces the maximum output when the Sun is directly overhead (at midday) and zero output when the Sun is on the horizon (at dawn or dusk). This is called a cosine response (or cosine correction), because the electrical signal from the pyranometer varies with the cosine of the angle between the Sun's rays and the vertical.

The instrument has a voltage output of approximately $9\,\mu V/W/m^2$. If the pyranometer is provided with occulting disc to prevent direct radiation reaching to the sensitive element, it can measure the diffused radiation only. Not all pyranometers use thermopiles. Considerably cheaper and less sophisticated *solar-cell pyranometers*, based on light-sensitive semiconductor chips that provide approximate measurements, are also available. Their main drawback of these is that they respond only to a limited range of wavelengths. This means that a solar-cell version responds to wavelengths in a much narrower band at about 300–1100 nm (with a peak in the infrared region at around 800–1100 nm).

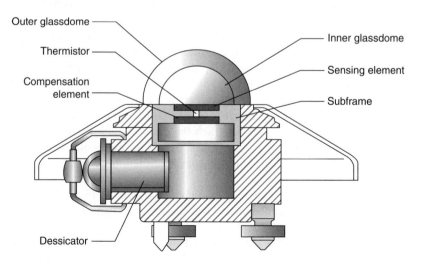

**Figure 5.3** Thermoelectric pyranometer.

### 5.3.3 Sources of Errors in Radiation Meters

Since the pyranometer output signal is influenced by the conditions at which the measurements are made, it needs to be properly calibrated from time to time in order to give accurate measurement results. The errors in meters are introduced as the conditions at the time of application may be different from the conditions at the time of calibration. The following errors are possible:

In actual conditions, the wavelength characteristic of radiometer differs slightly for different radiometer. Observation errors therefore occur when the energy distribution of solar radiation against wavelength varies with the Sun's elevation or atmospheric conditions.

Since thermo-electromotive force of a thermopile is non-linear and the heat conductivity inside a radiometer depends on temperature, the sensitivity of these instruments varies, and an error occurs when the ambient temperature and the temperature of the radiometer are not the same.

The field of view of a pyrheliometer is somewhat larger than the viewing angle of the Sun. Errors occur when the field of view is different, as the influence from diffuse sky radiation near the Sun differs. Pyrheliometers with different fields of view may make different observations depending on the turbidity of the atmosphere.

The characteristic may also deviate, and errors may occur because of the uneven thickness, curvature or material of the glass cover.

### 5.3.4 Sunshine Recorder

Sunshine duration is the length of time that the ground surface receives radiation by direct sunlight, that is, sunlight reaching the Earth's surface directly from the Sun. In 2003, WMO defined sunshine duration as the period during which direct solar irradiance exceeds a threshold value of $120 \, \text{W/m}^2$. This value is equivalent to the level of solar irradiance shortly after sunrise or shortly before sunset in cloud-free conditions. It was determined by comparing the sunshine duration recorded using a Campbell–Stokes sunshine recorder with the actual direct solar irradiance.

A sunshine recorder is a device that records the duration of sunshine in hours at a given location during the course of a day. The results provide information about the weather and climate of a geographical area. Campbell–Stokes sunshine recorders and Jordan sunshine recorders have long been used as instruments to measure sunshine duration. These instruments have the advantage of having no moving parts and require no electric power. But they have a disadvantage that the characteristics of the recording paper or photosensitized paper used in them affect the measurement accuracy. Since the threshold value for the definition of sunshine is set as a direct solar irradiance of $120 \, \text{W/m}^2$, sunshine recorders using photosensors as radiation detectors are also available.

Figure 5.4 shows a Campbell–Stokes sunshine recorder. It is the most widely used sunshine recorder in the world. A homogeneous transparent glass sphere is supported on an arc and is focused so that an image of the Sun is formed on the recording paper placed in a metal bowl attached to the arc. The glass sphere is concentric to this bowl, which has three partially overlapping grooves into which recording cards for use in the summer, winter or spring and autumn are set. Three different recording cards are used depending on the season. As the Sun moves across the sky, the spot moves across the

**Figure 5.4** Campbell's sunshine recorder.

card burning a trace; when the Sun is obscured, the trace is interrupted. The length of the burn trace left on the card represents the sunshine duration. The recording card is scaled with hour marks so that the exact time of sunshine occurrence can be ascertained. Measuring the overall length of burn traces reveals the sunshine duration for that day. For exact measurement, the sunshine recorder must be accurately adjusted for planar levelling, meridional direction and latitude. From sunshine duration, the amount of monthly average of daily total solar radiation on a horizontal surface can be obtained using the following empirical relation:

$$Q/Q_0 = a + b\left(\frac{N}{N_0}\right) \tag{5.1}$$

where $Q$ is the daily total amount of global solar radiation at the ground surface, $Q_0$ is the daily extraterrestrial radiation, $N$ is the sunshine duration, $N_0$ is the possible sunshine duration and $a$ and $b$ are constants for a particular location and depend on the latitude of the location. Values of $a$ and $b$ have been obtained for many cities in the world.

## 5.4 Solar Radiation on Different Surfaces

In order to determine the solar radiation on different surfaces of the Earth, it will be useful to understand certain basic definitions stated as follows.

### 5.4.1 Zenith and Zenith Angle

Zenith is a point directly overhead. Zenith angle $\theta_z$ is the angle between the solar incidence beam and vertical line on the horizontal surface.

### 5.4.2 Solar Time

Solar radiation calculations must be made in terms of solar time. The length of a solar day varies throughout the year because the Sun moves along the ecliptic at a varying rate throughout the year. It is more convenient to define time in terms of the average of local solar time. The difference between the true time and the mean time is known as *Time Equation.*

The time equation is written as

$$\text{Solar time} = \text{Standard time(true time or clock time)} \pm 4(L_{st} - L_{loc}) + E \quad (5.2)$$

where $L_{st}$ is the standard longitude used for measuring the standard time of the country and $L_{loc}$ is the longitude of the observer location. The positive sign is used if the standard meridian of the country lies in the western hemisphere, and negative for eastern hemisphere. $E$ is the correction used to take into account the variation in length of the solar day due to Earth's rotation. The value of $E$ is normally calculated using the following equation:

$$E = 9.87 \sin 2B - 7.53 \cos B - 1.5 \sin B \quad (5.3)$$

where $B = \left(\frac{360}{364}\right)(n - 81)$, where $n =$ day of the year, starting from January 1st.

### 5.4.3 Latitude (∅)

It is the angle made by the radial line joining the given location on Earth's surface to the centre of the Earth with the projection of line on the equatorial plane. By convention, the latitude is positive for the northern hemisphere and negative for the southern hemisphere.

### 5.4.4 Declination Angle (δ)

The declination angle, denoted by $\delta$, is defined as the angle between the equator and a line drawn from the centre of the Earth to the centre of the Sun. It varies seasonally due to the tilt of the Earth on its axis of rotation and the rotation of the Earth around the Sun. It is positive when measured in the northern hemisphere. If the Earth were not tilted on its axis of rotation, the declination would always be 0°. However, the Earth is tilted by 23.45°, and the declination angle varies plus or minus this amount depending on whether the measurement is made in the northern or southern hemisphere. Only in the spring and fall equinoxes $i$, the declination angle is equal to 0°. The declination angle can be approximately determined from the following equation:

$$\delta = 23.45 \times \sin\left[\frac{360}{365}(284 + n)\right]^{\circ} \quad (5.4)$$

### 5.4.5 Hour Angle (ω)

The hour angle measures the time before or after solar noon in terms of 15 degrees per hour. The time after solar noon is expressed using a positive hour angle, and the time before solar noon is expressed using a negative hour angle. Therefore, at 2 h before solar noon, the hour angle is −30°, and at 2 h after solar noon, it is +30°.

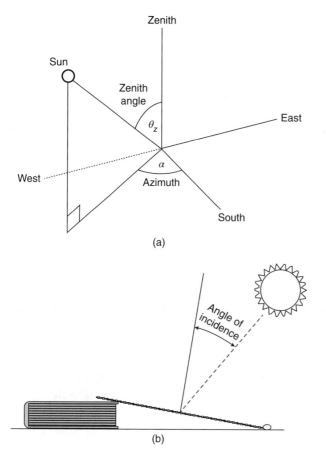

**Figure 5.5** (a) Azimuth angles, (b) incidence angle.

### 5.4.6 Surface Azimuth Angle ($Y$)

It is the angle made on a horizontal plane between the line at due south and the horizontal projection of the normal to the inclined plane surface (collector). Due south is taken as zero, and east of south as positive, Fig. 5.5(a).

### 5.4.7 Tilt Angle ($\beta$)

This is the angle between the inclined surface and the horizontal plane.

### 5.4.8 Angle of Incidence

The angle of incidence $\theta_i$ is the angle a ray of Sun makes with a line perpendicular to a surface (collector), as shown in Fig. 5.5(b).

The angle of incidence can be represented by

$$\cos \theta_i = (\cos \phi \cos \beta + \sin \phi \sin \beta \cos \gamma) \cos \delta \cos \omega + \cos \delta \cos \omega$$
$$+ \cos \sin \omega \sin \beta \sin \gamma + \sin \delta (\sin \phi \cos \beta - \cos \phi \sin \beta \cos \gamma) \quad (5.5)$$

### 5.4.9   Solar Radiation on an Inclined Surface

The solar collectors or solar panels are never placed in a horizontal surface [6]. These are always installed at an angle (TILTED) with the horizontal surface. It is therefore important to estimate the radiation on a tilted surface for designing and evaluating the performance of solar systems. However, in general, in national meteorological stations, global irradiance is measured only on horizontal surfaces. For determining the radiation on a tilted surface, the relation between beam radiation on the tilted surface with that on the horizontal surface $R_b$ is used. The daily total radiation incident on a tilted surface is composed of direct (beam) $I_b$, ground reflected $I_r$ and sky-diffuse $I_d$ components: the total radiation on a surface arbitrary orientation can be calculated as

$$I_T = I_b r_b + I_d r_d + (I_b + I_d) r_r \tag{5.6}$$

where $r, r_d$ and $r_r$ are known as tilt factors for beam, diffused and reflected radiations. Here $r_b$ is calculated as

$$r_b = \frac{I'_b}{I_b} \tag{5.7}$$

where $I'_b$ is beam radiation on the inclined surface and $I_b$ is the radiation on the horizontal surface. If $\theta_i$ is the angle at which the plane is inclined and $\theta_z$ is the zenith angle, then

$$r_b = \frac{\cos \theta_i}{\cos \theta_z} \tag{5.8}$$

$r_b$ is purely geometric quantity that converts instantaneous horizontal beam radiation to beam radiation on a tilted surface. Equation (5.8) cannot be used directly for the long-term beam radiation. Normally, it should be integrated over a month to find the correct value.

$r_d$, the tilt factor for diffused component, is given by

$$r_d = \frac{1 + \cos \beta}{2} \tag{5.9}$$

where $\beta$ is the slope of the tilted surface.

$r_r$ is the tilt factor for a reflected component which comes from the ground and surrounding objects. Since $r_d = \frac{1 + \cos \beta}{2}$, $r_r$ can be written as

$$r_r = \rho \left( \frac{1 - \cos \beta}{2} \right) \tag{5.10}$$

where $\rho$ is the reflection coefficient of the ground.

## 5.5   Utilization of Solar Energy

Solar energy can be utilized in a number of ways. In its most simple form, the solar energy can be used to heat water during winter using solar collectors. A solar panel as shown in Fig. 5.6 is used to collect heat from the sunlight. The pipes inside the solar panel heat the water flowing through the pipes. The hot water coming out of the pipes can be used for different purposes in a house or commercially in big establishments. Similarly, solar heat energy can be used to dry grains and many other agricultural products before being stored. Otherwise, insects and fungi will make them unusable. Examples include

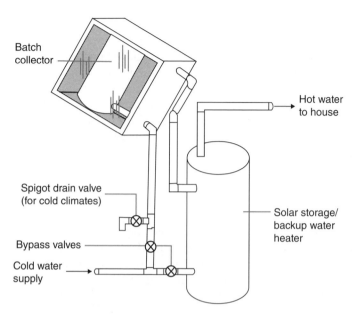

**Figure 5.6** Passive batch solar water heating.

wheat, rice, coffee, copra (coconut flesh), certain fruits and timber. In addition, concentrating collectors can be used to heat water to high temperatures for generating electricity. These are known as solar thermal systems.

In photovoltaic systems, solar cells made of semiconductor material convert solar energy directly into electrical energy. These devices produce electricity directly from electromagnetic radiation, especially light, without any moving parts.

In the photosynthesis process, the absorbed solar energy by plants is converted into chemical form of energy. In chlorophyll-containing plants, the process of photosynthesis splits water molecules ($H_2O$), releasing oxygen and storing the energy produced by that chemical reaction inside a carbohydrate molecule. The energy resident in each photon is transferred into the final organic compound product, which the plant stores as adenosine triphosphate.

Although the potential of solar energy is huge, the present market share of solar power generation is very low. Presently, developing and developed countries are steadily increasing their investments in solar power plants. The major reasons for slow deployment of solar technologies are as follows:

- The capital cost of per kW installation of solar energy is high compared to other fossil fuel and even renewable energy technologies.
- The intermittent nature of power generation means that there has to be energy storage capacity to keep the supply without interruptions. This also adds to the cost.

## 5.6 Solar Thermal Systems

Solar thermal systems convert solar energy into heat energy by absorbing it. These are mostly used in residential and industrial applications such as domestic water heating,

heating of swimming pools, space heating, water processes for industrial heating and agricultural drying. It is also possible to convert solar thermal energy to mechanical energy through heat engines using the Rankine cycle, Stirling cycle or Brayton cycle. The mechanical energy produced can be used to drive a turbine or water lifting or for refrigeration. Solar thermal collectors are the main component of a solar thermal system. Solar energy collectors are special kind of heat exchangers that transform solar radiation energy to internal energy of the transport medium. It absorbs the incoming solar radiation, converts it into heat and transfers this heat to a fluid (usually air, water or oil) flowing through the collector [7, 8].

The solar collectors are classified as follows:

- Non-concentrating type
- Concentrating type

A non-concentrating collector has the same area for intercepting and for absorbing solar radiation. These collectors absorb both beam and diffused radiation.

Concentrating solar collector usually has concave reflecting surfaces to intercept and focus the Sun's beam radiation to a smaller receiving area, thereby increasing the radiation flux and producing high temperature. They mainly focus beam radiation as it has unique direction.

Non-concentrating type is further categorized as follows:

i) Flat-plate collectors
ii) Evacuated tube collector

Concentrating type is further classified as (i) non-tracking type and (ii) tracking type.

- Non-Tracking Type
  Compound parabolic collectors
- Tracking Type
  1) Concentrating type with single-axis tracking:
     i) Linear Fresnel reflector (LFR)
     ii) Parabolic trough collector (PTC)
     iii) Cylindrical trough collector (CTC)
  2) Concentrating type with two-axis tracking
     i) Parabolic dish reflector (PDR)
     ii) Heliostat field collector (HFC)

### 5.6.1 Flat-Plate Collectors

Non-concentrating-type solar collectors do not have mirrors for concentration of solar light; hence, the concentration ratio is 1. Flat-plate collectors are the most widely used kind of collectors in the world for domestic solar water heating and solar space heating applications. A typical flat-plate collector is shown in Fig. 5.7. It has a black absorber surface insulated at the bottom and exposed to solar radiation. When solar radiation passes through a transparent cover and impinges on the blackened absorber surface, a large portion of this energy is absorbed by the plate and then transferred to the transport medium in the fluid tubes. The fluid delivers this heat to a thermal storage tank or use. The liquid commonly used is water; however, sometimes a mixture of water and ethylene glycol is also used. Tubes can be welded to the absorbing plate, or they

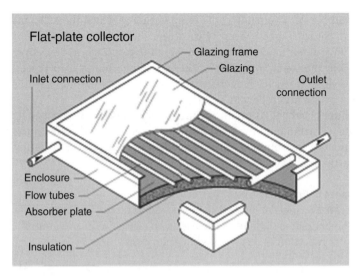

**Figure 5.7** Flat-plate collector.

can be an integral part of the plate. The liquid tubes are connected at both ends by a large-diameter header tubes. The transparent cover is used to reduce convection losses from the absorber plate by keeping the air stagnant. It also reduces radiation losses from the collector, as the glass is transparent to the short-wave radiation received by the Sun, but it is nearly opaque to long-wave thermal radiation emitted by the absorber plate (greenhouse effect). Flat-plate collectors are normally permanently fixed in position and do not have moving facility to track the Sun. The collectors face directly towards the equator, facing south in the northern hemisphere and north in the southern hemisphere. The optimum tilt angle of the collector is equal to the latitude of the location with angle variations of 10–15°, more or less depending on the application.

Commercial solar absorbers are made by electroplating, anodization, evaporation, sputtering and applying solar selective paints. Header pipes that are used to admit and discharge the fluid have slightly larger diameter typically 1.8–2.5 cm. The metal commonly used for absorber plate and tubes and header pipes is copper, but other metals and plastics have also been tried. In the bottom and along the side walls, thermal insulation by 2.5–8 cm thick layer of glass wool is provided to prevent heat loss.

The absorber is usually covered with one or more transparent or translucent cover sheets to reduce convective heat loss. In the absence of a cover sheet, heat is lost from the absorber as a result of not only forced convection caused by local wind but also natural convective air currents created because the absorber is hotter than ambient air.

The number of cover sheets on commercial flat-plate collectors may vary from none to three or more. Since the incoming energy is not lost by absorption or reflection by the cover sheet, collectors with no cover sheet have high efficiencies when operated at temperatures very near the ambient temperature. However, a considerable amount of the incident energy is lost in the collectors without cover at temperatures much above ambient or at low solar irradiance levels. A typical application for an uncovered flat-plate collector is for swimming pool heating, where temperatures not more than 10°C (18°F) above ambient are required.

With increase in the number of cover sheets, the temperature at which the collector can operate is increased considerably. One or two cover sheets are common, but triple-glazed collectors have been designed for extreme climates. Each added cover sheet increases the cost, but the collection efficiency at high temperatures is increased by reducing the convection loss. However, it decreases the efficiency at low temperatures because of the added absorption and reflectance of the cover.

Glass is preferred for outer cover sheets on most commercial collectors because of its superior resistance to the environment. The plastic sheet can be installed beneath the glass which protects it from the environment. Glass also does not transmit UV radiation and thus protects the plastic from UV radiation.

### 5.6.1.1 Thermal Performance of Collector

The performance of the solar thermal flat-plate collector depends on the amount of solar insolation absorbed by the plate. It is defined as the useful energy gain or collector efficiency. The amount of solar radiation $Q$ received by the solar collector is

$$Q = IA \tag{5.11}$$

where $I$ is the instantaneous intensity of solar radiation in $W/m^2$ as given by Eq. (5.6), and $A$ is the area of collector surface in $m^2$. Since a part of this radiation is reflected back to the sky, another component is absorbed by the glazing, and the rest is transmitted through the glazing and reaches the absorber plate as short-wave radiation. A conversion factor is required to get the actual flux absorbed by the collector. Basically, it is the product of the rate of transmission of the cover and the absorption rate of the absorber. Thus, Eq. (5.11) can be written as

$$Q = I\,(\tau\alpha)A \tag{5.12}$$

where $\tau\alpha$ is the transmissivity–absorptivity product for intensity of solar radiation incident on the collector.

As the heat is absorbed by the collector, its temperature increases, and the heat is lost to the atmosphere by convection and radiation. The rate of heat loss can be represented by the overall heat transfer coefficient $U_l$ as

$$Q_l = U_l A(T_c - T_a) \tag{5.13}$$

where $T_c$ is the average temperature of the collector and $T_a$ is ambient temperature.

The efficiency of the flat-plate collector can be expressed as

$$\eta = \frac{\text{rate of useful energy extracted by the collector } (Q_u)}{\text{Total solar radiation}}$$

here $Q_u = Q - Q_l$

$$\eta = \frac{I(\tau\alpha)A}{I(\tau\alpha)A - U_{lA}\,(T_c - T_a)} \tag{5.14}$$

### 5.6.2 Evacuated Tube Collector

Conventional simple flat-plate solar collectors were developed for use in sunny and warm climates. Their benefits, however, are greatly reduced when conditions become unfavourable during cold, cloudy and windy days. The performance of flat-plate collectors is improved by minimizing the convective heat loss from the collector by placing the solar radiation absorbing surface in a vacuum. These collectors are called evacuated tube collectors. These collectors have very low overall heat loss when operated at high temperatures. This is because they are essentially single-glazed collectors with the space between the glazing and absorber evacuated, thereby eliminating the convective loss. They are the most efficient solar collectors having conversion efficiency of over 90%.

A number of designs have been developed, and some of these are commercially available. Two main types of evacuated tube collectors are (i) direct-flow evacuated tube and (ii) heat-pipe evacuated tube collectors.

### 5.6.2.1 Direct-Flow Evacuated Tube Collector

Direct-flow evacuated tube collectors also known as "U" pipe collectors shown in Fig. 5.8(a) have two heat pipes running through the centre of the tube. One pipe acts as the flow pipe while the other acts as the return pipe. Both pipes are connected together at the bottom of the tube with a "U-bend", hence the name. The heat absorbing reflective plate acts as a dividing strip which separates the flow and the return pipes through the solar collector tubes. The absorber plate and the heat transfer tube are also vacuum sealed inside a glass tube providing exceptional insulation properties. The outer tube is transparent, allowing light rays to pass through with minimal reflection. Sunlight passing through an outer glass tube heats the absorber tube contained within it. The absorber can consist of either copper (glass–metal) or specially coated glass tubing (glass–glass). The glass–metal evacuated tubes are typically sealed at the manifold end, and the absorber is actually sealed in the vacuum; thus, the fact that the absorber and heat pipe are dissimilar metals creates no corrosion problems. The inner tube is coated with a special selective coating (Al–N/Al) which features excellent solar radiation absorption and minimal reflection properties. The tops of the two tubes are fused together, and the air contained in the space between the two layers of glass is pumped out. This "evacuation" of the gasses forms a vacuum, which is an important factor in the performance of the evacuated tubes. Since the fluid flows into and out of each tube, the tubes are not easily replaced. In addition, should a tube break, it is possible that all of the fluid could be pumped out of the system if a closed loop is used, or the water will flow out as in a broken pipe if an open loop is used. Direct-flow evacuated tubes can collect both direct and diffuse radiation and do not require solar tracking. However, various reflector shapes placed behind the tubes are sometimes used to usefully collect some of the solar energy, which may otherwise be lost, thus providing a small amount of solar concentration.

### 5.6.2.2 Heat-Pipe Evacuated Tube Collector

In heat-pipe evacuated tube collectors shown in Fig. 5.8(b), a sealed heat pipe, usually made of copper to increase the collector efficiency at cold temperatures, is attached to a heat absorbing reflector plate within the vacuum sealed tube. The hollow copper heat pipe within the tube is evacuated of air but contains a small quantity of a low-pressure alcohol/water liquid that turns into steam when heated and rises to the top of the pipe. The top part of the heat pipe, and therefore the evacuated tube, is connected to a copper heat exchanger called the "manifold". When the hot vapour still inside the sealed heat tube enters the manifold, the heat energy of the vapour is transferred to the water or glycol fluid flowing through the connecting manifold.

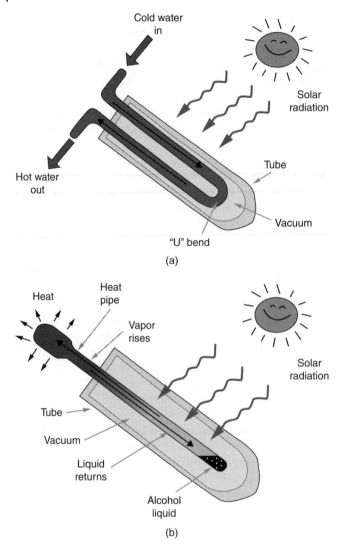

**Figure 5.8** (a) Direct flow evacuated tube collector. (b) Heat pipe type.

The main advantage of heat-pipe evacuated tube collectors is that there is a "dry" connection between the absorber plate and the manifold, making installation much easier than with direct-flow collectors. In addition, in the event of an evacuated tube cracking or breaking and the vacuum becoming lost, the individual tube can be exchanged without dismantling the entire system.

Evacuated tube collectors heat to higher temperatures and provide considerably more solar yield per square metre compared flat panels. However, they are more expensive than flat panels, but require less cost to repair in the event of damage. Evacuated heat tubes perform better than flat-plate collectors in cold climates because they only rely on the light they receive and not the outside temperature.

The efficiency of the evacuated tube solar collector depends mainly on the angle of inclination which should be approximately the latitude of the location. The collectors, however, cannot be installed at less than 20° angle. Evacuated tube solar collectors with internal adjustable collector surfaces allow this type of collector to be installed at angles ranging from 20° to 90°. The collector surfaces inside the evacuated tubes are set to the optimal angle during the installation. An evacuated tube collector is shown in Fig. 5.9.

### 5.6.3 Parabolic Collectors

Concentrating-type solar collectors are generally used along with a tracking mechanism for applications where high temperatures are required. These are devices that optically reflect and focus the incident solar energy onto a small receiving area. As a result of this concentration, the intensity of the solar energy is magnified. Because in this the temperature of the receiver can approach several hundred degrees.

The compound parabolic collector is a solar collector which is stationary and has no tracking system. This is a non-imaging concentrator that concentrates light rays and has the capability of reflecting to the absorber all of the incident radiation within wide limits.

The basic shape of the compound parabolic concentrator (CPC) is illustrated in Fig. 5.10. The name, compound parabolic concentrator, derives from the fact that the CPC is comprised of two parabolic mirror segments with different focal points. The focal point parabola 1 lies on parabola 2, whereas the focal point of parabola 2 lies on parabola 1. The two parabolic surfaces are symmetrical with respect to reflection through the axis of the CPC.

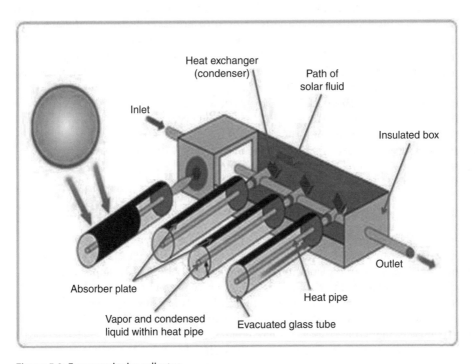

**Figure 5.9** Evacuated tube collector.

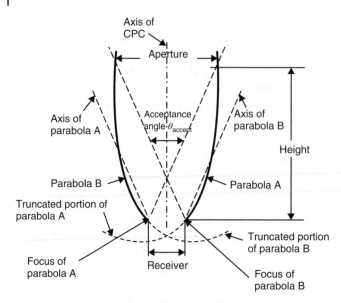

**Figure 5.10** Compound parabolic dish collector.

By using multiple internal reflections, any radiation that is entering the aperture, within the collector acceptance angle, is absorbed by the surface located at the bottom of the collector.

In particular, a working fluid at 500°C can drive a conventional heat engine to produce mechanical work and can also be used to produce electricity. The absorber can take a variety of configurations. But generally, tubes are used which are selectively coated and attached to the bottom, or it can be cylindrical as shown in Fig. 5.10. In the CPC shown in Fig. 5.10, the lower portion of the reflector is circular, while the upper portions are parabolic. As the upper part of a CPC contributes little to the radiation reaching the absorber, it is usually truncated, thus forming a shorter version of the CPC, which is also cheaper. CPCs are usually covered with glass to avoid dust and other materials from entering the collector and thus reducing the reflectivity of its walls. It is possible to attach a tracking system with CPC. In stationary type, radiation will only be received for the duration when the Sun is within the collector acceptance angle.

### 5.6.4 Linear Fresnel Reflector (LFR)

Linear Fresnel reflector is a concentrating-type solar collector with single-axis tracking mechanism. In 1822, a French physicist named Augustin Fresnel invented a lens of thin (low-weight and low-volume) for short focal length. Its first application was in lighthouses: to focus light horizontally and make it visible over a greater distance.

The principle of LFC is to concentrate the sunlight on an absorber tube that is fixed on a linear tower. However, unlike in parabolic trough collectors (PTCs), this is done by several linearly aligned mirrors as shown in Fig. 5.11. Each mirror with a characteristic curvature directs the beam radiation on to the absorber tube. In addition, each solar mirror is flat instead of expensive parabolic mirrors used in parabolic trough collector.

**Figure 5.11** Linear Fresnel reflector.

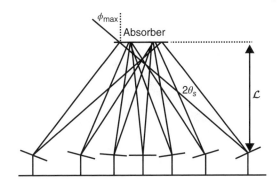

The entire optical system is enclosed in a sealed glazed casing. This concentrated energy is transferred through the absorber into the turbines for power generation. For generating sufficient power, many towers are required.

One problem with LFR system is avoidance of shading and light blocking by adjacent reflectors which means larger spacing between reflectors.

Compact linear Fresnel reflector (CLFR) technology is a new configuration of the Fresnel reflector field that overcomes the problem of reflector spacing [9]. Traditional LFR technology design is based around one absorber tower. The classical linear Fresnel system has only one linear receiver, and therefore, there is no possibility of changing the direction of orientation of a given reflector. However, if the size of the field is large, which is normal for generating electricity in the megawatt class, it is reasonable to assume that there will be many linear receivers in the system. If they are close enough, then individual reflectors have the option of directing the reflected solar radiation to at least two receivers. This additional possibility in reflector orientation provides the means for much more densely packed arrays and lower absorber tower heights, because patterns of alternating reflector orientation can be set up such that closely packed reflectors can be positioned without shading and blocking.

The avoidance of large reflector spacing and tower heights is an important cost issue when the cost of ground preparation, array substructure cost, tower structure cost, steam-line thermal losses and steam-line cost are considered.

### 5.6.5 Parabolic Trough Collector (PTC)

Figure 5.12 shows a typical parabolic trough collector (PTC), which is basically composed of a parabolic-trough-shaped concentrator that reflects direct solar radiation onto a metal black tube located in the focal line of the parabola. Since the collector aperture area is bigger than the outer surface of the receiver tube, the direct solar radiation is concentrated. The dish structure must fully track the Sun to reflect the beam into the thermal receiver. The concentrated radiation reaching the receiver tube heats the fluid that circulates through it, thus transforming the solar radiation into thermal energy in the form of sensible heat of the fluid. This fluid can be efficiently heated up to 400°C. Since a PTC is an optical solar concentrator, it has to be positioned at every moment in accordance with the Sun's position (i.e. Sun's vector) so that the incoming direct solar radiation is reflected onto the receiver tube. It is sufficient to use single-axis tracking of the Sun, and long collector modules can be used.

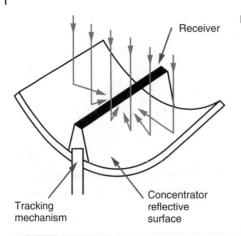

**Figure 5.12** Parabolic trough collector.

The collector can be oriented in the east–west direction, and tracking is from north to south. If the collector is oriented in the north–south direction, the tracking is in the east–west direction.

Parabolic trough technology is the most mature technology available for solar heating to about 400°C. The concentration ratio of a PTC is the ratio between the collector aperture area and the total area of the absorber tube. Usual values of the concentration ratio are about 20, although the maximum theoretical value is in the order of 70. High concentration ratios are associated with higher working temperatures.

A typical solar field with PTCs is composed of a number of parallel rows of collectors, with several collectors connected in series within every row. The number of PTCs connected in series within every row depends on the temperature required, while the number of rows connected in parallel depends on the required thermal output power of the system.

The selection of the working fluid for a solar field with PTCs is a very important. Thermal oil is commonly used in parabolic trough collectors for temperatures above 200°C, because if normal water is used, these operation temperatures would produce a high pressure inside the receiver tubes. This high pressure would require stronger joints and piping and thus increases the price of the collectors and complete solar field.

### 5.6.6 Cylindrical Trough Collector (CTC)

CTCs are similar to parabolic trough collector and are made by bending a sheet of a reflective material into a cylindrical shape. The receiver tube is covered with another glass tube (to reduce heat loss) and is placed at the focal line. When the collector is facing the Sun, some rays that hit the reflector surface are reflected on the receiver tube. The receiver tube is blackened at the outside surface to increase absorption. The reflected rays heat up the working fluid which is flowing in the receiver tube. For efficient performance, CTCs have to face the Sun at all time; hence, they are usually equipped with a single-axis tracking system. This means that at different times of the day, the reflector surface will have different orientations. Reflector orientation can have a significant impact on the structure and thermal efficiency of the collector.

### 5.6.7 Parabolic Dish Reflector

Large solar mirror collectors are a major subsystem of many solar energy systems, particularly for solar thermal generators. Thermal systems may use many collectors covering large sites. Parabolic dish concentrators offer the highest thermal and optical efficiencies of all the current concentrator options. A parabolic dish reflector, shown schematically in Fig. 5.13, uses parabolic-dish-shaped mirrors to focus the incoming solar radiation onto a receiver positioned at the focal point of the dish. These concentrators are mounted on active tracking systems to follow the Sun. It tracks the Sun in two axes. The dish structure must fully track the Sun to reflect the beam into the thermal receiver, and hence, the shape of the parabola needs to be precise. In order to focus the solar energy into the collector all the time, tracking mechanisms similar to the ones used in a single-tracking system are employed in double so that the collector is tracked in two axes. The receiver absorbs the radiant solar energy, and the fluid in the receiver is heated to high temperatures around 750°C.

This fluid is then used to generate electricity in a small Stirling or Brayton cycle engine attached to the receiver. Parabolic dish systems are the most efficient of all solar technologies and can have efficiency of about 25%. The heat transfer medium usually employed as the working fluid for an engine is hydrogen or helium. Alternate thermal receivers are heat pipes wherein the boiling and condensing of an intermediate fluid are used to transfer the heat to the engine. The heat engine system takes the heat from the thermal receiver and uses it to produce electricity, using an engine generator coupled directly to the receiver, or it can be transported through pipes to a central power-conversion system. Because the receivers are distributed throughout a collector field, as parabolic troughs, parabolic dishes are often called distributed receiver systems.

The main use of this type of concentrator is for parabolic dish engines. A parabolic dish engine system is an electric generator that uses sunlight instead of crude oil or coal to produce electricity. The main parts of this system are the solar dish concentrator and the power conversion unit.

**Figure 5.13** Parabolic dish.

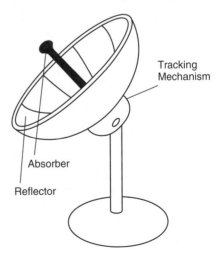

Tracking Mechanism

Absorber

Reflector

The advantages of this technology are as follows:

- The high fluid temperature obtained leads to high conversion efficiency of solar power to electricity.
- The Stirling engine system can be used either as a relatively small distributed power source, or by combining a lot of the units. MW levels of electricity can be generated.
- The Stirling engine system requires a very small quantity of water as the engine is air cooled. Condenser cooling normally required in steam power stations is not required.

### 5.6.8 Heliostat Field Collector (HFC)

Central receiver concentrating solar power plants offer significant performance advantages over line-focus systems. For extremely high inputs of radiant energy, a multiplicity of flat mirrors, or heliostats, using altazimuth mounts, can be used to reflect their incident direct solar radiation onto a common target as shown in Fig. 5.14. Central receiver consists of the following:

- Solar concentrator (heliostat field)
- Receiver
- Storage system
- Power generator

The reflecting element of a heliostat is typically a thin, back (second) surface, low-iron glass mirror. This heliostat is composed of several mirror module panels rather than a single large mirror. The thin glass mirrors are supported by a substrate backing to form a slightly concave mirror surface. Individual panels on the heliostat are also canted towards a point on the receiver. The heliostat focal length is approximately equal to the distance from the receiver to the farthest heliostat. Since the solar receiver is located in a fixed position, the heliostats must track the Sun. The mirrors are therefore mounted on individual frames that are tipped up and down and rotated east to west by small motors similar to those used in electric clocks. These motors are controlled by a computer which determines the position of each heliostat and directs the motors to position them so that its reflection hits the receiver throughout the year according to position of the Sun. Subsequent "tuning" of the closer mirrors is possible.

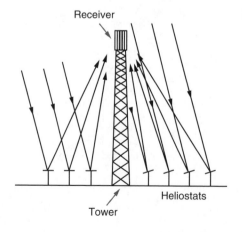

**Figure 5.14** Heliostat.

These are also known as central receiver systems. HFC has a large potential for mid-term cost reduction of electricity generation compared to parabolic trough technology [10]. The basic difference between the central receiver concept of collecting solar energy and the trough or dish collectors is that in this case, all of the solar energy collected in the entire field is transmitted optically to a small central collection region, instead of being piped around a field as hot fluid. Because of this characteristic, central receiver systems are characterized by large power levels (1–500 MW) and high temperatures 540–1000°C.

Plants with large-scale storage capability used molten nitrate as working fluids instead of steam and water in small plants. However, the high cost of the heliostat field remains a barrier to the widespread adoption of such plants.

## 5.7 Application of Solar Energy

Major applications of solar thermal energy in industrial and commercial sectors are as follows:

- Hot water usage for bathing and washing
- Pre-heated water up to 80°C to boilers
- Pasteurization, condensation and cleaning in milk dairies
- Drying and tanning in leather process industries
- Degreasing and phosphating in metal finishing industry
- Resin emulsification in the polymer industry
- Drying in food, wood, livestock and pharmaceutical industry
- Swimming pool water heating and so on
- Electric power generation
- Desalination of water

Solar energy can also be used in air conditioning and refrigeration.

### 5.7.1 Solar Water Heating

Solar energy can be used to heat liquids to low temperature of less than 80°C for domestic and service hot water [11, 12]. Most of the solar water heaters are used for domestic purpose only. When no pumping or blowing is involved, the solar system is known as passive. If the solar heat is collected in a fluid, usually water or air, which is then moved by pumps or fans for use, the solar system is said to be *active.*

The main part of a solar heating system is the collector, where solar radiation is absorbed, and energy is transferred to the fluid. Collectors used in water heating are either *flat-plate* or *evacuated tube collectors.*

### 5.7.2 Passive Systems with Thermosiphon Circulation

At the heart of a solar thermal system is the solar collector. It absorbs solar radiation, converts it into heat and transfers useful heat to the solar system. There are several different design concepts for collectors: besides simple absorbers used for swimming pool heating, more sophisticated systems have been developed for higher temperatures, such as integral storage collector systems, flat-plate collectors, evacuated flat-plate collectors and evacuated tube collectors.

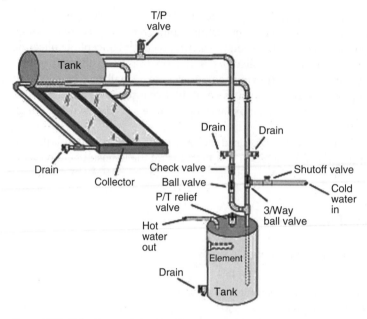

**Figure 5.15** Solar thermosiphon heating system.

Thermosiphon systems, shown schematically in Fig. 5.15, heat potable water or any other fluid and use natural convection to transport it from the collector to storage. For storing water overnight or on cloudy days, a storage tank is needed. The principle of the thermosiphon system is that cold water has a higher specific density than warm water and so, being heavier, will sink down. Since the driving force is only a small density difference, larger than normal pipe sizes must be used to minimize pipe friction. A thermosiphon system's storage tank must be positioned well above the collector; otherwise, the cycle can run backwards during the night, and all the water will cool down. Therefore, the collector is always mounted below the water storage tank, so that cold water from the tank reaches the collector via a descending water pipe. If the collector heats up the water, the water rises again and reaches the tank through an ascending water pipe at the upper end of the collector. The cycle of tank–water pipe–collector ensures the water is heated up until it achieves an equilibrium temperature. The consumer can then make use of the hot water from the top of the tank, with any water used being replaced by cold water at the bottom. The collector then heats up the cold water again. Due to higher temperature differences at higher solar irradiances, warm water rises faster than it does at lower irradiances. Therefore, the circulation of water adapts itself almost perfectly to the level of solar irradiance. Furthermore, the cycle does not work properly at very small height differences. In regions with high solar irradiation and flat roof architecture, storage tanks are usually installed on the roof. Thermosiphon systems operate very economically as domestic water heating systems, and the principle is simple, needing neither a pump nor a control. However, thermosiphon systems are usually not suitable for large systems, that is, those with more than $10\,\text{m}^2$ of collector surface. Furthermore, it is difficult to place the tank above the collector in buildings with sloping roofs, and single-circuit thermosiphon systems are only suitable for frost-free regions.

The main disadvantage of thermosiphon systems is that they are tall units, which makes them aesthetically unattractive. Usually, a cold water storage tank is installed on top of the solar collector, supplying both the hot water and the cold water needs of the house, thus making the collector unit taller. Additionally, extremely hard or acidic water can cause scale deposits that clog or corrode the absorber fluid passages. For direct systems, pressure-reducing valves are required when the city water is used.

### 5.7.3   Integrated Collector Storage Systems (Passive)

An integrated collector storage system combines collection and storage of thermal energy in a single unit. The basic configuration of these systems is that there is a tank or a set of interconnected tanks with their exposed surfaces as energy absorbing surfaces enclosed in an insulated box with a transparent cover on the top. In the morning, water is poured into these tanks which are heated by the Sun during daytime. The hot water moves to the top of tank. The hot water is drawn from the top of the tank, and the cold make-up water enters the bottom of the tank on the opposite side.

### 5.7.4   Active Solar Systems

Active solar systems are ideal for heating in hotels, hospitals, apartments, dormitories and other commercial applications. In contrast to thermosyphon systems, in active solar systems, an electrical pump can be used to move water through the solar cycle of a system by forced circulation. Collector and storage tank can then be installed independently, and no height difference between the tank and the collector is necessary.

There are two types of active solar heating systems:

1) Direct circulation (open-loop) systems
2) Indirect circulation (closed-loop) systems

#### 5.7.4.1   Direct Circulation Systems

Direct circulation systems are most suitable for homes in climates where the temperature rarely dips below the freezing point. Direct circulation systems use pumps to circulate potable water through the collectors. These systems are appropriate in areas that do not freeze for long periods and do not have hard or acidic water. These systems may require a recirculation freeze protection (circulating warm tank water during freeze conditions), and this in return requires electrical power for the protection to be effective. An open-loop system operates at atmospheric pressure. A pool uses this method as it is generally not operated in freezing months.

In direct circulation systems, the solar collector is separate from the storage cylinder and the potable water is directly heated in the solar collector. It consists of a solar collector, a circulation pump and a combined preheat tank and backup heating system. The water that will be used as domestic hot water is circulated directly into the collectors from the storage tank.

There are two types of direct systems – drain-down and re-circulating. In both systems, a controller will activate a pump when the temperature in the collectors is higher than the temperature in the storage tank. The drain-down system includes a valve that will allow the water in the collector loop to drain into a reservoir tank when the pumps stop when the outdoor temperature reaches 38°. Drain-back systems must be carefully installed to ensure that the piping always slopes downwards, so that the water will completely drain from the piping. When the temperature is higher than 38° and the collectors are hotter than the storage tank, the valve allows the system collectors to refill, and the heating operation resumes. The re-circulating system will pump heated water from the storage tank through the collectors when the temperature drops to 38 degrees.

These two systems have serious drawbacks. The cycling of air and water in a drain-down system collector as a result of periodically draining down (thereby emptying the collectors) can cause a buildup of mineral deposits in the collectors and reduce their efficiency. The re-circulating system circulates buildup from potable water heated from the storage tank through collectors during potential freeze conditions and effectively cools the water (wasting energy).

### 5.7.4.2 Indirect Circulation (Closed-Loop) Systems

It uses a heat transfer fluid (water or a diluted anti-freeze fluid) to collect heat and a heat exchanger to transfer the heat to the potable water indirectly. Heat exchangers transfer the heat from the heated fluid to the potable water (or other fluid). Some indirect systems have "overheat protection" by-pass which removes the heat that cannot be used. This protects the collector and the glycol fluid from becoming super-heated when the load is low and the solar intensity of incoming solar radiation is high. The heat transfer fluid is usually a glycol–water mixture with the glycol concentration depending on the expected minimum temperature. The glycol is usually food-grade propylene glycol because it is non-toxic. This system is typically used in hot water heating and radiator or in floor home/commercial heating. Because the system is closed to the atmosphere, pressure can build up as the temperature rises. Such systems will incorporate pressure relief valves and solar expansion tanks for safety purposes.

An indirect system that exhibits effectiveness, reliability and low maintenance is the drain-back system (see Fig. 5.16). The drain-back system typically uses distilled water as the collector circulating fluid. The collectors in this system will only have water in them when the pump is operating. This means that in case of power failure as well as each night, there will be no fluids in the collector that could possibly freeze or cool down and delay the start-up of the system when the Sun is shining. The fluids that are circulated into the collectors are separated from the heated water that will be used in the home by a double-walled heat exchanger. A heat exchanger is used to transfer the heat from the fluids circulating through the collectors to the water used in the home. The heat exchangers should be double-walled to prevent contamination of the household water. The heat exchanger may be separate from the storage tank or built into it. The controller in these systems will activate the pumps to the collectors and heat exchanger when design temperature differences are reached.

This system is very reliable and widely used. The main requirement is that the collectors must be mounted higher than the drain-back tank/heat exchanger. This may be impossible to do in a situation where the collectors are to be placed on the ground. An indirect or direct system can be used for heating swimming pools and spas.

### 5.7.5 Air Heating Systems

Most solar air heaters work on the same principle as a greenhouse wherein sunlight is converted to heat within a glass or plastic-covered enclosure. The heat is then trapped in the enclosure by the glass or plastic. Solar heating is an economical way for persons to augment the heating of homes and other buildings. Solar air heating systems absorb

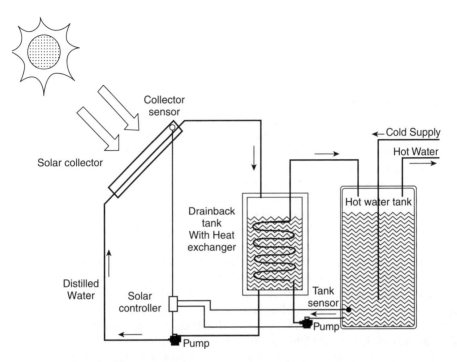

**Figure 5.16** Drain-back hot water system.

thermal energy from direct sunlight to heat air; this heated air can then be circulated through buildings to provide heat. Heating domestic water often constitutes a large part of the household energy bill, and designing a solar air collector to help offset this expense is a viable option. Air heating systems built for both space and domestic water heating are useful year-round and not just during the winter space-heating season.

Systems that use air to heat water need an air-to-water heat exchanger. These are indirect water heating systems that circulate air via ductwork through the collectors to an air-to-liquid heat exchanger. In the heat exchanger, heat is transferred to the potable water, which is also circulated through the heat exchanger and returned to the storage tank. Normally, double storage tank type system is used, because air systems are mostly used for domestic hot water heating.

Air from inside the house is drawn by a fan into a series of channels in a space behind the absorber where it is heated by the hot absorber plate. The heated air is then utilized for process in the home directly. Excess heat is stored in a storage medium (such as rocks) so the heat will be available during the night when solar heating is not available.

A simple controller is used to turn on the fan(s) in this type of system. The controller uses sensors in the collector to activate the system when it is hotter in the collector than in the house interior or storage medium.

Air collectors can be mounted vertically on the south wall of a building if used for space heating only. In that location, properly designed overhang will prevent them from heating up in the summer.

Air collectors are more practical in climates with longer and colder winters. The investment in storage systems for air collectors is substantial. The use of air collectors to add heat into the house directly can be readily achieved with properly oriented windows in the area.

## 5.8   Solar Thermal Power Generation

Solar energy is being used worldwide in solar thermal power plants with a total capacity of about 2.5 GW currently in operation; about 1.5 MW is currently under construction. These plants are particularly suitable for generating electricity in regions with high direct irradiation. Concentrating solar power plants concentrate solar rays to heat a fluid, which then directly or indirectly runs a turbine and an electricity generator. Concentrating the Sun's rays allows for the fluid to reach working temperatures high enough to ensure fair efficiency in turning the heat into electricity, while limiting heat losses in the receiver. This can supplement or completely replace fossil-fuel-operated power plant. The collectors are required to track the Sun in order to achieve a sufficient concentration. The sunlight is concentrated by mirrors that bundle the light onto a heat exchanger, which transfers the absorbed energy to a heat transfer fluid.

At the moment, the collected thermal energy is predominantly transformed into electricity in steam power plants. These are suitable for capacity sizes from 10 MW upwards and for temperatures of up to about 600°C and can be coupled with parabolic trough, linear Fresnel and solar tower systems. Two common designs of CSP plants – parabolic troughs and power towers – concentrate sunlight onto a heat transfer fluid (HTF), which is used to drive a steam turbine. Stirling engines are suitable for smaller capacities up to several tens of kilowatts, which are typical for dish concentrators. Gas turbines can be used for larger capacities and higher temperatures.

In order to use solar energy for reliable power generation, solution to the problem for reduced or curtailed energy production when the Sun sets or is blocked by clouds must be found [13]. Thermal energy storage (TES) can provide a workable solution to this challenge. Thermal energy storage can be used to provide electric supply generation in the night and to provide backup energy during periods with reduced sunlight caused by cloud cover. Since 2006, plants have been built with thermal storage systems into CSP plants, almost exclusively using sensible heat storage in a mixture of molten salts. The concept of thermal storage is simple: throughout the day, excess heat is diverted to a storage material (e.g. molten salts). When production is required after sunset, the stored heat is released into the steam cycle and the plant continues to produce electricity.

Adding TES provides several advantages to a CSP plant. First, unlike a plant without storage that must sell electricity when solar energy is available, a CSP plant with TES can shift electricity production to periods of highest prices. Second, plant with TES may completely replace a conventional power plant as opposed to just supplementing its output. Finally, the dispatch ability of a CSP plant with TES can provide high-value ancillary services such as spinning reserves.

## 5.9   Desalination of Water

Desalination is the process of removing salt and other minerals from saline water mainly obtained from the sea. The desalination recovery ratio is the ratio of the desalinated water volume to the seawater volume. The energy cost is 0.86 kWh/m$^3$ for conversion of seawater with saline content of 34,500 ppm at a temperature of 250°C.

Desalination processes have been used for many decades, but their high-energy requirement and, therefore, their prohibitive costs have prevented their widespread

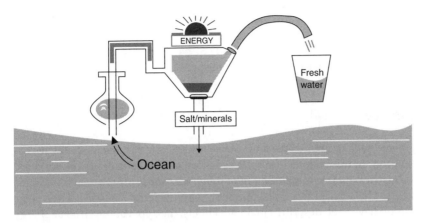

**Figure 5.17** Desalination of water.

adoption across countries, However, because of the necessity, many of the countries in the Middle East make extensive use of desalination for freshwater. For example, the Kingdom of Saudi Arabia and the Gulf States are currently almost completely dependent on desalination for much of their water needs, and this incurs considerable use of non-renewable energy.

Thermal desalination involves distillation processes where saline feed water is heated to vaporize, causing freshwater to evaporate and leave behind a highly saline solution, namely the brine. Freshwater is then obtained from vapour cooling and condensation. The Multi-stage flash (MSF) process is divided into sections or stages. Saline water is heated at the boiling temperature between 90 and 110°C, with a decreasing pressure through the stages. Part of the water quickly vaporizes at each stage while the rest continues to flow through the following stages (Fig. 5.17).

Concentrated solar–thermal desalination plants are solar power plants that make use of solar radiation primarily in the infrared (IR) range to power the desalination of saltwater to freshwater. The most modern solar–thermal desalination systems generally produce concentrated heat energy, which is used to create pressurized steam, which is used to power reverse osmosis desalination systems.

## 5.10   Steam Pressurization Systems Using Heat Energy

The technology for producing pressurization from heat energy is very well established due to its use many prior industrial applications. For example, in a steam engine, the pressure vessel of the steam engine boiler is heated from externally applied heat energy, and as a consequence, the steam within the pressure vessel is pressurized (in the case of a steam engine, this pressurized steam is subsequently released to generate mechanical energy). As another example, for solar–thermal–electrical generators, the pressure vessel is heated from heat energy obtained from a solar concentrator. Both of these example steam pressurization systems also include a cooling cycle to cool and return the steam. In contrast, for solar desalination applications of interest here, the pressurization to drive desalination can make use of a pressure vessel that is heated using heat energy obtained from a solar concentrator; the pressurized steam is then released to drive the (reverse osmosis) desalination process. Again, the system also needs to include a cooling cycle to cool and return the steam. As noted earlier, saltwater reverse osmosis desalination requires only moderate pressure of approximately 55 bar, and such use of conventional heated pressure vessels can be used to achieve this pressure.

## 5.11  Summary

Solar energy is available in abundance and can be utilized for various purposes resulting in less pollution. In this chapter, the solar radiation and its spectral distribution are described. The instruments for measurement of solar radiation are important for designing solar systems. The solar power collectors with various designs as used in the solar energy application have been described in detail. Finally, the applications of solar energy in heating and cooling are discussed.

## References

1 Boyle, G. (ed.) (2012) *Renewable Energy - power for a sustainable future*, Third edn, Oxford university Press.

2 Duffie, J.A. and Beckman, W.A. (1991) *Solar Energy of Thermal Process*, second edn, Wiley & Sons, New York.

3 IEA (2004) *World Energy Outlook 2004*, International Energy Agency, Paris, IEA/OECD.

4 Perez, R. *et al.* (1990) Modeling Daylight Availability and Irradiance Components from Direct and Global Irradiance. *Solar Energy*, **44** (5), 271–289.

5 Paulescu, M. *et al.* (2013) *Weather Modelling and Forecasting of PV Systems Operation*, Chapter 2, Springer.

6 Mehleri, E.D. *et al.* (2010) Determination of the optimal tilt angle and orientation for solar photovoltaic arrays. *Renewable Energy*, **35** (11), 2468–2475.

7 Mills, D. (2004) Advances in solar thermal electricity technology. *Solar Energy*, **76**, 19–31.

8 Soteris, K.A. (2004) Solar thermal collectors and applications. *Progress in Energy and Combustion Science*, **30**, 231–295.

9 Mills, D.R. and Morrison, G.L. (1999) Compact linear Fresnel reflector solar thermal power plants. *Solar Energy*, **68**, 263–283.

10 Schell, S. (2011) Design and evaluation of Esolar's heliostat field. *Solar Energy*, **85**, 614–619.

11 Norton, B. (1992) *Solar Energy Thermal Technology*, Springer, London.

12 Kreider, J.F. (1982) *The Solar Heating Design Process*, McGraw-Hill, New York.

13 Madaeni, S.H., Siosanshi, R. and Denholm, P. (2012) Estimating the capacity value of concentrating solar power plants with thermal energy storage: a case study of the southwestern United States. *IEEE Transaction on Power System*, **27** (2), 1116–1124.

# 6

# Photovoltaic Systems

## 6.1 PV Solar Cells and Solar Module

In photovoltaic energy systems, the solar energy (mainly sunlight) is directly converted to electric energy using photovoltaic solar cell modules. They have no moving parts, therefore require little maintenance. The basic principle of photovoltaic effect was discovered by Becquerel in 1839 using selenium. He discovered photovoltaic effect at a junction between a semiconductor (selenium) and an electrolyte. But it was not considered for power generation until 1954 when silicon semiconductors became available, and PV solar cells based on semiconductor technology were developed. Selenium and silicon are both semiconductor materials, but it was only when the three researchers, Gerald Pearson, Daryl Chapin and Calvin Fuller, at Bell Laboratories discovered a silicon solar cell, which was the first material to directly convert enough sunlight into electricity to run electrical devices. The efficiency of the silicon solar cell, which was produced at Bell Laboratories, was only 4%.

In most of today's solar cells, the absorption of photons, which results in the generation of the charge carriers, and the subsequent separation of the photo-generated charge carriers take place in semiconductor materials. A semiconductor is a material that has an electrical property which lies in between that of a conductor and an insulator. The semiconductor materials commonly used now are silicon and germanium. However, the availability of silicon in abundance has made it more commonly used semiconductor material. Solar cells using silicon were first produced in 1954. They found their immediate application in space activities. Their first use was to power a small radio transmitter in the US space satellite Vanguard I. Now almost in all spacecraft, the power source is PV cells.

Photovoltaic power is one of the fastest growing renewable energy technologies. Annual production of PV cells grew tenfold from about 50 MW in 1990 to more than 500 MW by 2003. Rapid progress in increasing the efficiency and reducing the cost of PV cells has been made over the past few decades [1]. Their uses are now widespread in domestic, commercial and industrial buildings.

A brief description of semiconductors and their properties is given here to understand the basics of energy conversion.

*Operation and Control of Renewable Energy Systems*, First Edition. Mukhtar Ahmad.
© 2018 John Wiley & Sons Ltd. Published 2018 by John Wiley & Sons Ltd.

## 6.1.1   Semiconductor Technology

In semiconductors, there is an energy gap between the valence band and the conduction band but the energy gap is not very large. These materials therefore have properties between those of a conductor and an insulator. Due to a number of favourable characteristics, silicon is the solar industry's most common semiconductor material. However, in its pure crystalline form, silicon is a poor conductor of electricity as it has equal number of free electrons and holes. In order to make the silicon material suitable for an electronic or solar application, impurity atoms are added to crystalline silicon. This process is known as doping. The doped semiconductor is called *extrinsic semiconductor*. The impurities that are added are boron, gallium, aluminium or phosphorous, arsenic or antimony. Silicon is a group IV element on the periodic chart and has four electrons in its outermost shell. When silicon is doped with group V element such as arsenic (As), or phosphorous, its atoms replace silicon atoms at a small number of points on the crystal lattice. Since their atom has five electrons in its outer shell, an *extra* electron is added to the crystal. This extra electron (often called a *donor* electron) is easily excited into the conduction band as a freely moving current carrier. Silicon can also be doped with an element from group III of the periodic table, such as gallium (Ga), boron or aluminium. In case silicon is doped with gallium or aluminium, the impurity has only three electrons in its outermost shell, so there is a deficiency of one electron at every point where a gallium/aluminium atom replaces a silicon atom. The gallium often takes an electron from a neighbouring silicon atom leaving a "hole" or empty state in the valence band of the silicon. This "hole" is free to roam around in the valence band and effectively acts as a positive charge carrier. The electrons move forward to fill up the spaces (holes) in front of them and create another space behind them. The holes move in the direction opposite of electrons; hence, they behave as a positive charge. Thus, doping with group III elements produces a p-type semiconductor because the effective charge carriers (holes) are positive. On the other hand, doping with group V elements results in an n-type semiconductor since the charge carriers (electrons) are negative.

A PV cell consists primarily of two layers of semiconductor material, which create a p–n junction. In a p–n junction, one layer of negatively doped material is connected physically to another layer of positively doped material. The two layers are connected together by a flat surface, allowing as much surface area contact as possible. When a junction between p- and n-type materials is formed, the carriers (free electrons and holes) diffuse from a higher concentration side to the lower concentration side. For generating electricity, the negative side, or the side having excess electrons, is faced towards the source of light while the positive side is faced away. When light strikes the surface of the negative side of the PV cell, the photons begin striking the electron shells of the PV-cell material. The semiconductor material absorbs photons with energies equal to or greater than its band gap. Due to photon absorption, an electron of the valence band is promoted to the conduction band and is free to move through, becoming a free-roaming electron. The positive layer of the material being so close in proximity combined with the abundance of electrons in the negative layer forces the electron towards the positive side of the PV cell. However, once on the positive side, the electron is unable to return to its original parent atom and, as a result, is stuck on the positive side of the cell. In order for this free charge to be captured for current generation, decay to the lower energy state, that is, recombination with the hole in the valence band, has to be prevented through charge separation. Thus, an electron–hole pair is produced. The energy available in the photon is given by

$$E = h\nu = \frac{hc}{\lambda} \tag{6.1}$$

where $h$ is Planck's constant $= 6.63 \times 10^{-34}$ J s, $c$ is the speed of light, $\lambda$ is the wavelength of photons in metres. For energies of photons less than the band gap energy, no absorption takes place. For photons having energy much more than the band gap energy, still only one electron–hole pair is generated and the remaining energy is wasted as heat. One of the most critical requirements for a single-junction cell is that the band gap energy must be optimized to transfer maximum energy from the incident light to the photo-generated electron–hole pairs.

Because the efficiency depends on the product of open-circuit voltage ($V_{OC}$) and short-circuit current ($I_{SC}$), there is an optimum energy band gap for producing maximum efficiency of PV cells. It has been found that the optimum energy band gap of 1.5 eV is ideal for solar power generation. In practice, semiconductors with energy band gaps in the range of 1.0–1.7 eV are used.

The band gap energy of silicon (1.12 eV) is almost ideal and allows absorption of photons in the near-infrared (NIR), visible and ultraviolet spectrum [2, 3]. Silicon has an indirect energy band gap of approximately 1.1 eV at room temperature. The indirect energy band gap results in a low optical absorption coefficient ($\alpha \leq 104$ cm$^{-1}$) for photons with energies below 3.4 eV, and this means that the silicon must be 100–300 μm thick for achieving efficient light absorption. The electron–hole pairs generated by the light should be able to diffuse up to that distance to reach the electric field in the depletion region of the junction if they are to contribute to the photocurrent. (Increasing w to more than a few microns by lowering the doping concentrations is at the expense of increasing the device series resistance to the point where it severely degrades the device efficiency.)

Today's most successful materials for thin-film photovoltaics are α-Si, where the optical absorption is increased by impurity scattering. Cadmium telluride (CdTe), with a band gap of 1.48 eV, and copper indium gallium selenide (CIGS), whose band can be tuned around the nominal value of 1.04 eV by controlling its composition and which has the highest absorption constant of 3-6 $\times 10^5$ cm$^{-1}$ reported for any semiconductor, are now favourite materials.

## 6.2 Solar Cell Characteristics

An ideal solar cell can be modelled as a current source in parallel with a diode. When there is no light present to generate any current, the PV cell behaves as a diode. As the intensity of incident light increases, current is generated by the PV cell, as illustrated in Fig. 6.1. The diode current (when the junction is not illuminated) is given by

$$I_d = I_0 \left\{ \exp\left(\frac{V}{V_T}\right) - 1 \right\} \tag{6.2}$$

where $I_0$ is the reverse saturation current; $V_T$ is $1.38 \times 10^{-23}$ T, where $T$ is the cell temperature in Kelvin; and $V$ is the measured cell voltage.

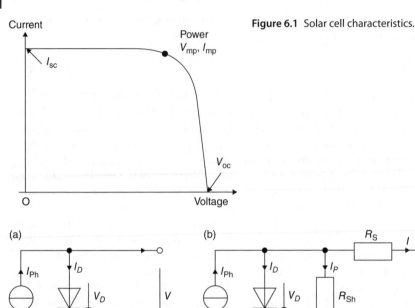

**Figure 6.1** Solar cell characteristics.

**Figure 6.2** Equivalent circuit of solar cell under light; (a) ideal cell, (b) real cell.

When the p–n junction is illuminated, the cell current is given by, the total current $I$ equal to the current $I_{Ph}$ generated by the photoelectric effect minus the diode current $I_d$, according to Eq. (6.2):

$$I = I_{Ph} - I_d = I_{Ph} - I_0 \left\{ \exp\left(\frac{V}{V_T}\right) - 1 \right\} \tag{6.3}$$

As shown in Fig. 6.2, the short-circuit current $I_{SC}$ corresponds to the short-circuit condition when the impedance is low and is calculated when the voltage equals 0.

$$I(\text{at } V = 0) = I_{SC} \tag{6.4}$$

The open-circuit voltage ($V_{OC}$) occurs when there is no current passing through the cell.

$$V(\text{at } I = 0) = V_{OC} \tag{6.5}$$

$V_{OC}$ is also the maximum voltage difference across the cell for a forward-bias power quadrant.

$V_{OC} = V_{max}$ for forward-bias power quadrant.

$I_{SC}$ occurs at the beginning of the forward-bias sweep and is the maximum current value in the power quadrant. For an ideal cell, this maximum current value is the total current produced in the solar cell by photon excitation.

$I_{SC} = I_{max} = I_{Ph}$ for forward-bias power quadrant.

The power produced by the cell in Watts can be easily calculated by the equation $P = IV$. At the $I_{SC}$ and $V_{OC}$ points, the power will be zero and the maximum value for power will occur between the two. The voltage and current at this maximum power point (MPP) are denoted as $V_{MP}$ and $I_{MP}$, respectively.

## 6.2.1 Equivalent Circuit

The solar cell can be seen as a current generator which generates the current $I_l$. The dark current $I_d$ flows in the opposite direction and is caused by a potential between the +ve and −ve terminals. In addition, there are two resistances: one in series $(R_s)$ and the other in parallel $(R_{sh})$. The series resistance is caused by the fact that a solar cell is not a perfect conductor. The parallel resistance is caused by leakage of current from one terminal to the other due to poor insulation. In an ideal solar cell, $R_s = 0$ and $R_{sh} = \infty$. The equivalent circuit of an ideal cell and practical cell (including all resistances) is shown in Fig. 6.2.

The equation for load current of the practical cell from the equivalent circuit is

$$I_l = I_0 \left[ \exp \left\{ \frac{V + IR_s}{V_T} \right\} - 1 \right] - (V + IR_s)/R_{sh} \tag{6.6}$$

A shunt resistance $R_{sh}$ of a few hundred ohms does not reduce the output power of the solar cell appreciably. In reality, $R_{sh}$ is much larger than a few hundred ohms and can in most cases be neglected. The series resistance, however, can drastically reduce the output power. For example, a series resistance of only $5\,\Omega$ can reduce the output power by 30%.

## 6.2.2 Solar PV Module

Since the total power available from single solar cell is very small, it cannot be utilized for any useful work. Individual solar cells are usually manufactured and combined into modules that consist of 36 cells (12 V) to 72 cells (24 V) depending on the output voltage and current of the module. When a module is connected to a system, an important consideration is the behaviour of the module when it is not illuminated. During night time, when none of the cells is generating photocurrent, the module may be considered as a series connection of diodes that may be forward biased by the storage batteries. This will result in leakage current through the diodes and discharging of batteries.

Another consideration in series connection of solar cells relates to the shading of individual cells. If any cell of the module is shaded, it will not produce any current. Since the cells are connected in series, it may become forward biased if other unshaded modules are connected in parallel. This will result in heating of the shadowed cell and may damage the module.

When cells are mounted in modules, the unit is fixed on a durable back cover with a transparent cover on the top and hermetically sealed to make it suitable for outdoor applications. The modules vary in size according to the manufacturers but are typically between 0.5 and 1 m² and generate around 100 W/m² of energy under peak solar conditions for a 10% efficient module.

## 6.2.3 Series and Parallel Connections of Cells

In order to obtain higher current and voltage output, cells are connected together in a module. To increase the voltage of the module, cells are connected together in series, and to increase the current, cells are connected in parallel. How many cells are to be connected in series or in parallel is determined by the desired string or module output. When cells are connected in a series, the voltage of the entire string is the sum of the individual cells' voltages as shown in Eq. (6.7) for two cells. The total current of the string is equal to the current of one cell in the string, as shown in Eq. (6.8).

$$V_o = V_{OC1} + V_{OC2} = 2V_{OC} \tag{6.7}$$

or

$$I_{SC1} = I_{SC2} = I \tag{6.8}$$

Alternatively, when cells are connected in parallel, the currents add and the voltages are equal as seen in Eq. 6.9 and 6.10. The *I-V* curve of two cells connected in parallel is shown in Fig. 6.3. The current of the combined curve is equal to the sum of the two currents at a specific voltage.

$$V_o = V_{OC1} = V_{OC2} \tag{6.9}$$
$$I_{SC1} + I_{SC2} = I \tag{6.10}$$

Unfortunately, it is very difficult to get two identical cells in reality. In addition, cell mismatch occurs when a cell in a series connected string is shaded or damaged preventing it from generating current equal to the other cells in the string. This mismatch results in degradation to the module and a lower power output.

When two cells with mismatched characteristics are connected in series, the composite characteristics can be obtained by using the individual output voltage of the cell corresponding to common current, for all operating points as shown in Fig. 6.4. Cell mismatch has a substantial effect on the *I-V* curve of the whole module. A cell with a current output lower than that of the rest of the cells in the string is referred to as a "weak cell". When cells are connected in series, the total current is limited by the weakest cell while the voltages add as normal. The combined output of the cells is dependent on the reverse voltage behaviour of the weak cell. The effects of shading one cell in a series connection can be seen in Fig. 6.5. One cell in a series string of three cells is shaded 50%. This causes the current produced by this cell to drop by about 50%, and the current of the resultant curve drops similarly. The voltages of the three cells add while the current is limited by the weak shaded cell.

The short circuit is the worst-case condition for mismatched cells since it results in the reverse biasing of the weak cells. In open-circuit conditions, the cells are forward biased

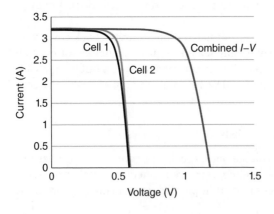

**Figure 6.3** *I-V* curve of two cells connected in parallel.

**Figure 6.4** Series-connected cell.

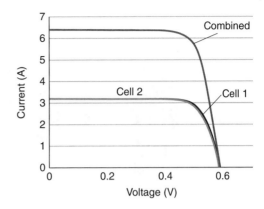

**Figure 6.5** The effect of mismatched series string of three cells with one cell shaded.

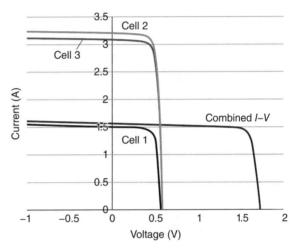

by their photo-generated currents. The shading of cell 2 results in it having a lower $V_{OC}$ which ultimately lowers the $V_{OC}$ of the combined cells. However, over a large number of cells, typically 36 in a module, this drop in Voc is not significant. A mismatched cell lowers the performance of the entire module by limiting the current output and slightly decreasing the $V_{OC}$ of the module.

### 6.2.4 Solar PV Panel

In a solar PV panel, several solar modules are connected in series/parallel as shown in Fig. 6.6. The panel contains a group of modules that can be wired and packaged off-site. When modules are connected in series, it is desirable to have each module produce maximum power at the same current. For parallelly connected modules, it is desirable to have each module produce maximum power at the same voltage.

In parallel connection, blocking diodes are connected in series with each series string of modules as shown in Fig. 6.7. In a failed string, the voltage is lower than the normal string, and when the two are connected in parallel, the current will flow from the normal string to the failed string. The blocking diode will not allow the power absorption by the failed string from other strings. In addition, bypass diodes are connected across each

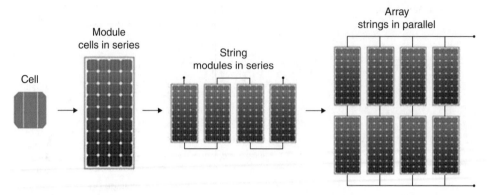

**Figure 6.6** Solar PV panel connected in series and parallel.

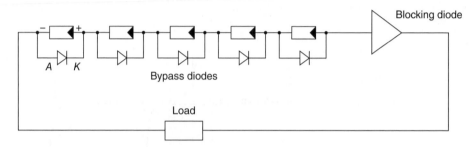

**Figure 6.7** Solar PV panel with blocking diodes.

module, so that if one module fails, the output of the remaining modules in a string will bypass the failed module.

### 6.2.5 PV Array

Because a single solar PV panel can only produce a limited amount of power, several panels are connected to form an array. A PV array is an interconnected system of PV modules that function as a single electricity-producing unit. In smaller systems, an array may consist of a single panel or for large systems may contain many panels connected together. The panels should be so installed that the shade of one panel must not fall on other panel throughout the year.

#### 6.2.5.1 Design of PV System
The main aspect of interconnecting:

Suppose that the PV array comprises 16 modules of 160 W peak. The array peak power is $16 \times 160 = 2.56$ kW.

Taking the derating of array as 0.77, and daily sunshine hours of 6, the dc energy output of the array is $2.56 \times 0.77 \times 6 = 11.82$ kWh.

If the power loss in dc cable is 3%, the total dc power output $= 0.97 \times 11.82 = 11.47$ kWh, then the inverter power rating may be 11.5 kW.

Matching array voltage to the maximum and minimum inverter operating voltages:

Many of the inverters available will have a voltage operating window. When the temperature is at maximum, then the maximum power point (MPP) voltage of the array should never fall below the minimum operating value of the inverter.

If the ambient temperature is 35°C, the maximum temperature of the solar cell is 70°C. Now if the module used has rated MPP voltage of 35.4 V and voltage coefficient of –0.170 V/C, then the voltage at maximum temperature will be $= 35.4 - (35 \times 0.17) = 29.05$ V. If voltage drop in dc cable is 3%, then the input voltage to the inverter at maximum temperature is $0.97 \times 29.05 = 28.17$ V.

If the minimum voltage window of the inverter is 140 V, taking the factor of safety as 1.1, the minimum inverter voltage should be 154 V.

The minimum number of modules in a string can be $154/28.17 == 5.46$, or 6 modules may be used. As the voltage of the cell increases, if the temperature falls from 25°C, the maximum voltage of the inverter window should be obtained for the place on the basis of the minimum temperature. If the effective temperature of the cell is taken as 25°C, then no voltage increase is required. The maximum number of modules for a 400 V inverter will be $400/29.05 = 13.76$, or it can be 14.

Total number of modules required is 16 which may be connected in two strings of 8 modules.

## 6.3 Maximizing Power Output of PV Array

The output powers of photovoltaic (PV) panels have non-linear curves and depend on solar irradiance, degradation level and temperature. Their MPP changes non-linearly with environmental conditions and load profile. Since the photovoltaic electricity is expensive compared to the electricity from conventional sources, photovoltaic systems should be designed to operate at their maximum output power in any environmental conditions [4].

Two methods are generally used to increase the solar panel output power in photovoltaic systems. In the first one, the system will increase the incident solar irradiance on the solar panel. The system requires a sun tracker to track the Sun's position for increasing the received solar irradiance by the panels for different positions of the Sun. In that case, the environmental conditions define the quality of the generated power for each load. This method is used for medium- to high-scale photovoltaic systems and has a high cost. In the second method, solar panels are fixed and load profile is changed to maximize the power output under changing conditions of insolation and temperature. The maximum power is transferred to the load when the impedance of the source matches the load resistance. Therefore, in order to transfer maximum power from the panel to the load, the internal resistance of the panel has to match the load resistance as seen by the PV panel. For a fixed load, the equivalent resistance seen by the panel can be adjusted by changing the power converter duty cycle. To accomplish this objective, a switching converter is placed between the PV source and the load.

Tracking the MPP of a photovoltaic (PV) array is usually an essential part of a PV system. As such, many MPP tracking (MPPT) methods have been developed and implemented. With an MPPT control, it is possible to reach the output panel characteristics around the optimal power point. This method is utilized in small-scale photovoltaic applications. Combination of sun tracking or load matching and MPPT approaches can raise the energy output of a solar panel in photovoltaic systems.

### 6.3.1 Solar Tracking

Employing a tracking system with solar panel is one of the methods to increase the overall efficiency of the solar panel. Solar trackers are used to orient photovoltaic panels, reflectors, lenses or other optical devices towards the Sun. Since the Sun's position in the sky changes with the seasons and the time of day, trackers are used to align the collection system to maximize the power output. Solar tracking systems are of several types and can be classified according to several criteria. They can be classified depending on the number of rotation axes. Thus, solar tracking systems are with a single rotation axis or with two rotation axes. Since solar tracking involves moving parts which require a control, these are expensive. A single-axis tracking system is therefore the best solution for small PV power plants. Single-axis trackers will usually have a manual elevation (axis tilt) adjustment on the second axis depending on the different months in a year.

### 6.3.2 Design of Simple Automatic Solar Tracker

A simple cost-effective solar tracker can be constructed with the help of the following components.

- Photo transistor
- Voltage comparators
- dc motor

To track the Sun, the system should work in closed-loop form, and the controller needs to sense the light through a light sensor. Cadmium sulfide (CdS) photoresistor is used for tracking the Sun. The CdS photoresistor is a passive element that has a resistance inversely proportional to the amount of light (LDR) incident on it. To utilize the photoresistor, it is placed in series with another resistor. A voltage divider is thus formed at the junction between the photoresistor and another resistor shown in Fig. 6.8. The heart of the aforementioned circuit are two voltage comparators made using LM358 Dual Op-Amp. It is well known that when the intensity of light falling on an LDR increases, its resistance decreases. Here LDR is connected with a series resistor (R3 and R4); hence,

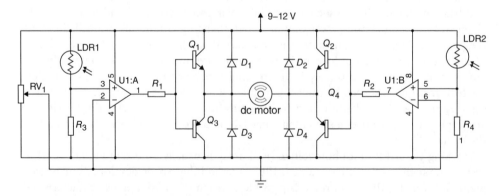

**Figure 6.8** Simple automatic solar detector.

**Table 6.1** Rotation based on light.

| Input A | Input B | Output C |
|---------|---------|----------|
| 0 | 0 | STOP |
| 0 | 1 | CLOCKWISE |
| 1 | 0 | ANTI CLOCKWISE |
| 1 | 1 | STOP |

when the intensity of light falling on an LDR increases, voltage across the corresponding resistor (R3 or R4) increases.

The output of the voltage comparator will be high when the voltage at the non-inverting terminal (+) is higher than the voltage at the inverting terminal (−). Inverting (−) terminals of both comparators are shorted and connected to a variable resistor (RV1), which is used to set the reference voltage. Thus, the sensitivity of both LDRs can be adjusted by varying the 10 K pot shown on the left side of the circuit diagram. When the light falls on an LDR increases, voltage at the non-inverting (+) terminal of the corresponding comparator increases and its output goes HIGH, as shown in Table 6.1.

The direction of motor rotation is controlled by the H-bridge formed by the complimentary symmetry transistors BC547 and BC557. Consider the case when the output of the first comparator (U1:A) is high and the output of the second comparator (U1:B) is low. In this case, transistors Q1 and Q4 will turn on and the resulting current rotates the motor in clockwise direction. When the output of the first comparator is low and the output of the second comparator is high, transistors Q2 and Q3 will turns on and the resultant current rotates the motor in anticlockwise direction. If the outputs of both comparators are low, transistors Q3 and Q4 turn on, but no current will flow through the motor. Similarly, if the outputs of both comparators are high, transistors Q1 and Q2 turn on, but no current will flow through the motor. The dc motor should be connected to the panel in such a way that the rotation of the motor rotates the panel in the direction of movement of the Sun.

### 6.3.3 Load Matching for Optimal Operation

In order to obtain maximum power from an electrical circuit, the source impedance must be matched to the load impedance. A dc/dc converter acts as an interface between the load and the module.

By changing the duty cycle, the load impedance as seen by the source is varied and matched at the point of the peak power with the source so as to transfer the maximum power. MPPT techniques are used to always maintain a maximum power output from the module.

## 6.4 Maximum Power Point Tracking Algorithm

As discussed earlier, the power output of a PV system depends on irradiance and cell temperature. Therefore, the operating current and voltage which maximize the power

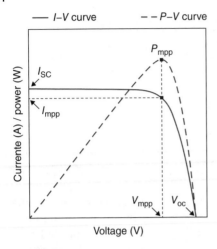

— *I–V* curve    – – *P–V* curve    **Figure 6.9** PV and IV curves.

output will change with environmental conditions. In order to maintain efficient operation despite environmental variations, one approach is to use a MPPT algorithm to dynamically tune either control current or voltage to the maximum power operating point. Tracking the MPP of a photovoltaic array is an essential part of a PV system. As such, many MPPT methods have been developed and implemented. The methods vary in complexity, sensors required, cost, range of effectiveness, implementation hardware, popularity and so on. [5].

The power curve of a PV array is shown in Fig. 6.9. The MPPT techniques are developed to automatically find the voltage or current at MPP at which the PV array should operate to obtain the maximum power output under a given temperature and irradiance. In order to continuously obtain maximum power from the solar panels, it is necessary to control the optimal impedance between the storage device or load and PV cell. MPPT is an adaptation of the dc-dc switching technique used in voltage regulators. Most MPPT techniques respond to changes in both irradiance and temperature, but some are specifically more useful if the temperature is approximately constant. Here few common MPPT techniques are described.

### 6.4.1 Constant-Voltage Method

The Constant-Voltage (CV) algorithm is the simplest MPPT control method. The operating point of the PV array is kept near the MPP by regulating the array voltage and matching it to a fixed reference voltage equal to the $V_{MPP}$ of the PV panel. This method assumes that individual insulation and temperature variations on the array are insignificant and that the constant reference voltage is an adequate approximation of the true MPP. The CV method does not require any input. However, measurement of the voltage of the panel is necessary in order to set up the duty cycle of the dc/dc converter. For low insolation conditions, the constant-voltage technique is more effective than either the perturb and observe or the incremental conductance algorithm.

### 6.4.2 Hill-Climbing/Perturb and Observe Techniques

Hill-climbing techniques are the most popular MPPT methods due to their ease of implementation and good performance when the irradiation is constant. The advantages of both methods are the simplicity and low computational power needed. The shortcomings are also well known: oscillations around the MPP, and they can get lost and track the MPP in the wrong direction during rapidly changing environmental conditions.

### 6.4.2.1 Perturb and Observe

The P&O algorithm is also called "hill climbing", but both names refer to the same algorithm depending on how it is implemented. The PV array is connected to a power converter that can vary the current of the PV array. Hill climbing involves a perturbation on the duty cycle of the power converter, and P&O involves a perturbation in the operating voltage of the dc link between the PV array and the power converter. In the case of the hill climbing, perturbing the duty cycle of the power converter implies modifying the voltage of the dc link between the PV array and the power converter, so both names refer to the same technique.

In this method, the sign of the last perturbation and the sign of the last increment in the power are used to decide what the next perturbation should be. From Fig. 6.9, it is clear that by incrementing the voltage on the left-hand side of MPP, the power will increase, and decrementing the voltage will decrease the power. On the right-hand side, the opposite will happen. Therefore, if after perturbation, the power increases, then this perturbation is to be kept, and if power decreases, the perturbation should be reversed. The process is repeated periodically until the MPP is reached. The algorithm is shown in Table 6.2.

The oscillation can be minimized by reducing the perturbation step size. However, a smaller perturbation size slows down the MPPT. A solution to minimize the oscillation is to have a variable perturbation size that gets smaller towards the MPP. Fuzzy logic control can also be applied to optimize the magnitude of the next perturbation.

Normally, two sensors are required to measure the PV array voltage and current from which power is computed. However, depending on the power converter topology, only a voltage sensor might be needed. In this case, the PV array current from the PV array voltage is estimated, eliminating the need for a current sensor. DSP or microcomputer control is more suitable for hill climbing and P&O even though discrete analog and digital circuitry can also be used.

### 6.4.3 Incremental Conductance (IC)

The disadvantage of the perturb and observe method to track the peak power under rapidly varying atmospheric condition is overcome by the incremental conductance method [6]. The incremental conductance algorithm is based on the fact that the slope

**Table 6.2** Perturb and Observe Algorithm.

| Perturbation | Change in power | Next perturbation |
| --- | --- | --- |
| Positive | Positive | Positive |
| Positive | Negative | Negative |
| Negative | Positive | Negative |
| Negative | Negative | Positive |

of the curve power versus the voltage (current) of the PV module is zero at the MPP, positive (negative) on the left of it and negative (positive) on the right, as can be seen in Fig. 6.9.

The IC can determine when the MPP has been reached and stop perturbing the operating point. If this condition is not met, the direction in which the MPPT operating point must be perturbed can be calculated using the relationship between $dI/dV$ and $-I/V$. At MPP, $\frac{dI}{dV} = -\frac{I}{V}$. This relationship is derived from the fact that $dP/dV$ is negative when the MPPT is to the right of the MPP and positive when it is to the left of the MPP. The MPPT regulates the PWM control signal of the dc-to-dc boost converter until the condition $(dI/dV) + (I/V) = 0$ is satisfied.

This algorithm has advantages over P&O in that it can determine when the MPPT has reached the MPP, whereas P&O algorithm oscillates around the MPP. In addition, incremental conductance can track rapidly increasing and decreasing irradiance conditions with higher accuracy compared to perturb and observe. One disadvantage of this algorithm is the increased complexity when compared to P&O.

## 6.5  Types of Solar Cells and Technologies

Solar cells are broadly classified into three types:

1) Crystalline
    i) Monocrystalline
    ii) Polycrystalline
2) Thin Film
    i) Amorphous silicon
    ii) Cadmium telluride (CdTe)
    iii) Copper indium gallium selenide (CIGS)
    iv) copper indium selenide (CIS)
3) Concentrating Photovoltaic

### 6.5.1  Crystalline Solar Cells

Crystalline silicon PV cells are the most common photovoltaic cells in use today. Bell Laboratories fabricated the first crystalline silicon solar cells in 1953, achieving 4.5% efficiency, followed in 1954 with devices with 6% efficiency. In the next 10 years, the efficiency of crystalline silicon cells was improved to around 15%. The improvements in research-cell efficiencies were achieved for various kinds of solar cells over the past 30 years. The majority of solar cells representing about 90% of the world's total production are manufactured from silicon wafers which may be monocrystalline or polycrystalline [7, 8]. The main reason for use of silicon technology is that it is non-toxic and abundantly available in the Earth's crust. PV modules produced from crystalline silicon have demonstrated their long-term stability over decades.

The silicon atom has 14 electrons; however, the orbital arrangement has solely four valence electrons to be shared by alternative atoms. These valence electrons play a crucial role in photovoltaic effect. Large numbers of silicon atoms bond with one another by means of their valence electrons to form a crystal. In a crystalline solid, each silicon atom normally shares one of its four valence electrons in a covalent bond with each of four neighbouring silicon atoms. The crystal therefore consists of basic units of five silicon atoms: the original atom plus the four atoms with which it shares the valence electrons.

The energy conversion efficiency and cost of solar cells are important issues in the selection. The efficiency of the cell influences the PV system, from material production to system installation. The solar cell efficiency is limited mainly because of photon losses; minority carrier loss due to recombination in the silicon bulk and at the surface; and heating loss due to series resistance in the gridlines and busbars. The crystalline PV cells provide the highest energy conversion efficiencies of all commercial solar cells and modules. Standard cells are produced using one of two different boron-doped p-type silicon substrates:

i) monocrystalline and
ii) polycrystalline.

Both types of cells are typically 125 mm (5 in) or 156 mm (6 in) square. The solar cells are assembled into modules by soldering and laminating to a front glass panel using ethylene vinyl acetate as an encapsulant. These cells are shown in Fig. 6.10.

### 6.5.1.1 Monocrystalline Solar Cells

Monocrystalline solar cells are manufactured from crystals of very pure silicon. In monocrystalline silicon crystal, there is a perfect periodic arrangement of atoms. The more perfectly aligned silicon molecules make better solar cells in terms of converting solar energy (sunlight) into electricity (the photovoltaic effect). The efficiency of solar panels depends on the purity of silicon; more pure silicon produces more electricity, but the production of enhanced quality of silicon increases the cost.

The process of making solar cells is similar to making semiconductors. These cells are fabricated using thin wafers cut from large cylindrical monocrystalline ingots prepared by the exacting Czochralski (CZ) crystal growth process and doped to about one part per million with boron during ingot growth. The silicon dioxide of either quartzite gravel or crushed quartz is placed into an electric arc furnace and heat is applied which results in production of carbon dioxide and molten silicon. This simple process yields silicon with 1% impurity, which is not suitable for solar cell manufacturing and must be further purified. This is accomplished by passing a rod of impure silicon through a heated zone

**Figure 6.10** (a) Monocrystalline silicon PV cell, (b) polycrystalline.

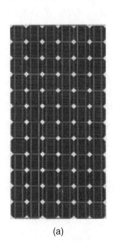

(a)  (b)

several times in the same direction. This procedure results in moving the impurities towards one end with each pass. After certain number of passes, the silicon is deemed pure, and the impure end is removed.

Wafers are then sliced out of the ingot, then sealed back to back and placed in a furnace to be heated to slightly below the melting point of silicon (1410°C) in the presence of phosphorous gas. In the early era of terrestrial PV cell production, small 2–5-inch-diameter CZ ingots were used, the small size and high cost of which obstructed cost reduction for monocrystalline cells.

Much research and development has been devoted to reducing the production costs for CZ ingots and wafer processing over the past 20 years. CZ wafers with side lengths of 125 and 156 mm, sliced using special saws from 6- and 8-in.-diameter ingots and 3–5 ft in length, respectively, are now widely used for PV cell fabrication. The fabrication of monocrystalline cells and modules using wafers of the same size as those used for poly-crystalline cell production has improved the competitiveness of monocrystalline cells against their polycrystalline counterparts in terms of manufacturing cost per output watt. Monocrystalline cells represented 38% of all solar cells manufactured in 2008.

The cells are wired together to create a solar panel. Under standard conditions, their conversion efficiency is more than those of polycrystalline cells. Monocrystalline panels are superior in quality and are more efficient but expensive compared to polycrystalline cells and are often used where high reliability is needed.

### 6.5.1.2 Polycrystalline Silicon Cells

The first solar panels based on polycrystalline silicon, which also is known as polysili-con (p-Si) and multi-crystalline silicon (mc-Si), were introduced to the market in 1981. Unlike monocrystalline-based solar panels, polycrystalline solar panels do not require the Czochralski process. Polycrystalline cells, on the other hand, are made from square silicon substrates cut from polycrystalline ingots grown in quartz crucibles. The front surface of the cell is covered with micrometre-sized pyramid structures (textured sur-face) to reduce reflection loss of incident light.

Solar cell wafers are made from polycrystalline silicon using different technologies. Polycrystalline silicon ingots and wafers were developed as a means of reducing the production costs for silicon ingots. Polycrystalline cells are currently the most widely produced cells, making up about 48% of world's solar cell production in 2008. The effi-ciencies of typical commercial crystalline silicon solar cells with standard cell structures are in the range of 16–18% for monocrystalline substrates and 15–17% for polycrys-talline substrates. These efficiencies are considerably lower than the 25% efficiency levels obtained in the laboratory.

The efficiencies of polycrystalline cell modules, however, are almost the same as those for monocrystalline cells (14%) due to the higher packing factor of the square polycrys-talline cells; monocrystalline cells are fabricated from pseudo-square CZ wafers and have relatively poor packing factors.

## 6.6 Thin-Film Solar Cells

The complete process of producing crystalline cells involves more than 20 separate steps to prepare and process ingots, wafers, cells and circuit assemblies. Thin-film technologies have the potential to reduce the cost of silicon cells by eliminating wafer slicing and reducing material consumption as much as by a factor of 10. The absorber layer can be deposited at the required thickness, greatly reducing the wastage of material.

Solar cells made from thin-film technology have efficiency less than that of crystalline solar cells (8–10%). As these cells are much less expensive, these types of cells can reduce the overall cost of solar modules [9]. The main thin-film materials currently being investigated are CdTe, CuIn, CIS, CIGS, amorphous silicon (a-Si) and crystalline silicon (c-Si). These thin-film solar cell technologies are described here.

### 6.6.1 Amorphous Silicon Solar Cells (a-Si)

Amorphous silicon is an alloy of silicon with hydrogen in the non-crystalline form and is the most developed of the thin-film technologies [10]. Amorphous silicon solar cells have been used in consumer products such as calculators and digital watches since the early 1980s. Now these are also used to provide power to some private homes, buildings and remote facilities. In the early studies of amorphous silicon, it was determined that plasma-deposited amorphous silicon contained a significant percentage of hydrogen atoms bonded into the amorphous silicon structure. These atoms were discovered to be essential to the improvement of the electronic properties of the material. Amorphous silicon is therefore known as "hydrogenated amorphous silicon" or a-Si:H.

Amorphous silicon panels are formed by vapour-depositing one or several layers of silicon material – about 1 micrometre thick – on a substrate material such as glass or metal. Amorphous silicon can also be deposited on inexpensive materials such as glass, plastic or stainless steel. The technology of a-Si:H for PV is based on two types of device design: a single-junction and a multi-junction p–i–n structure. Although major progress has been made in recent years in improving the deposition and other processes, the improvement in efficiency is only possible with multi-junction devices.

The fundamental photodiode inside an amorphous-silicon-based solar cell has three layers deposited in either the p–i–n or the n–i–p sequence. The three layers are a very thin (typically 20 nm) p-type layer, a much thicker (typically a few hundred nanometres), undoped intrinsic (i) layer and a very thin n-type layer. As illustrated in Fig. 6.11, in this

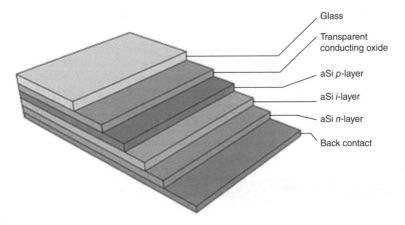

Glass

Transparent conducting oxide

aSi *p*-layer

aSi *i*-layer

aSi *n*-layer

Back contact

**Figure 6.11** Amorphous silicon.

structure, excess electrons are actually donated from the n-type layer to the p-type layer, leaving the layers positively and negatively charged (respectively) and creating a sizable "built-in" electric field (typically more than $10^4$ V/cm). Sunlight enters the photodiode as a stream of photons that pass through the p-type layer, which is a nearly transparent layer. Most of the photons are absorbed in the much thicker intrinsic layer. Each photon that is absorbed generates one electron and one hole photocarrier. The photocarriers are swept away by the built-in electric field to the n-type and p-type layers, respectively – thus generating solar electricity.

However, single-layer cells suffer from significant degradation in their power output (in the range of 15–35%) when exposed to the Sun. The mechanism of degradation is called the Staebler–Wronski effect, named after their discoverers. The conversion efficiency can be significantly improved by depositing two or three such diodes one on top of another to create a multi-junction device. These multi-junction cells are produced by layering the different materials on top of each other, with the shortest wavelengths on the "top" and increasing through the body of the cell. As the photons have to pass through the cell to reach the proper layer to be absorbed, transparent conductors need to be used to collect the electrons being generated at each layer. In this way, the absorption of light can be split into two or three separate layers. Such cells are more stable and have higher efficiency compared to the single-junction solar cells.

### 6.6.2 Cadmium Telluride (CdTe)

Cadmium telluride, a thin-film technology, offers improved performance in that it operates close to its maximum efficiency, particularly in hot, humid conditions. Although thin-film cells are cheaper to make, their efficiency has lagged behind that of conventional ones. Still, they have shown more improvement [11]. The material is very well suited for solar cell applications as it has a high optical absorption coefficient in the visible (high intensity) portion of the solar spectrum. Furthermore, it has a band gap of around 1.45 eV, which is close to the optimum (1.5–1.6 eV) for a terrestrial solar cell. CdTe films are polycrystalline and can be fabricated by physical vapour deposition, sputtering and electrodeposition. CdTe cell shown in Fig. 6.12 is composed of four layers and a transparent and conducting oxide (TCO) which acts as a front contact. Next, a layer of cadmium sulfide is deposited from solution onto a glass sheet coated with a transparent

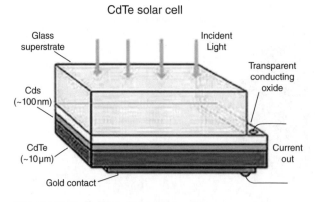

**Figure 6.12** Cadmium telluride solar cell.

conducting layer of tin oxide. This is followed by the deposition of the main cadmium telluride cell by a variety of techniques including close-spaced sublimation. The fourth layer is the back contact on top of the CdTe layer. The efficiency of CdTe solar cells is 16% on a 1-cm$^2$ laboratory cell. The main problem with CdTe modules is due to the toxicity of cadmium. On one hand, CdTe is a very stable compound and probably non-toxic. There are, however, definite environmental hazards and safety issues related to the production of CdTe modules: the release of cadmium into the atmosphere in the case of fire and the recycling of CdTe modules.

### 6.6.3  Copper Indium Gallium Diselenide (CIGS)

Copper indium diselenide (CIS) and copper indium gallium diselenide (CIGS) are direct-gap polycrystalline semiconductors with very high optical absorption coefficients and are presently being widely studied for application in solar cells. The CIGS can be designed to have a band gap value of 1.04 eV by controlling its composition and is the most suitable material for solar cells. Its conversion efficiency on a glass substrate is now approaching 20%. CIGS material has the advantage that it can be deposited on many substrates: glass, plastic, stainless steel and aluminium.

Advances in fabrication technology have been made for large-scale production of these cells. Recent developments in the field of CIGS have seen a trend towards flexible devices, with polyamide or metal foil substrates. Their flexible nature, resistance to solar radiation and high specific power have led CIGS collar cells to becoming increasingly used for space applications.

The structure of a CIGS solar cell is quite complex because it contains several compounds as stacked films that may react with each other. Fortunately, these reactions are either thermodynamically or kinetically inactive at ambient temperature. CIGS can be made by any of the following processing techniques: sputtering, evaporation, electrochemical deposition, nanoparticle printing and ion-beam deposition. Soda lime glass is commonly used as a substrate, because it contains Na, which has been shown to yield a substantial open-circuit voltage increase. A MO (Molybdenum) layer is deposited (commonly by sputtering) which serves as the back contact and to reflect most unabsorbed light back into the absorber. Following Mo deposition, a p-type CIGS absorber layer is grown by one of several unique methods. A thin n-type buffer layer is added on top of the absorber. The buffer is typically CdS deposited via chemical bath deposition. The buffer is overlaid with a thin, intrinsic ZnO layer (i-ZnO) which is capped by a thicker, Al-doped ZnO layer. The i-ZnO layer is used to protect the CdS and the absorber layer from sputtering damage while depositing the ZnO. The Al-doped ZnO serves as a transparent conducting oxide to collect and move electrons out of the cell while absorbing as little light as possible.

### 6.6.4  Copper Indium Selenide (CIS)

CIS technology gives the best performance in the laboratory with 19.5% efficiency achieved for small cells but has proved difficult to commercialise. Unlike the other thin-film technologies, which involve deposition onto a glass superstrate, CIS technology generally involves deposition onto a glass substrate. It is a direct-gap material and can be either p-type or n-type. Although a p-n junction of CIS is possible, it is unstable. However, the CIS of p-type when combined with CdS of a n-type material to form a p–n junction is stable. Unlike the a-Si cells, the CIS cells have shown good long-term outdoor stability, with interesting recovery behaviour for cells that have been away from light. Exposure to more intense light has actually caused efficiencies to increase, whereas exposure to elevated temperatures has resulted in loss of efficiency.

### 6.6.5 Crystalline Silicon (c-si) Thin-Film Solar Cells

Crystalline silicon (c-Si) solar cells currently account for about 87% of the worldwide PV modules produced. Solar cells made from a thin layer of crystalline silicon have high efficiency and lower costs when compared to conventional thick crystalline silicon solar cells. There are four types of c-Si solar cells: single-crystal, polycrystalline, ribbon and silicon film deposited on low-cost substrates.

Crystalline silicon thin-film solar cells deposited by plasma-enhanced chemical vapour deposition (PECVD) can be easily combined with amorphous silicon solar cells to form tandem cells; the band gaps involved (1.1 eV for crystalline silicon and 1.75 eV for amorphous silicon) are very close to the theoretically ideal combination. It has resulted in the production of stabilized tandem cells with efficiencies of 12%. *PECVD is a process by which thin films of various materials can be deposited on substrates at lower temperatures compared to standard chemical vapour deposition (CVD).*

## 6.7 Concentrating Photovoltaic Systems

The important factor in a solar electric power system based on photovoltaic conversion is the cost of solar cells. There are two alternatives for the reduction of the cost of photovoltaic systems: a cell fabrication technology which substantively reduces fabrication costs; or use of concentrating collectors in order to maximize the cell output, while minimizing the cell area.

Concentrating photovoltaic systems use lenses or mirrors to concentrate sunlight onto high-efficiency solar cells. These solar cells are typically more expensive than conventional cells used for flat-plate photovoltaic systems. However, the concentration decreases the required cell area while also increasing the cell efficiency. Concentrating PV systems have several advantages over flat-plate systems. First, concentrator systems reduce the size or number of cells needed that allow the use of more expensive semiconductor materials which would otherwise be very expensive if used without concentrator. Secondly, the efficiency of solar cell increases when it is placed under concentrated light. The increase in efficiency depends largely on the design of the solar cell and the material used to make it. Third, a concentrator can be made of small individual cells. This is an advantage because it is harder to produce large-area, high-efficiency solar cells than it is to produce small-area cells. CPV systems must track the Sun to keep the light focused on the PV cells. Therefore, they require highly sophisticated solar tracking system. The concentrating system is already described in Chapter 5.

## 6.8 New Emerging Technologies

In recent years, the solar photovoltaic market is dominated by products based on silicon wafers. The cost of wafer in solar module is about 50% of the total cost. The cost of the solar modules can be reduced by replacing the silicon wafers with other materials. Research is going on to find new semiconductor materials which have optimal band gap, inactive grain boundaries, stability properties and easy processing. Spectrum splitting through *multi-junction cells* with band gap energies designed to match the solar spectrum is also being developed [12]. Many configurations and materials have been investigated for tandem and multi-junction cell concepts.

The dye-sensitized solar cells (DSCs) provide a technically and economically credible alternative concept to present-day p–n junction photovoltaic devices. DSC cells are photoelectrochemical devices that convert solar energy into electricity. Their name is due to the presence of a particular molecule, similar to the organic pigment of green leaves – chlorophyll – which determines their colour and performs the task to capture the photons of solar light. Several organic dyes other than chlorophyll can also do the job; at present, the best-known natural dyes are the purple-red dyes found in, for example, blackberries, raspberries, hibiscus flowers and blackcurrants.

In contrast to the conventional systems where the semiconductor assumes both the task of light absorption and charge carrier transport, the two functions are separated here. Light is absorbed by a sensitizer, which is anchored to the surface of a wideband semiconductor. Charge separation takes place at the interface via photo-induced electron injection from the dye into the conduction band of the solid. Carriers are transported in the conduction band of the semiconductor to the charge collector. The use of sensitizers having abroad absorption band in conjunction with oxide films of nanocrstalline morphology permits to harvest a large fraction of sunlight. Nearly quantitative conversion of incident photon into electric current is achieved over a large spectral range extending from the UV to the near-IR region. Overall solar (standard AM 1.5) to current conversion efficiencies (IPCE) over 10% have been reached.

Among the most interesting approaches using silicon are the following:

1) The amorphous silicon–germanium alloys (a-Si,Ge:H) where the band gap can be varied from 1.75 eV down to below 1.3 eV. Triple-junction amorphous silicon alloy solar cells have achieved normal efficiency of 145 and 13% stable conversion efficiency.
2) Micromorph tandem solar cells consisting of a microcrystalline silicon bottom cell and an amorphous silicon top cell are considered as one of the most promising new thin-film silicon solar cell concepts. These cells ($\mu$c-Si:H (1.12 eV)/$\alpha$-Si:H (1.75 eV), with enhanced stability properties against light-induced degradation, have normal and stable efficiencies of 14.7% and 10.7%, respectively.

Multi-junction solar cells in tandem configuration are fabricated from a-Si/$\mu$-Si (amorphous Si/microcrystalline Si) and organic or inorganic semiconductors. Multi-junctions incorporate material alloys such as amorphous or polycrystalline silicon carbide ($\alpha$-Si:C) and silicon germanium ($\alpha$-Si:Ge). III–V materials have ideal band gap energies for highly efficient photon absorption (e.g. 1.0–1.1 eV for InGaAsN, 1.4 eV for GaAs).

In addition, fine-tuning of both lattice constant and band gap can be achieved by modifying the alloy composition, resulting in a large flexibility that is exploited for growing multi-junction cells. Lattice-matched and metamorphic three-junction GaInP/GaInAs/Ge cells can have maximum efficiency under concentrated sunlight.

Combinations of PV and solar thermal systems are predicted to have efficiencies greater than 60%. In concentrated solar power designs, the PETE module will receive the direct incident light, convert a fraction to electricity and transfer its waste heat to a solar thermal cycle. Theoretical efficiencies of combined photon-enhanced thermionic emission PETE/solar thermal cycles reach above 50% and may provide affordable renewable energy on the utility scale.

Apart from semiconductor solar cells, currently organic materials have become attractive for possible application in photovoltaic devices due to their potential application in large-area, printable and flexible solar panels [13]. Either low-molecular-weight molecular semiconductors, such as merocyanine, or conjugated polymeric semiconductors have been demonstrated to show photovoltaic effect in various types of p–n or Schottky-type devices. OPVs are light, can have tandem structures and can be fabricated on plastic substrates, with flexibility for potential applications in consumer electronics. Organic solar cells can be coloured, transparent and applied to flexible, light films. They generate electricity even under cloudy skies. It will soon be possible to generate solar energy anywhere: in cars and windows or inside buildings.

The field of organic photovoltaics (OPVs) has progressed quite significantly in the last 10 years. The highest efficiency obtained from (OPVs), such as bulk heterojunction polymer:fullerene solar cells, has risen from 2.5 to 11%. This rapid progress suggests that the commercialization of OPVs should be realized soon.

## 6.9 Solar PV Systems

Solar PV systems can be classified based on the end-use application of the technology. There are two main types of solar PV systems.

- Stand-alone PV solar system
- Grid-connected or grid-tied solar system

When the grid supply is not available, stand-alone solar PV systems can provide much needed electricity, eliminating the other methods of power generation such as diesel generators. First stand-alone systems for power generation were developed to supply loads requiring small amount of power but difficult to supply from the main grid. Such systems include space satellites, marine warning lights and telecommunication repeater stations. Gradually, the markets have shown remarkable growth in PV stand-alone power for homes ("solar homes"), villages and medical and educational facilities in developing countries, especially for lighting, water pumping, radio and television. Except for water pumping applications, nearly all stand-alone systems include batteries for energy storage.

Schematic diagram of stand-alone system for dc load and ac load is shown in Fig. 6.13. As shown in the figure, the basic stand-alone dc system consists of a PV array, charge controller, battery for energy storage, dc distribution board and load. The purpose of charge controller is to control the current and voltage inputs to the batteries to protect them from damage at either end of the charging cycle. In the beginning of the cycle, if the batteries are almost discharged, a large amount of charging current if not regulated may damage them because of overheating. In the end of the cycle, the overcharged batteries will generate hydrogen gas and will dehydrate the batteries. The battery storage is required to collect energy from solar arrays during high sunshine and make it available during night or low sunshine.

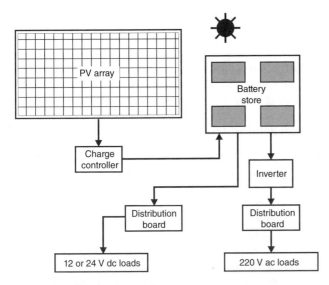

**Figure 6.13** Stand-alone system.

For ac/dc loads, the system requires an inverter to convert dc into ac. Hybrid ac-dc systems are especially suitable for connecting mid-range ac power consumers with dc generators. With such systems, the battery on the dc side can be simultaneously charged via a generator which acts as an additional source of power. These systems are described in detail in Chapter 13. Here only a brief description of these systems is presented.

### 6.9.1 Grid-Connected PV System

*A grid-connected PV system is the least expensive and lowest-maintenance option for a home solar electric system.* As the cost of PV system continues to decrease and awareness about environment concerns has increased, grid-connected interactive PV systems are now being installed in many countries. The inverters in a grid-connected solar PV system perform two major tasks. One is to ensure that the PV module(s) is operated at the MPP. The other is to inject a sinusoidal current into the grid. Since the inverter is connected to the grid, it must supply power with the voltage and harmonic content standards given by the utility companies.

Two types of grid-connected photovoltaic systems are considered: (i) grid-connected PV systems without battery storage and (ii) grid-connected PV systems with battery storage.

### 6.9.2 Grid-Connected System without Battery Storage

A grid-connected PV system consists of five main components: (i) a PV array, (ii) an inverter, (iii) the main service panel or breaker box, (iv) safety disconnects and (v) meters. The PV array converts the solar energy into dc current. The inverter converts the dc power produced by the PV array into ac power consistent with the voltage and power quality requirements of the utility grid. A bidirectional interface is provided between the PV system ac output circuits and the electric utility network. This allows the ac power produced by the PV system to either supply local electrical loads or feed the grid when the PV system output is greater than the onsite load demand. At night and during the periods when the local electrical power demand is more than the PV system output, the balance of power required by the loads is received from the electric utility. When the utility grid is down, these systems automatically shut down and disconnect from the grid. This safety feature ensures that the PV system will not continue to operate and feed back onto the utility grid when the grid is down for service or repair.

### 6.9.3 Grid-Connected System with Battery Storage

This type of system is extremely popular for homeowners and small businesses where backup power is required for critical loads such as refrigeration, water pumps, lighting and other necessities. Grid-connected photovoltaic power systems comprise photovoltaic panels, MPPT, solar inverters, power-conditioning units and grid connection equipment. These systems may or may not have batteries. Under normal circumstances, the system operates in a grid-connected mode, supplying the on-site loads and charging the battery or sending excess power back onto the grid while keeping the battery fully charged [14]. In the event when grid power is not available, control circuitry in the inverter opens the connection with the utility through a bus transfer mechanism and operates the inverter from the battery to supply power to the dedicated critical load circuits only. In this configuration, critical loads are supplied from a dedicated load sub-panel. Non-essential loads are not energized in this condition.

A grid-connected system can be as small as a couple of kilowatts and can be as large as hundreds of megawatts. Residential grid-connected photovoltaic power systems which have a capacity less than 10 kilowatts can meet the load of most consumers. It can feed excess power to the grid, which in this case acts as a battery for the system.

## 6.10 Design and Control of Stand-Alone PV System

An example of designing a PV system for a typical house in a rural area is described here. The first step in the design process is to calculate the power consumption of the house. To calculate the load of the appliances, their power consumption and their use over a period in a day are tabulated in Table 6.3.

Thus, the total power consumption in 24 h is 8740, or the average power consumption per day is $8740/24 \cong 365$ W. However, a system cannot be designed on the basis of average demand. Assuming that the washing machine and the electric press are not used simultaneously, the maximum demand is calculated as follows:

$$\text{Electric press or washing machine } 1000\,\text{W} + \text{Refrigerator } 100\,\text{W}$$
$$+ \text{ computer } 160\,\text{W} + 2\text{ fans } 140\,\text{W} + \text{television } 150\,\text{W} = 1550\,\text{W}$$

The power will be supplied through PV modules connected to dc bus through dc-to-dc converter. The energy storage system consists of 12 V lead–acid batteries connected to the same dc bus through a bidirectional battery charger. Battery is charged during daytime and supplies power during night or when there is no sunlight. Since all the appliances used in the house are ac, these are connected to a single-phase ac bus. The interfacing between the dc bus and the ac bus is through a single-phase 12/220 V voltage source inverter.

**Table 6.3** Typical home appliance and their use per day.

| Appliance | Power (watts) | Hours/day | Watt hour/day |
|-----------|---------------|-----------|---------------|
| Fans (3) | $70 \times 3 = 210$ | 10 | 2100 |
| Lights (4) | $40 \times 4 = 160$ | 5 | 800 |
| Computer | 160 | 4 | 640 |
| Refrigerator | 100 | 12 | 1200 |
| Television | 150 | 10 | 1500 |
| Washing machine | 1000 | 1.5 | 1500 |
| Electric press | 1000 | 1.0 | 1000 |
| | | **Total** | 8740 |

**Table 6.4** PV module parameters.

| | |
|---|---|
| Power | 175 W |
| Area | $1.38 \, \text{m}^2$ |
| $V$ (MPP) | 23.9 V |
| $V_{OC}$ | 30.02 V |
| $I$ (MPP) | 7.32 A |
| $I_{SC}$ | 7.963 A |
| Efficiency | 15% |

Now in designing the solar PV system, the location of the village must be considered. Since the irradiation during winter and summer may be different at a particular location, the design is generally based on the worst month. First, we select a commercially available module with the following parameters as given in Table 6.4.

Using MATLAB model of PV module for that particular location, the average solar irradiation is found. Suppose that the average solar radiation obtained is 530 W/m². The average power obtained is 55% of maximum power of 175 W which is 96.5 W. Since the average sunshine hours in a day is 7 h, the total power in a day is

$$P_{av} = 96.5 \times 7 = 676 \, \text{Wh}$$

Total number of modules required to supply $8740/676 = 13$ taking a factor of safety (for system losses, etc.) as 1.2, the number of modules required $= 15.6$ or 16. Now 16 modules can supply a maximum load of $16 \times 175 = 2800 \, \text{W}$, which is sufficient to take care of peak load as well. The dc bus voltage can be taken as voltage of module 23.9 or 24 V which will be adjusted by a dc-to-dc converter. The dc-to-dc converter should have 10% higher rating, and therefore, it will be rated at $2800 \times 1.10 = 3000 \, \text{W}$.

### 6.10.1 Battery Rating

Two batteries in series connection will yield 24 V, and the number of batteries required to be connected in parallel will be calculated as follows:

If each battery is 17 A h capacity from its discharge characteristic as available from the manufacturer is that it can supply 0.85 A continuously for 20 h with final voltage drop of 1 V, then for average demand of 365 W, the number of batteries required = average power/$V_{battery} \times I_{battery}$.

OR number of batteries required = $\frac{365}{12 \times 0.85}$ = 36 batteries.

| First | $15 \times 40 = 600$ | $10 \times 80 = 800$ | $08 \times 150 = 1200$ | $4 \times 2000 = 8000$ | 0 |
|---|---|---|---|---|---|
| Total | 2200 W | 5600 W | 9500 W | 36000 W | 30000 W |
| Use in hours | 8 | 8 | 6 | 6 | 3 |
| Total kWh | $8 \times 2.2 = 17.6$ | $8 \times 5.6 = 44.8$ | $6 \times 9.5 = 57$ | $6 \times 3.6 = 21.6$ | 30 |

## 6.11 Summary

In this chapter, the basic principle of solar energy conversion using PV cells is presented. The characteristic of PV solar cells including their connections to form arrays is described. Next, various types of solar cells and their properties are discussed. Advances in solar PV technology and newer solar cells are also described. In addition, solar power tracking and MPPT techniques are described. Finally, PV solar systems in isolated and grid-connected modes along with examples are presented.

## References

1 Chalmers, R. (1976) *The Photovoltaic Generation of Electricity*, Scientific American.
2 Chapin, D.M. *et al.* (1954) A new silicon p-n junction photocell for converting solar radiation into electric power. *Journal of Applied Physics*, **25**, 676.
3 Khan, B.H. (2011) *Non-Conventional Energy Technologies*, 2nd edn, McGraw-Hill.
4 Dave, F. (2010) *Introduction to Photovoltaic Systems Power Point Tracking*, Application notes Texas Instruments, Texas Instruments.
5 Esram, T. and Chapman, P.L. (2007) Comparison of Photovoltaic array maximum power point tracking techniques. *IEEE Transactions On Energy Conversion*, **22** (2), 439–449.
6 Lee, J.H. *et al.* (2006) *Advanced Incremental Conductance MPPT Algorithm with a Variable Step Size*, EPE-PEMC, pp. 603–607.
7 Green, M.A. (1993) Silicon solar cells: evolution, high-efficiency design and efficiency enhancements. *Semiconductor science and Technology*, **8** (1), 1.
8 Saga, T. (2010) Advances in crystalline silicon solar cell technology for industrial mass production. *Asia Materials*, **2**, 96–102.
9 Beucarne, G. (2007) *Silicon Thin Film Solar Cells, Advances in Optoelectronics*, Hindawi Publication.
10 Carlson, D.E. (1989) Amorphous Silicon Solar Cells. *IEEE Transaction on Electron Devices*, **32**, 2775–2780.
11 Poortmans, J. and Arkhipov, V. (eds) (2006) *Thin Film Solar Cells Fabrication, Characterization and Application*, John Wiley.

**12** R. McConnell and M. Symko-Davies (2006) Multijunction Photovoltaic Technologies for High-Performance Concentrators, IEEE 4th World Conference on Photovoltaic Energy Conversion (WCPEC-4) Waikoloa, Hawaii May 7–12.

**13** Su, Y.W., Lan, S.C. and Wei, K.H. (2012) Organic Photovoltaics. *Materials Today*, **15** (12), 554–562.

**14** Mriun, B. *et al.* (2005) Performance Parameters for Grid-Connected PV Systems NREL conference paper February 2005.

# 7

# Wind Energy

## 7.1 Wind as Source of Energy

The wind power is basically nothing but another form of solar energy. Approximately 1% of the total solar energy absorbed by the Earth is converted to kinetic energy in the atmosphere, in the form of wind. Since early recorded history, people have realized the potential of wind energy and utilized it for various applications. It was used to propel boats along the Nile River as early as 5000 BC, and it was used to pump water and grind grain between 500 and 900 BC. By the 11th century, windmills were used in food production in the Middle East. The windmills were further improved by the Dutch and others and were adapted for industrial applications such as sawing wood, making paper and draining lakes and marshes. In the late 19th century, the wind power was used in windmills to pump water for farms and ranches. However, due to industrialization and rural electrification, in the 20th century, there was gradual decline in the use of windmills for mechanical applications.

First large-sized automatically operating wind turbine for generation of electricity was built by Charles Brush of the United States in 1888. From 1920 to 1940, propeller-type horizontal-axis wind turbines (HAWTs) with two or three blades were used to supply electricity in rural areas where supply of electricity from the grid was not available. However, the use of wind turbines to generate electricity at a commercial scale started in the 1970s as a result of technical advances in the field of turbine and mainly due to escalating oil prices because of OPEC crisis in 1971. During the last few years, the three-blade upwind horizontal-axis large turbines on monopole tower have become the standard. In terms of total number of wind turbines in use at present, they are of small capacity of the order of 10 kW or less. But in terms of total generating capacity, the turbines that are used in large wind farms have capacity in the range of 1.5–5 MW.

### 7.1.1 Origin of Wind

The unequal heating of Earth's surface due to its tilt, rotation and difference in insolation results in the wide distribution of pressure over the Earth's surface. Difference in temperature causes pressure difference which in turn results in air movement called wind. The regions which receive higher amount of heat cause heating of air which rises in the atmosphere. The upward movement of hot air induces cooler air from the surroundings to rush in. Thus, horizontal air currents and wind patterns are created on the Earth's surface. The Earth's rotation, geographic features and temperature gradients

*Operation and Control of Renewable Energy Systems*, First Edition. Mukhtar Ahmad.
© 2018 John Wiley & Sons Ltd. Published 2018 by John Wiley & Sons Ltd.

affect the location and nature of the resulting winds [1]. Two major forces that determine the speed and direction of wind on a global basis are the differential heating between the equator and the poles and the rotation of the Earth. The primary force for global winds is developed because there is a difference in heating of the Earth at equatorial and polar regions. Polar regions have lower temperatures compared to the tropical regions. Due to this difference in temperature, the heat is transported by the winds from tropical regions to polar regions. About 30% of the total global heat transfer takes place due to atmospheric currents along with ocean currents. But due to the spinning of the Earth about its own axis, a force is produced which is responsible for deviation of air currents towards the west due to *Coriolis* force. The westerly motion causes the north-east winds in the northern hemisphere and south-east winds in the southern hemisphere.

Apart from these global currents, local winds are created due to localized uneven heating of small area of the Earth. Land masses are heated by the Sun more quickly than sea during daytime. The hot air from the land rises and flows towards the sea, creating a low pressure at the ground level. The cold air rushes from the sea towards the land. This is called sea breeze. That is why wind is more pronounced in coastal areas during daytime. At night, the direction of wind is reversed, but as the difference in temperatures of the Earth and sea is not very large, this wind speed is low.

The local winds are also produced in mountain regions due to difference in heating of mountain slope and neighbouring air of the ground. As the density of air due to heating of the slope decreases, the air moves towards the top following the slope. At night, the air flows in the reverse direction as the mountains cool down faster than the low land. As a result, the production of electricity using wind is highly sensitive to local wind conditions and the ability of wind turbines to reliably extract energy from the wind.

### 7.1.2 Wind Power Potential

Approximately 1% of the total solar energy absorbed by the Earth is converted to kinetic energy in the atmosphere, in the form of wind which is ultimately dissipated by friction at the Earth's surface [2]. If this energy is assumed to be dissipated uniformly over the entire surface area of the Earth, it means that there is an average power source of $3.4 \times 10^{14}$ W for the total land area of the Earth. This power is equivalent to an annual supply of energy of 10,800 EJ (1 EJ is $10^{18}$ J), which is 22 times the present global annual consumption of commercial energy. But the wind energy is not distributed uniformly over the Earth. Regional patterns of dissipation depend not only on the wind source available in the free troposphere but also on the frictional properties of the underlying surface. Estimates of global wind energy potential range from a low of 70 EJ/year (19,400 TWh/year) (onshore only) to a high of 450 EJ/year (125,000 TWh/year) (onshore and near-shore).

Wind power is used to generate mechanical power by rotating turbines and the generators to produce electricity. This turbine power can also be used for doing some mechanical work (such as grinding grain or pumping water). Wind energy is a clean energy which can reduce greenhouse gas (GHG) emissions. A number of different wind energy technologies are available for a wide range of applications. However, the primary use of wind energy mitigating climate change is to generate electricity from larger, grid-connected wind turbines. These turbines may be deployed either "onshore" or "offshore". At present, a number of wind energy installations mostly onshore are working in various countries. The wind power capacity installed by the end of 2009 was

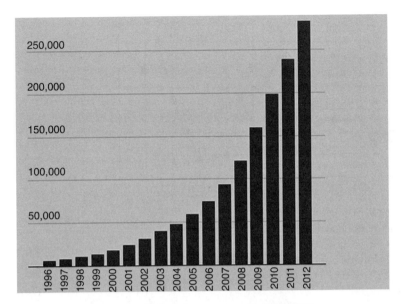

**Figure 7.1** Cumulative installed wind power capacity 1980–2013.

roughly 1.8% of the total electricity demand, and if the present trend of deployment continues, it is likely to grow to more than 20% by 2050. At the end of 2013, the wind farms installed in more than 85 countries had a combined generating capacity of 318,000 MW. New data from the Global Wind Energy Council shows that new wind generating capacity of about 35,000 MW was added worldwide in 2013 (Fig. 7.1).

Power generation from onshore wind energy is already being integrated into electricity supply system without any problem. Although average wind speeds vary considerably by location, it is possible to extract significant amount of wind energy in most regions of the world. In some areas with good wind resources, the cost of wind energy is already competitive with the current energy market prices, even without considering relative environmental impacts. Nonetheless, in most regions of the world, continued advances in onshore and offshore wind energy technologies are required for further reducing the cost of wind energy and improving GHG emission reduction potential.

## 7.2 Power and Energy in Wind

The wind power depends on the mass of airflow rate (density) $m$, velocity of air $v$, passing through an area of interest $A$ [3, 4].

The kinetic energy by definition is given by

$$\text{KE} = \frac{1}{2}mv^2 \tag{7.1}$$

here $m$ is given by

$$m = \rho A v \tag{7.2}$$

where $\rho$ = density of air.

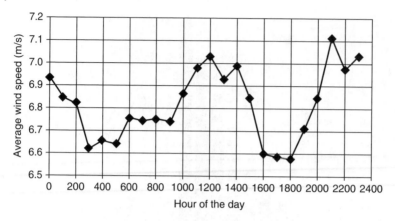

**Figure 7.2** Wind energy distribution curve.

Power in wind is kinetic energy per unit time,

$$P = \frac{1}{2}mv^2 = \frac{1}{2}\rho A v^3 \tag{7.3}$$

From the aforementioned expression, it is clear that the power available in wind is proportional to the cube of wind speed. The wind turbines are used to convert this kinetic energy of moving air into mechanical energy. The energy that a wind turbine will produce depends on both its wind speed–power curve and the wind speed frequency distribution at the site. Histogram showing the number of hours for which the wind blows at different wind speeds during a given period of time is used to determine the wind speed frequency distribution. From the histogram, operating range of the wind speed of the turbine (the speed between shutdown speed and cut-in speed) is obtained. The energy produced by the turbine at a particular speed is obtained by multiplying the number of hours of its duration by the turbine power at that speed. This data is used to plot *wind energy distribution curve* as shown in Fig. 7.2. The total energy produced in a given period of time can be calculated by adding the energy produced at all the wind speeds within the operating range of the turbine. In order to obtain the wind speed–power curve, maps giving estimates of the mean wind speeds are available for various locations for most of the countries.

## 7.3 Aerodynamics of Wind Turbines

A wind turbine extracts kinetic energy from the wind and converts it into mechanical energy. The turbine is then used to drive a generator for the production of electricity [5, 6]. Since the wind turbine power production depends on the interaction between the rotor and the wind, the major aspects of wind turbine performance such as power output and loads are determined by the aerodynamic forces generated by the wind. It is therefore desirable to use a simple model to know the power developed by an ideal turbine rotor and the effect of the rotor operation on the local wind field. The simplest model of a wind turbine, considered by Betz, is the so-called actuator disc model where the turbine is theoretically replaced by a circular disc through which the airstream flows with a velocity $V$ and across which there is a pressure drop from $P_1$ to $P_2$. An actuator disc concept, representing a turbine rotor with an infinite number of blades, is used to derive the one-dimensional momentum equation. It was originally intended to provide an analytical means for evaluating ship propellers.

### 7.3.1 Momentum

Momentum theory assumes the following:

- The airflow is homogeneous and incompressible.
- There is no frictional drag.
- It is a steady-state time-invariant fluid flow.
- The thrust over the disk is uniform.
- The static pressure far upstream and far downstream of the disk (rotor) is equal to the undisturbed ambient static pressure.

The analysis assumes a control volume as shown in Fig. 7.3, in which the control volume boundaries are surface of the stream tube and two cross sections of the stream tube. According to the basic momentum theory, when fluid flows through an obstacle such as an actuator disc, it produces an axial thrust which is a consequence of the pressure drop in the fluid at sections adjacent immediately before and after the disc, in the flow stream. Similarly, since the energy is given up by the fluid as it moves from high pressure to low pressure, its speed also decreases.

Applying the conservation of linear momentum to control volume enclosing the whole system, the air mass flow rate must remain same throughout the stream tube. Therefore,

$$\dot{m} = \rho A_1 V_1 = \rho A_2 V_2 = \rho A_{43} 3 = \rho A_4 V_4 \tag{7.4}$$

The thrust $T$ is equal and opposite to change in momentum of air stream. Thus,

$$T = \dot{m}(V_4 - V_1) \tag{7.5}$$

or

$$T = (\rho A_1 V_1)V_1 - (\rho A_4 V_4)V_4 \tag{7.6}$$

The thrust can also be calculated from pressure conditions as

$$T = A_2(P_2 - P_3) \tag{7.7}$$

**Figure 7.3** Actuator disk model of wind turbine.

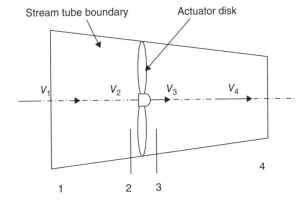

Since the thrust is positive, $V_4$ is less than $V_1$. No work is done on either side of the rotor. The Bernoulli's equation therefore can be used in the two control volumes on either side of the disk. In upstream flow of the wind turbine, the Bernoulli's equation is

$$\frac{1}{2}\rho V_1^2 + P_1 = \frac{1}{2}\rho V_2^2 + P_2 \tag{7.8}$$

and in the downstream of wind turbine, it is

$$\frac{1}{2}\rho V_3^2 + P_3 = \frac{1}{2}\rho V_4^2 + P_4 \tag{7.9}$$

where it is assumed that $P_1 = P_4$ and $V_2 = V_3$.

Using the value of $P_2 - P_3$ from Eqs (7.8) and (7.9) in Eq. (7.4), the following expression is obtained for thrust.

$$T = \frac{1}{2}\rho(V_1^2 - V_4^2) \tag{7.10}$$

From Eqs (7.4) and (7.10), one gets

$$V_2 = \frac{V_1 + V_4}{2} \tag{7.11}$$

This means that the velocity of the wind at the rotor plane is the average speed of upstream and downstream wind.

Now if the axial interference (induction) factor $a$ is defined as the fractional decrease in the wind velocity between the free stream and rotor plane, then $a$ is given by

$$a = \frac{V_1 - V_2}{V_1} \tag{7.12}$$

or

$$V_2 = (1 - a)V_1 \tag{7.13}$$

and

$$V_4 = (1 - 2a)V_1 \tag{7.14}$$

From Eq. (7.14), it can be concluded that if the rotor absorbs all the wind energy, that is, $V_4 = 0$, and $a = 0.5$. The limit $0 < a < 0.5$ is wind turbine state of propeller operation. The power output $P$ is equal to thrust times the velocity at the disk; the power $P$ is

$$P = \frac{1}{2}\rho A_2(V_1^2 - V_4^2)V_2 \tag{7.15}$$

or

$$P = 2\rho A V^3 a(1 - a)^2 \tag{7.16}$$

here the control volume area is replaced by $A$, the rotor area, and free stream velocity $V_1$ is replaced by $V$.

The maximum power will be obtained when

$$\frac{dP}{dA} = 0 \tag{7.17}$$

$$2\rho A V^3(1 - 4a + 3a^2) = 0 \tag{7.18}$$

or

$$(1 - 4a + 3a^2) = 0 \qquad (7.19)$$

or

$$a = 1 \text{ or } 1/3$$

since $a$ cannot be greater than 1/2, or $a = 1/3$. Substituting the value of $a = 1/3$ in Eq. (7.18)

$$P_{max} = \frac{16}{27} \left( \frac{1}{2} \rho A V^3 \right) \qquad (7.20)$$

The factor $\frac{16}{27}$ is known as Betz's coefficient or *Betz Limit Cp*.

Power coefficient, $C_p$, is defined as the ratio of power extracted by the turbine to the total contained in the wind resource

$$C_p = \frac{P_T}{P_W} \qquad (7.21)$$

Turbine power output

$$P_T = \frac{1}{2} \rho A v^3 C_p \qquad (7.22)$$

Or 59% efficiency is the *BEST* a conventional wind turbine can do in extracting power from the wind.

This result $\left( a = \frac{1}{3} \right)$ indicates that if an ideal rotor is designed and operated in such manner that the wind speed at the rotor is 2/3 of the free stream wind speed, then the turbine will be operating at maximum power generation point.

Similarly, the maximum axial thrust is given by

$$\frac{1}{2} \rho A V^2 [4a(1-a)] \text{ for } a = 1/3 \text{ or } T_{max} = 8/9 \rho A V^2 \qquad (7.23)$$

The variation of power coefficient $C_p$ with interference factor $a$ is shown graphically in Fig. 7.4. As shown in Fig. 7.4, when there is no load on the turbine, the blades just

**Figure 7.4** Variation of $C_p$ with $a$.

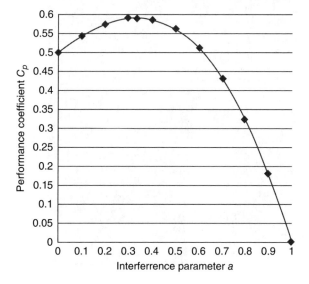

freewheel. As there is no reduction in speed, the value of $a = 0$, also $C_p = 0$ because the turbine is not generating any power. The maximum value of power occurs at $a = 1/3$ when $C_p = 0.593$ and when $a = 1$, $C_p$ is zero.

## 7.4 Types of Wind Turbines

Wind turbines can be classified into two general types: *horizontal axis* and *vertical axis*. A horizontal-axis turbine has its blades rotating on an axis parallel to the ground as shown in the Fig. 7.5. A vertical-axis machine has its blades rotating on an axis perpendicular to the ground. There are a number of available designs for both types of turbine, and each type has certain advantages and disadvantages. However, the horizontal-axis type machines are the most common, and very few vertical-axis machines are running commercially.

### 7.4.1 Horizontal-Axis Wind Turbines

HAWTs have the main rotor shaft and electrical generator at the top of a tower. HAWT should be pointed towards the wind to capture the maximum power. It has blades that resemble a propeller that spin on the horizontal axis. The dominant driving force in this turbine is the lift. Depending on the different relative position of the rotor and tower, the HAWT can be divided into *upwind wind turbine* and *downwind wind turbine*. If the rotor is in front of the tower, it is known as the upwind turbine, while if the rotor is installed behind the tower, it is called the downwind wind turbine. Upwind wind turbine

**Figure 7.5** Horizontal-axis wind turbine.

requires steering installation (yaw mechanism) to ensure that the rotor faces the wind during working which is a disadvantage.

Since electrical generators require speed higher than the speed of turbine, most large wind turbines have a gearbox, which turns the slow rotation of the rotor into a faster rotation. The basic advantage of upwind designs is that one avoids the wind shade behind the tower. Although there is some wind shade in front of the tower, that is, the wind starts bending away from the tower before it reaches the tower itself. By far, the vast majority of wind turbines have this design. The basic drawback of upwind designs is that the rotor has to be rather inflexible and placed at some distance from the tower.

The downwind wind turbine does not require steering mechanism as it can face the wind automatically. However, in the downwind wind turbine, because a part of wind blows to the blades across the tower, the tower disturbs the wind flow which should go across the blades. In this way, it causes the tower shadow effect, causing a drop in power each time a blade passed behind the mast, decreasing the efficiency. An important advantage of this type of turbine is that rotor can be made flexible. Thus, at higher speeds, the blades can bend, thus reducing the load on tower. Downwind turbines are generally noisier (additional aerodynamic noise), and the blades are subject to more forces than those of upwind turbines.

The HAWT can also be divided into the *lift-type wind turbine* and the resistance-type or *drag-type wind turbine*. For the drag design, the wind literally pushes the blades out of the way. Drag-powered wind turbines are characterized by slower rotational speeds and high torque capabilities. They are useful for pumping, sawing or grinding work. The lift type has a high rotating speed. Most of the HAWTs are downwind type that have the steering device and can rotate with the wind. For the small-sized wind turbine, the steering device employs the tail vane, while for the large-sized wind turbine, it often adopts the gear consisting of wind sensors and servomotor. Most commonly used HAWTs are upwind, lift-type, three-blade machines as shown in Fig. 7.5. The advantages and disadvantages of HAWT are as follows:

Advantages
- High efficiency since the blades are always perpendicular to wind.
- Higher stability because blades are to the side of the turbines' centre of gravity.
- Ability to wing warp, which gives the turbine blades the best angle of attack.
- Variable blade pitch minimizes damage due to high wind.
- Tall tower allows access to stronger wind.
- Tall tower allows placement on uneven land or in offshore locations.
- Good starting performance.

Disadvantages
- Difficult to transport because of taller tower and blades (20% of equipment costs)
- Difficult to install (require tall cranes and skilled operators)
- Effect radar in proximity
- Difficulty in maintenance

### 7.4.1.1 Horizontal-Axis Wind Turbines with Wake Rotation

In Section 7.3, it is assumed that there is no rotation of the flow. However, in any wind turbine, the rotor is not always stationary but rotates when the wind flows through it. When the rotor rotates, it generates angular momentum which affects the rotor torque.

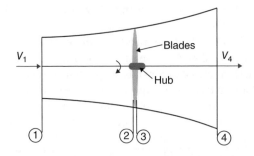

**Figure 7.6** Steam tube model for rotating wake.

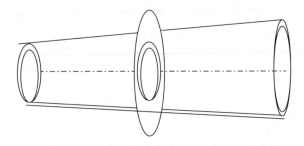

The wind imparts a torque on the wind turbine, and a thrust is created. The airflow behind the rotor is in the opposite direction to the rotor because the blades exert a torque on the wind. Thus, the flow in the wake has two components, axial and tangential. This tangential flow is referred to as wake rotation. Due to rotation, there is loss of kinetic energy of the wind rotor. The loss is high for low-speed rotors and low for high-speed rotors.

For the analysis of momentum with the wake rotation, an annular stream tube model as shown in Fig. 7.6 is used. The blade wake rotates with an angular velocity $\omega$, and the blades rotate with an angular velocity of $\Omega$. In this analysis, it is assumed that the angular velocity imparted to flow stream $\omega$ is small compared to the angular velocity $\Omega$ of the wind turbine rotor. It is also assumed that the pressure in the far wake is equal to the pressure in free stream.

If the ring radius of the annular stream tube is $r$ and thickness is $dr$, then the cross-sectional area of the ring is $2\pi r dr$.

Applying the energy equation before and after the blade, we get

$$P_2 - P_3 = \frac{1}{2}\rho(\Omega + \omega)^2 r^2 - \frac{1}{2}\rho\Omega^2 r^2 \tag{7.24}$$

or

$$P_2 - P_3 = \rho\left(\Omega + \frac{1}{2}\omega\right)\omega r^2 \tag{7.25}$$

The angular velocity of the air relative to the blade increases from $\Omega$ to $\Omega + \omega$, whereas the axial velocity remains constant. Therefore, the thrust on the annular element of the rotor is

$$dT = \rho\left\langle \Omega + \frac{1}{2}\omega\right\rangle \omega r^2 2\pi r dr \tag{7.26}$$

If $a' = \frac{\omega}{2\Omega}$, and $V_2 = V(1 - a)$, the thrust becomes

$$dT = 4\,a'(1 - a)\rho V\Omega r^3 2\pi dr \tag{7.27}$$

From the conservation of momentum, the torque exerted must be equal to the angular momentum $dQ$ of the wake.

$$\text{Or}\ \ dQ = 4\,a'(1 - a)\rho V\Omega r^3 \pi dr \tag{7.28}$$

The power output from the rotor is

$$P = \int \Omega\,dQ \tag{7.29}$$

*Tip Speed Ratio*   The Tip speed ratio (TSR) is used by wind turbine designers to properly match and optimize a blade set to a particular generator. TSR is defined as the speed of the blade at its tip divided by the speed of the wind. TSR is important in designing a wind energy conversion system (WECS). For a particular generator, if the blade set spins too slowly, then most of the wind will pass by the rotor without being captured by the blades. If the blades spin too fast, then the blades will always be cutting the turbulent wind.

This is because the blades will always be travelling through a location that the blade in front of it just travelled through (and used up all the wind in that location). In short, if the blades are too slow, they are not capturing all the wind they could, and if they are too fast, then the blades are spinning through used/turbulent wind. For this reason, TSRs are employed when designing wind turbines so that the maximum amount of energy can be extracted from the wind using a particular generator.

Calculation of TSR

Find the speed of the blade at its tip.

If $r$ is the length of blade, the distance travelled by the tip of the blade in one revolution $= 2\pi r$ m. In addition, if the tip of the blade is rotating at a speed of n rpm, the distance travelled by the tip in 1 min is $2\pi nr$ m.

Thus, thespeed of the tip $= 2\pi nr/60\,\text{m/s}$.

And TSR $\lambda = 2\pi nr/60v$, where v is the wind speed in m/s.

### Example 7.1

Determine the tip speed ratio for HAWT rotating at 60 rpm, for wind speed of 15 m/sec and length of blade is 10 m.

The speed of the tip $= 2 \times 3.14 \times 10 \times \dfrac{60}{60} = 62.8$

TSR $\lambda = \dfrac{162.8}{15} \cong 4$

*Optimal Tip Speed Ratio*   The optimal tip speed ratio for maximum power extraction is obtained by relating the time taken for the disturbed wind to re-establish itself $t_w$ to the time taken for a rotor blade of rotational frequency $\omega$ to move into the position occupied by its predecessor $t_b$. For a turbine having $n$ blades rotating at angular speed $\omega$

$$t_b = \frac{2\pi}{n\omega}$$

If the length of the strongly disturbed air stream upwind and downwind of the rotor is s, then the time period for the wind to return to normal is given by

$$t_w = \frac{s}{V}$$

If $t_b > t_w$, then some wind is unaffected. If $t_b < t_w$, then some wind is not allowed to flow through the rotor. The maximum power extraction occurs when $t_b \approx t_w$.

Or $\omega_{opt} \approx \dfrac{2\pi V}{ns}$

The tip speed ratio for maximum power extraction is

$$\lambda_{opt} \approx \frac{2\pi Vr}{nsV} = \frac{2\pi}{n}\left(\frac{r}{s}\right)$$

From practical observation $\frac{r}{s} \approx 2$

$$\lambda_{opt} \approx \frac{4\pi}{n}$$

## Example 7.2

A HAWT has the following data:

Blade length = 40 m. Number of blades = 3, wind speed = 15 m/sec. Determine the speed of the turbine for maximum power output.

**Solution:**

$$\text{Tip speed ratio for optimum power} = \lambda_{opt} \approx \frac{4\pi}{n} = \frac{4\pi}{3} = 4.188$$

$$\text{TSR } \lambda = 2\pi nr/60v = 4.188 = \frac{2\pi n \times 40}{60 \times 15} = 15\,\text{rpm}$$

## Example 7.3

A HAWT has the following data:

Blade length l = 52 m, wind speed = 12 m/sec at 1 atm, air density $\rho = 1.23\,\text{kg/m}^3$, power coefficient $C_p = 0.45$. Calculate the power developed by the turbine:

**Solution:**

Radius of blade $r = l = 52$ m

$$\text{Area of rotor} = \pi r^2 = 3.14 \times 52 \times 52 = 8495\,\text{m}^2$$

The power available from wind from Eq.7.23 is $= P_T = \frac{1}{2}\rho A v^3 C_p$

$$= \frac{1}{2} \times 1.23 \times 8495 \times 12 \times 12 \times 12 \times 0.45 = 4.05\,\text{MW}.$$

### 7.4.2 Vertical-Axis Wind Turbines

Currently, HAWTs dominate the wind energy market due to their large size and high power generation characteristics. However, vertical-axis wind turbines (VAWTs) are capable of producing a lot of power and have many advantages. The main advantage of it is that it can receive wind from any direction; therefore, there is no need of a steering device to change the direction of the rotor to face the wind. This is a big advantage on sites where the wind direction is highly variable or has turbulence. Since there is no need of the steering device, the structure of the vertical wind turbine is simple. It has another advantage that the gearbox and the generator can be directly coupled to the axis on the ground. The wind turbine itself is also near the ground, unlike the horizontal-axis turbine where everything is to be placed on a tower. Similarly to HAWTs, the VAWTs are also of two types: lift-based and drag-based. Lift-based designs are generally much more efficient than drag or "paddle" designs.

The first aerodynamic VAWT was developed by Georges Darrieus in France and was first patented in 1927. It has a rotor with two or three thin curved blades with an airfoil section. Its principle of operation depends on the fact that its blade speed is a multiple of the wind speed.

The most common type of lift-type turbine is the Darrieus-type wind turbine. It has many different models, such as the Φ structure, Δ structure, Y structure and H structure. But as its wind area is small and the starting wind speed is high, it is not available on large scale. The maximum torque in this turbine occurs when blades are moving across the wind at a speed greater than wind speed. The popular type of turbine is the H structure type shown in Fig. 7.7. During working, the rotor drives the generator to send power to the controller and output the required power to the electrical equipment.

The drag-based type uses aerodynamic resistance of the wind, and the most typical structure is S-type rotor, which consists of two semi-cylindrical blades whose axes are staggered. A drag-type wind turbine cannot rotate at speed higher than the wind speed.

**Figure 7.7** Vertical-axis wind turbine.

The main advantage of this type is the high starting torque, but the shortcoming is the asymmetrical gas flow around the rotor, which forms lateral thrust. As for the relatively large-sized wind turbine, it is difficult to employ this structure. The ratio of utilization of the wind is lower than that of the high-speed VAWT or HAWT.

Due to the reason that the gas flow of the vertical axis is more complex than that of the horizontal axis, VAWTs were developed at a later stage, and the theory is not yet mature. But the structure is simple. It can start smoothly with low wind speed and low noise. The disadvantages of the VAWT, on the other hand, are as follows:

- Most of them are only half as efficient as HAWTs due to the dragging force.
- Airflow near the ground and other objects can create a turbulent flow, introducing issues of vibration.
- VAWTs may need guy wires to hold them up (guy wires are impractical and heavy and cannot be used in farmed areas).

### 7.4.3 Main Components of Wind Turbine

There are four main components in a wind turbine. These are *turbine blades, nacelle, tower* and *control system*.

*Turbine blades*: Turbine blades of HAWTs are made of high density wood, PVC, aluminium alloy or glass fibre. These blades have airfoil type of cross section. The blades are also twisted from the outer tip to the rotor to reduce the chances of stalling. Larger rotors with longer blades sweep a greater area, increasing energy capture. But longer blades are heavier and incur greater structural load. Special designs with advanced materials such as carbon fibres can be used to reduce the weight and load. However, the advanced materials are expensive and are used only for large power turbines.

Unlike the HAWTs where the blades exert a constant torque about the shaft as they rotate, a VAWT rotates perpendicular to the flow, causing the blades to produce an oscillation in the torque about the axis of rotation. VAWT blades are designed such that they exhibit good aerodynamic performance throughout an entire rotation at the various angles of attack they experience, leading to a high time-averaged torque. The blades are therefore curved and thin which resemble an egg beater. The pitch of the blades cannot be changed. The diameter of the rotor is less than the tower height.

*Nacelle:* Nacelle is the name given to a streamlined enclosure of an aircraft. In wind energy systems, the nacelle holds all the turbine machinery. Since it is required to rotate so that it always follows the wind direction, it is connected to the tower via bearings. Turbine is the most important part for a wind turbine system. It consists of more than 10 components including the generator, rotor, gear, shaft, bearing and others. The most important part in the turbine is its rotor. The trend is to build large rotors so that a large amount of wind is swept for the same or lower loads. Wind turbines are available in a variety of sizes and power ratings. Depending on the number of blades, wind speed and type of turbine, HAWT or VAWT, rotors have been developed in various shapes and sizes.

*Towers*: In HAWT, the towers support the nacelle and rotor hub at its top. These are made from tubular steel, concrete or steel lattice. Large wind machine towers are usually made of steel, and most of these are of the tubular or conical type. Some towers have been built out of reinforced concrete sections. Lattice or truss towers, common in the early days, are rarely used except for very small machines in the range of 100 kW and below. Guyed pole towers are used for small wind machines. Towers must be designed to resist the full thrust produced by an operating windmill or a stationary wind machine in a storm.

Height of the tower is an important parameter in the design of HWAT. Because wind speed increases with height, taller towers enable turbines to capture more energy and generate more electricity. Generally, output power of the wind system increases with increase in height and also reduces the turbulence in wind.

The tower of VAWT has a hollow vertical rotor shaft which can rotate freely supported by bearings on top and bottom. Since all mechanical components are located at the ground, there is no load on the top of the tower. Thus, in VAWT, a very strong tower is not required.

*Control system*: Control of wind turbine is based on many factors [7]. These are as follows:

- To extract as much energy from the wind as possible
- Speed regulation so that noise is within limits
- Safety of turbine

The control system includes the turbine controller and inverter control of the generator. A turbine can be controlled by controlling the generator speed, blade angle adjustment and rotation of the entire wind turbine. Blade angle adjustment and turbine rotation are also known as *pitch and yaw* control, respectively. Inverter control is used to control the active power supplied by the generator. These are described later.

### 7.4.3.1 Drive Train

The mechanical power received from the rotor is transmitted to the electrical generator by means of mechanical transmission which contains all the moving parts known as drive train. In HAWT, this equipment is on top of the tower. The main criterion for the design of this equipment is therefore low maintenance. A typical HAWT drive train consists of a rotor shaft assembly (low speed shaft), a speed-increasing gearbox (not required in a direct drive system), a generator drive shaft, a rotor brake, an electrical generator and control equipment. The rotor shaft connects the rotor to the gearbox. The braking used in wind turbines is either aerodynamic braking or mechanical braking.

The coupling between the low-speed shaft and the generator shaft can be either rigid or flexible. At present, three most dominant drive train technologies used are as follows:

- High-speed drive train having overall speed ratio of 100:1
- Medium speed drive train having a speed ratio of 10:1 to 40:1
- Direct drive

The VAWT has its drive train on the ground. The blades of VWAT also do not require orientation towards the wind.

## 7.5 Dynamics and Control of Wind Turbines

Normally, large wind turbines are placed in group in wind farms. The monitoring and control of wind turbines separately as well as overall control of the wind farm are necessary for proper operation [8]. Control of HAWT is dependent on the type of machine:

whether it is upwind or down wind or a fixed pitch or variable pitch type. Moreover, a wind turbine can be of fixed speed or variable speed. A large wind turbine connected to the grid has generally four levels of controls:

- Wind farm control
- Supervisory control
- Operational control
- Subsystem control

*Wind farm control* is used for coordinated control of numerous wind turbines in a farm. It communicates with supervisory control of each turbine.

*Supervisory control* is at the top of hierarchy which decides when to start and stop the turbine depending on the wind speed and also monitors the condition of the turbine. The operational control is used in variable-speed turbines. Its main purpose is to run the machine at speed between 6 m/s and 11.7 m/s which is the region of maximum power coefficient. The purpose of this control is to operate the turbine at a constant tip speed ratio, corresponding to the maximum power coefficient and use pitch control to operate the turbine at maximum power output point. In this region, the turbine rotor runs at variable speed. For speeds above 11.7 m/s and less than about 25 m/s, the turbine is operated by using pitch control to keep it running at a constant speed.

*Operational control* determines how the turbine achieves its control objectives. *The subsystem controller* is used to control the generator, associated power electronics, yaw drive and pitch drive to achieve desirable performance.

### 7.5.1 Pitch Control

In case of stronger winds, it is necessary to waste part of the excess energy of the wind in order to avoid damage of the wind turbine. All wind turbines are therefore designed with some sort of power control. There are different ways of doing this safely on modern wind turbines. Pitch control, yaw and tilt control and stall control are the methods used for this purpose.

The pitch control system is one of the most widely used control techniques to regulate the output power of a wind turbine generator. The pitch of a turbine blade is controlled by rotating it from the root where it is connected to the hub. As the pitch angle is changed, the power captured by the turbine is also changed. Hydraulic actuators are used to control the pitch angle.

Pitch control can be active or passive. In active-pitch-controlled wind turbine, the turbine's electronic controller checks the power output of the turbine several times per second. When the power output becomes too high, it rotates the blade of the turbine through pitch mechanism which immediately tilts the blades slightly out of the wind. The blades are turned back into the wind whenever the wind speed becomes safe again. The pitch-controlled wind turbine will generally pitch the blades a few degrees every time the wind changes in order to keep the rotor blades at the optimum angle in order to maximize the output for all wind speeds.

In passive pitch control, the blades and their hub mountings are designed to twist to limit the load due to higher wind speeds. This scheme is not easy to implement and is generally used for stand-alone wind turbines.

## 7.5.2   Yaw Control

In yaw control, the entire wind turbine is rotated in the horizontal axis. Yaw control ensures that the turbine is constantly facing the wind to maximize the effective rotor area and, as a result, power. Because wind direction can vary quickly, the turbine may misalign with the oncoming wind and cause power output losses. Yaw control is used in small horizontal-axis turbines. It turns the nacelle so that the rotor always faces the wind. For small wind turbines, the same yaw control is used to control the power. The purpose of yaw control here is to yaw the turbine out of wind to limit the power during high wind.

Yaw control mechanism uses electric motors and gearboxes to orient the rotor.

## 7.5.3   Passive and Active Stall Power Control

The basic advantage of using stall control is that it avoids the moving parts in the rotor and a complex control system. However, stall control represents a very complex aerodynamic design problem and related design challenges in the structural dynamics of the whole wind turbine, for example, to avoid stall-induced vibrations. Around two-thirds of the wind turbines currently being installed in the world are stall-controlled machines.

The stall control of wind turbine can be active or passive.

### 7.5.3.1   Passive Stall Control

The blades of passive stall-controlled wind turbines are bolted onto the hub at a fixed angle. The geometry of the rotor blade profile, however, is designed aerodynamically to ensure that, when the wind speed becomes too high, it creates turbulence on the side of the rotor blade which is not facing the wind. As the actual wind speed in the area increases, the angle of attack of the rotor blade will increase, until at some point it starts to stall. Rotor blades for a stall-controlled wind turbine are twisted slightly in its longitudinal axis. This is partly done in order to ensure that the rotor blade stalls gradually rather than abruptly when the wind speed reaches its critical value.

### 7.5.3.2   Active Stall Control

An increasing number of larger wind turbines (with 1 MW and more capacity) are equipped with active stall power control mechanisms. The active stall machines are pitch-controlled machines, since they have pitch-controlling blades. At low wind speeds, the machines will usually be programmed to pitch their blades much as a pitch-controlled machine. However, when the machine reaches its rated power, and the generator is about to be overloaded, the machine will pitch its blades in the opposite direction from what a pitch-controlled machine is supposed to do. Thus, it will increase the angle of attack of the rotor blades in order to make the blades go into a deeper stall, thus wasting the excess energy in the wind.

One of the advantages of active stall is that one can control the power output more accurately than with passive stall, so as to avoid overshooting the rated power of the machine at the beginning of a gust of wind. Another advantage is that the machine can be run almost exactly at rated power at all high wind speeds. A normal passive stall-controlled wind turbine will usually have a drop in the electrical power output for higher wind speeds, as the rotor blades go into deeper stall.

## 7.6   Wind Turbine Condition Monitoring

There is a constant need for the reduction of operational and maintenance costs of WECSs through the adoption of reliable and cost-effective condition monitoring techniques that allow for the early detection of any degeneration in the system components. Wind turbines are different from traditional rotating machines as they usually operate in remote locations, rotate at low and variable speed and work under constantly varying loads. For these reasons, a reliable condition monitoring system (CMS) for wind turbines is essential to avoid catastrophic failures and to minimize the requirement of costly corrective maintenance.

Among different techniques available for condition monitoring, vibration analysis and oil monitoring are the most predominantly used for wind turbines. Vibration analysis can pinpoint the crack locations, and oil monitoring can detect lubricant deterioration. Gearbox fault in the drive train is considered to be most common fault occurring in wind turbines.

CMS is generally composed of electronic data acquisition module, and further for vibration-based condition monitoring, there is a number of accelerometers and other transducer installed on the turbine drive train, main shaft, generator and main bearing. The standard CMS can be divided into three main components; this includes data gathering or data collection followed by the data acquisition and finally fault diagnostic algorithm or fault diagnosis. In the first phase, data is collected and then converted into digital format for further processing. In the next stage, the data acquisition receives data from the data collection module for calculating the statistical function and values in frequency and time domains. This block further applies different techniques for analyzing of data pattern for further processing. Fast Fourier transformation is generally used to convert a time-domain signal into frequency-domain one. Vibration-based condition monitoring is one of the most effective techniques that can detect gearbox failure of wind turbine especially on the high speed side rather than on the low speed side. This is attributed to the inherent sensitivity of the accelerometers to high-frequency vibration.

The third block which is more critical and complex, because of working with diagnosing fault, due to its functionality and importance in machine fault diagnosis, different methods and techniques are developed, and still a major work is under development. Newly developed analysis techniques such as wavelet analysis, fuzzy logics, case-based reasoning and neural networks for vibration or fault pattern recognition are being successfully applied and examined.

Vibrations may not be evident while faults are developing, but analysis of the oil can provide early warnings. Oil analysis is mainly applied to the gearbox, as it is the only oil-lubricated component in the drive train. The objective of oil monitoring is to detect oil contamination and degradation. Oil analysis is mostly executed off-line by taking samples, although on-line sensors are available and used rarely. The typical parameters sought in an oil sample analysis include particle counts, water content, total acid number, viscosity and particle element identification.

It is possible to detect wind turbine drive train faults through the terminals of the associated generator [9]. Since wind power generation systems mostly use induction generators, the techniques and tools available, to monitor the condition of induction machines, can be used. Some of the technologies used for monitoring include sensors, which may measure the speed, output torque, vibrations, temperature, flux densities and so on. These sensors are together coupled with algorithms and architectures, which allow for efficient monitoring of the machine condition.

## 7.7  Wind Energy Conversion Systems (WECS)

WECSs convert wind energy into usable form of electrical energy using power generation systems. There are a number of ways by which WECSs can be classified [10].

### 7.7.1  Based on Capacity of Power Generation

These are classified as small (less than 2 kW), medium (2–100 kW) and large (more than 100 kW). Second classification is based on the speed of the turbine which may be fixed speed or variable speed. Third method is based on whether it is grid-connected which may be in the form of microgrid or isolated system feeding only local loads. Fourth method of classification is based on the type of electric generator along with its control used. Several types of generators are available for power generation using wind turbines. The overall wind power conversion technologies can be divided into three categories.

- Systems without power electronics
- Systems with partially rated power electronics (converter rating less than the total power rating of the system)
- Systems with full-scale power electronics

### 7.7.2  Systems without Power Electronics

Systems without power electronics are fixed-speed power generation systems using squirrel-cage induction generator (SCIG). The generator shaft is driven by the wind turbine, and its stator is connected to the grid in a grid-connected system or to the load in an isolated system. In the grid-connected system, the speed of the generator and therefore that of turbine are fixed by the grid frequency. The excitation is provided by the power supply from the grid. A capacitor bank may be connected across the stator of the induction generator to limit the reactive power absorption from the grid. In stand-alone systems, the capacitors are connected to provide excitation.

Initially, the induction machine runs as an induction motor till its speed is equal to the synchronous speed. At synchronous speed, the machine neither delivers power nor takes power from the grid except the losses. When the speed of the machine driven by wind turbine exceeds the synchronous speed, it works as a generator and starts supplying power to the grid or load as the case may be. The maximum power supplied by the generator occurs at a slip of 1–2%. This is why it is known as fixed-speed WECS. SCIGs are preferred because they are simple, require low maintenance, are robust and stable. Their main drawback is that in order to get more active power production, more reactive power is required. SCIGs used in fixed-speed turbines can cause local voltage collapse due to fault. During a fault, grid voltage is reduced which results in less power delivery. The machine therefore accelerates due to the imbalance between the mechanical power from the wind and the electrical power that can be supplied to the grid. When the fault is cleared, these machines absorb reactive power, further decreasing the network voltage. If the voltage does not recover quickly, the wind turbines continue to accelerate and

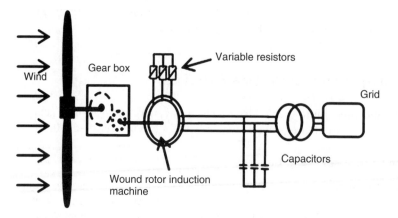

**Figure 7.8** Wound rotor induction generator with rotor resistance control.

to consume larger amounts of reactive power. This eventually leads to voltage and rotor speed instability.

Systems with partially rated power electronics and systems with full-scale power electronics are variable-speed power generation systems. These systems use pitch control in the turbine to extract wind power. In partially rated power electronic system, the performance is improved by using slip ring induction machine with power electronic control. Figure 7.8 shows a wound rotor induction generator with power electronic control of rotor resistance. This dynamic slip controller can provide a speed control in the range of 2–10%. The power converter requirement for control of rotor resistance is high current at a low voltage. It is also possible using this control to keep the output power fixed at higher speeds.

Another solution with partial power electronic control is by using doubly fed induction generator (DFIG). Here power electronic converter is connected in the rotor circuit to extract the slip energy from rotor and supply it to the grid. As shown in Fig. 7.8, the stator is directly connected to the grid and the rotor windings are connected to the grid through PE converter. When the machine is running at super-synchronous speed, the power is delivered by both the rotor and the stator to the grid. If the machine is running at sub-synchronous speed, still the power can be supplied to the grid through rotor. A speed variation of ±30% around the synchronous speed can be obtained by the converter having only 30% of total capacity of the machine.

The converter in the rotor circuit can provide control of both active and reactive power which provides better grid performance. This method of control is slightly more expensive than rotor resistance control, but it can save the money spent on reactive power controller, and it can extract more wind energy compared to the rotor resistance control scheme. Since the power rating of the converter is only about 30% of total power rating, the overall cost is less compared to systems with full-scale power converter.

Full-scale converters are rated equal to the total capacity of the system and therefore are expensive and produce more losses but have advantage in technical performance. These systems employ a DFIG, conventional synchronous generator or permanent magnet synchronous machine. A DFIG wind energy system is shown in Fig. 7.9. At very low wind speeds, the rotational speed of the system will be fixed at maximum allowable slip to prevent over-voltage of generator. For speeds where power production is below the

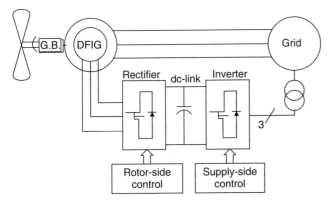

**Figure 7.9** Wind energy conversion system with DFIG.

maximum power, the wind turbine will vary the rotational speed proportional to the wind speed and keep the pitch angle fixed. When the turbine power is above the nominal power, a pitch angle controller is used to limit the power by suitably rotating the blades. The total electrical power of the WECS is regulated by controlling the DFIG through rotor-side converter. The grid-side converter is used simply to keep the dc-link voltage fixed.

The types of synchronous generators used in the wind turbine industry are as follows:

1) Wound rotor synchronous generator (WRSG)
2) Permanent magnet synchronous generator (PMSG).

The synchronous generator with a large number of poles can be used for direct-drive applications without any gearbox. PMSGs do not require external excitation current, require no slip rings, require less maintenance and have lower losses.

A multi-pole PMSG with wind turbine control is shown in Fig. 7.10. An advantage of this WECS is that the dc link provides decoupling between the turbine and the grid. The dc link can also be used to connect it to energy storage systems (if present) for better power control to the grid. As shown in Fig. 7.10, the generated active power is controlled by the generator-side converter, and the reactive power is controlled by the grid-side converter. A dc chopper is normally introduced to prevent overvoltage of dc link in case of grid faults.

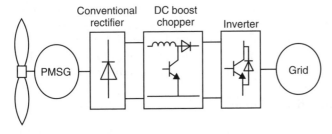

**Figure 7.10** WECS with PM generator.

## 7.8   Offshore Wind Energy

Offshore wind speeds are generally higher and more consistent than onshore winds. It is therefore possible to extract about 20–30% more energy using offshore wind installations. Offshore wind turbines are generally built with larger capacity because there are fewer constraints on the component and assembly equipment transportation.

The first offshore wind farm was built in Denmark in 1991. First large-scale offshore wind farm was also built by Denmark in 2002 with a capacity of 80.2 MW. Offshore wind technology consists of five key components and processes: the wind turbine, the foundation, the electrical connection, their installation and their operation and maintenance. The technologies for offshore wind farms are not developed especially for this purpose, but the known technologies already developed for other industries [11] are used.

### 7.8.1   Offshore Wind Turbines

Offshore wind turbines have, to date, largely use modified versions of the largest onshore wind designs that are more suitable for high-capacity factors [12]. Offshore turbines require technical modifications and substantial system upgrades for adaptation to the marine environment. These modifications include strengthening the tower to cope with loading forces from waves or ice flows, pressurizing the nacelles to keep corrosive seawater from critical electrical components and adding brightly coloured access platforms for navigation safety and maintenance access. Change in design is required to obtain several other benefits. Technology is already being developed for direct-drive, gearless nacelles, improved generators (e.g. fixed magnets, increased generator coils) and to increase condition monitoring.

Offshore turbines have fewer constraints than onshore turbines in terms of visual impact and noise (particularly in regard to planning), but there are greater costs associated with reliability and servicing. For this reason, offshore turbines may have automatic greasing systems to lubricate bearings and blades as well as heating and cooling systems to maintain gear oil temperature within a specified range.

### 7.8.2   Foundation

The offshore foundation system depends on the water depth. Most of the projects installed so far have been in water less than 22 m deep, with a demonstration project in Scotland at a depth of 45 m. Shallow water technology at present uses monopile foundations for about 20 m depth. These are tried and tested technologies used in marine construction. Basically, these are simple steel tubes, hammered into the seabed. The industry is also considering the use of concrete gravity-based structures, adaptations of monopiles (tripods and triples) and jacket structures (as used for oil and gas platforms). For depths beyond 60 m, floating foundations are being developed. However, floating foundations still need reliable solutions, including advanced control systems to deal with wind, ocean waves, tides, ice formation and water currents simultaneously.

### 7.8.3   Electrical Connection and Installation

Installation of foundations and turbines is currently achieved with the use of standard jack-up barges and some custom-built vessels. Typically, these have 4–6 legs that extend into the seabed and lift the vessel completely out of the water. Normal HVAC subsea cables are used if the distance is not large. Cable installation uses a "cable plough" that digs a shallow trench in the seabed and buries the cables. Jack-up vessels can currently install in depths up to 35 m. Much beyond this, special floating installation vessels with hydraulics and jet thrusts might be needed. In future, when the wind farms will be located far away from the shore, HVDC will be used.

### 7.8.4   Operation and Maintenance

In the area of operation and maintenance, the priority is to increase the reliability of wind turbines and therefore minimize unscheduled repairs. Remote condition monitoring may reduce the need for repairs.

## 7.9   Advantages of Offshore Wind Energy Systems

Offshore wind has three inherent cost advantages over onshore wind. These are as follows:

1) *Larger capacity*: Offshore wind farms could be as large as 1.5 GW or even more due to the availability of almost unlimited space allowing a higher number of turbines. In addition, turbines can be larger in size as transportation by sea is not a major problem.
2) *Higher wind speed*: Offshore wind speeds are generally higher and more consistent than onshore winds.
3) *No requirement of land*: These can be located in the coasts near load centres.

## 7.10   Environmental Impact of Wind Energy Systems

There are many advantages of using wind power for the generation of electricity in terms of emission of carbon dioxide or other pollutants, but there are certain negative aspects which also must be looked into [13].These are as follows:

- Noise
- Electromagnetic interference
- Effect on wildlife
- Effect on ecosystem

### 7.10.1   Impact of Noise

As with any moving machinery, wind turbines also generate noise during operation. There are two main sources of noise in wind turbines. *Mechanical noise* caused by gearbox, bearings, generator and pitch control mechanism. The other noise known as *aerodynamic noise* is created due to interaction of airflow with the blades of turbine. The sound thus produced can be described as a *swishing sound.* This noise tends to increase with the speed of rotation, that is, it is more at higher wind speeds. Although the noise at higher speeds in more dominant, usually it is not noticeable because of the background noise produced by the winds. The nuisance due to noise is therefore more at low wind speeds which may look strange.

### 7.10.2 Electromagnetic Interference

The presence of tall wind turbine towers near radio, television or microwave towers can sometime reflect some of EM radiation in such a way that the reflected wave may interfere with the original signal at the receiving end. This can cause the received signal to be distorted significantly. The electromagnetic scattering properties of wind turbines are not simple to describe. The extent of electromagnetic interference due to wind turbines depends mainly on the materials used to make the blades and on the shape of the tower.

## 7.11 Combining the Wind Power Generation System with Energy Storage

Most of the renewable energy systems including wind energy are intermittent in nature. Since it is difficult to predict and control the output of wind generation, its potential impacts on the electric grid are different from the traditional energy sources. At a high penetration level, an extra fast-response reserve capacity is needed to cover the shortfall of generation when a sudden deficit of wind takes place. Various storage systems such as pumped hydro and battery are possible to integrate the wind energy to the present network. The pumped hydro is difficult to be constructed because of the geographic restrictions and environmental regulation. The batteries are not suitable to stabilize the output of the large wind energy farm because of their cost. WTES, which employs low-cost thermal energy storage system and light and low-cost heat generator, could be a better solution than the combination of wind power and thermal plant [14, 15].

## 7.12 Summary

Wind energy systems are now considered one of the important renewable energy sources. They now contribute to the energy mix in more than 70 countries of the world, with Denmark and Germany taking the lead. Wind turbines are classified into two general types: *horizontal axis* and *vertical axis*. A horizontal-axis turbine has its blades rotating on an axis parallel to the ground, whereas a vertical-axis machine has its blades rotating on an axis perpendicular to the ground. However, the horizontal-axis type machines are the most common, and very few vertical-axis machines are running commercially.

The future trend of WECSs is the increase in the power capacity of wind turbines and generators so that the cost of generated electricity may be reduced.

Offshore WECSs are also being considered and installed in a few countries.

## References

1 Burton, T. (2011) *Wind Energy Handbook*, 2nd edn, Wiley.
2 Xi, L.U. *et al.* (2009) Global potential for wind-generated electricity, Proceedings of the National Academy of Sciences of the United States of America.
3 Golding, E.W. (1955) *Generation of Electricity by Wind Power*, E. & E. N. Spon Ltd.

 **4** Strategies Unlimited (1988) *The Potential for Wind Energy in Developing Countries*, Strategies Unlimited.
 **5** Manwell, J.F. *et al.* (2002) *Wind Energy Explained: Theory, Design, and Application*, John Wiley & Sons.
 **6** Walkewr, J.F. and Jenkins, N. (1997) *Wind Energy Technology*, John Wiley & Sons.
 **7** Laks J.H. *et al.* (2009) Control of wind turbines: past, present, and future, American Control Conference .
 **8** Pao, L.Y. and Johnson, K.E. (2009) A tutorial on dynamics and control of wind turbines, American Control conference.
 **9** Lu, B., Li, Y., Wu, X., and Yang, Z. (2009) A review of recent advances in wind turbine condition monitoring and fault diagnosis, EEE pp. 1–7, Jun. 2009.
 **10** Amirat, Y. *et al.* (2009) A brief status on condition monitoring and fault diagnosis in wind energy conversion systems. *Renewable and Sustainable Energy Reviews*, **13** (9), 2629–2636.
 **11** Li, H. and Chen, Z. (2008) Overview of different wind generator systems and their comparisons. *IET Renewable Power Generation*, **2**, 13–138.
 **12** Olimpo, A.L. *et al.* (2014) *Offshore Wind Energy Generation: Control, Protection, and Integration to Electrical Systems*, John Wiley & Sons.
 **13** Jaber, S. (2013) Environmental impacts of wind energy. *Journal of Clean Energy Technologies*, **1** (3), 251–254.
 **14** Toru, O. *et al.* (2015) Concept study of wind power utilizing direct thermal energy conversion and thermal energy storage. *Renewable Energy*, **83**, 332–338.
 **15** Ramteen, S. and Paul, D. (2013) Benefits of co-locating concentrating solar power and wind. *IEEE Transaction sustainable energy*, 877–885.

# 8

# Biomass Energy Systems

## 8.1 Biomass Energy

The term *biomass* is used for materials derived from plants and animals, including their wastes and residues. Biomass is produced by green plants that use the energy of sunlight to convert carbon dioxide and water into simple sugar and oxygen. The energy obtained from biomass is known as *biomass energy*. Photosynthesis is fundamental to the conversion of solar radiation into stored biomass energy. The initial energy in the biomass system is obtained from solar radiation in the form of photosynthesis. In plants, algae and certain types of bacteria, the photosynthetic process results in the release of molecular oxygen and the removal of carbon dioxide from the atmosphere that is used to synthesize carbohydrates (oxygenic photosynthesis). This conversion process consists of a series of chemical reactions that require carbon dioxide ($CO_2$) and water ($H_2O$) and store chemical energy in the form of sugar. Light energy from the Sun drives these reactions. Oxygen ($O_2$) is a by-product of photosynthesis and is released into the atmosphere. The following equation summarizes photosynthesis:

$$6CO_2 + 6H_2O + \text{light energy} \rightarrow C_6H_{12}O_6 \text{ (glucose)} + 6O_2 \qquad (8.1)$$

The reaction produces $C_6H_{12}O_6$ which is glucose and oxygen which is released into atmosphere. The theoretical achievable efficiency of conversion is limited by both the limited wavelength range applicable to photosynthesis and the quantum requirements of the photosynthetic process. Because of these limitations, the theoretical maximum efficiency of solar energy conversion is approximately 11%. In practice, however, the magnitude of photosynthetic efficiency observed in the field is further decreased by factors such as poor absorption of sunlight due to its reflection, respiration requirements of photosynthesis and the need for optimal solar radiation levels. The net result is an overall photosynthetic efficiency of between 3% and 6% of total solar radiation [1]. When these carbohydrates are burnt, they are converted back into carbon dioxide and water and release the energy they captured from the Sun. The biomass can be used to produce energy using any of the following methods:

- Burned in power plants to produce heat or electricity, with fewer harmful emissions than coal.
- Fermented to produce fuels, such as ethanol, for cars and trucks.
- Digested by bacteria to create methane gas for powering turbines.

*Operation and Control of Renewable Energy Systems*, First Edition. Mukhtar Ahmad.
© 2018 John Wiley & Sons Ltd. Published 2018 by John Wiley & Sons Ltd.

- Heated under special conditions, or "gasified", to break down into a mix of gases that can be burned for electricity or used to make a range of products, from diesel to gasoline to chemicals.

In this way, biomass can be considered as a sort of natural battery for storing solar energy. Biomass takes carbon out of the atmosphere while it is growing and returns it as it is burnt. If it is managed on a sustainable basis, biomass is harvested as part of a constantly replenished crop such as fast-growing willow trees. This maintains a closed carbon cycle with no net increase in atmospheric $CO_2$ levels. If the consumption does not exceed its natural level of production, the burning of biomass will not generate more heat or carbon dioxide than would have been formed by the natural process.

As described in this section, a wide variety of fuels can be considered as "biomass". These include animal manure, landfill waste, wood pellets, vegetable oil, algae, crops such as corn, sugar, switchgrass and other plant material – even paper and household garbage – can be used as a biomass fuel source. Various biofuel resources are shown in Fig. 8.1.

Fossil fuels such as coal, oil and gas are also derived from biological material; however, they absorbed $CO_2$ from the atmosphere many millions of years ago. The vital difference between biomass and fossil fuels is one of time scale. That is why fossil fuels are not considered as renewable because they are out of carbon cycle for a very long time. Fossil fuels offer a high energy density when burnt. By doing this, the stored carbon is oxidized into carbon dioxide and the hydrogen into water (vapour). If these emissions are not captured and stored, they are released back into the atmosphere, thus returning the carbon captured millions of years ago and thus adding to increased amounts of $CO_2$ into the atmosphere.

There are a number of conversion technologies used for the production of power and heat as well as transportation fuels. Biomass fuel can be converted directly into

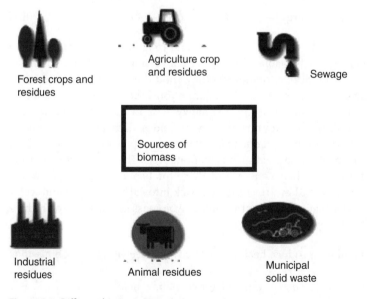

**Figure 8.1** Different biomass resources.

heat energy through combustion, similarly to the burning of a log. However, in most cases, biomass is converted into a more convenient form, by chemical or biological process, to produce *biofuels*. Examples of biofuels include methane gas, liquid ethanol, oils and solid charcoal. The biofuels react with oxygen to release heat, and the energy is dissipated.

## 8.2 Biomass Production

There is a wide variety of biomass feedstocks available, and they can be produced anywhere in the world. Every region has its own locally generated biomass feedstocks from agriculture, forest and urban sources Furthermore, most feedstocks can be made into liquid fuels, heat, electric power and/or biobased products.

The two groups of feedstocks primarily responsible for producing biomass fuels are as follows:

Group I Sources:
- Forest industries
- Agriculture.

Group II Sources:
- Food waste
- Industrial waste and co-products

In the first category, the biomass may consist of sawdust which is produced when a log is cut to make lumber in lumber mills and logging by-products. A lot of biomass fuels are available as by-product from other activities, such as saw milling and manufacturing of plywood and particle board.

Energy crops (switchgrass), crop residue and so on are also included in the first group. Forest industry provides biomass materials as by-products such as logging residues and as main products which are cut directly from trees and forests. The planted forests are usually thinned out to maintain the growth space among stumps. The woody material thus cut is utilized as biomass.

Dedicated energy crops are another source of woody biomass for energy. Short-rotation (3–15 years) techniques for growing poplar, willow, Eucalyptus or even non-woody perennial grasses (e.g. Miscanthus) have been developed over the past two to three decades.

In the second category, the biomass may come from construction or organic municipal waste, food waste and even chicken litter. Food wastes are generated in restaurants, supermarkets, residential blocks, cafeterias, airline caterers, food processing industries and so on. In most of the cases, it is dropped in landfill sites and is left to rot, thereby releasing greenhouse gases in the atmosphere. Anaerobic digestion (AD) can be used as described later to convert this waste into energy. Industrial waste energy recovery and recycling come under the technology of cogeneration or combined heat and power (CHP) generation. Energy recovery from municipal wastes known as MSW and from some of the industries is also possible and is being utilized.

Since the rapid expansion of biomass energy today relies largely on wood from forests, the energy produced by the combustion of biomass from forest wood and woody residue as well as other technologies for the generation of energy is discussed here.

## 8.2.1    Forest Industries

Wood is the first biomass resource which has been used by human beings as fuel for cooking for the last few centuries. Wood is one of the oldest energy sources and has been in constant use throughout the modern era, despite the widespread adoption of other types of fuels. Even now, more than 2 billion people depend on wood energy for cooking and/or heating, in households in developing countries. It represents one-third of the global renewable energy consumption, making wood the most decentralized energy in the world. Forests are a major source of woody biomass. However, woody materials can also be sourced from agriculture. For example, fast-growing tree species such as hybrid willow (Salix) and poplar have been developed for the production in agricultural settings (i.e. grown as row crops on farms).

Today, wood energy has become an important source of green energy because of climate change and energy security concerns. Wood energy is considered as a climate neutral and socially viable source of renewable energy, provided the trees harvested as biomass are re-planted as fast as the wood is burnt. In such condition, the new trees absorb the carbon dioxide produced by the combustion of old trees, and the carbon cycle theoretically remains in balance. Thus, no extra carbon is added to the atmospheric balance sheet. Biomass can only maintain carbon balance if fast-growing crops are grown on otherwise unproductive land; in this case, the re-growth of the plants offsets the carbon produced by the combustion of the crops. But cutting or clearing forests for energy, either to burn trees or to plant energy crops, releases carbon into the atmosphere that would have been remained stored if the trees had remained untouched.

Biomass for fuel can be gathered as waste or grown in fields. The most common biomass resources are therefore from agricultural and forestry. Forest residues and wood wastes represent a large potential resource for energy production and include forest residues, forest thinnings, and primary mill residues [2]. But the energy potential of waste biomass is relatively small compared to the virgin biomass as an energy resource. The key to the large-scale production of energy, fuels, from biomass is to grow suitable virgin biomass species in an integrated biomass-production conversion system (IBPCS) at costs that enable the overall system to be profitable.

Even though the costs for these fuels are usually greater than that of coal, they reduce the fuel price risk by diversifying the fuel supply. These also result in significantly lower sulfur dioxide and nitrogen oxide emissions compared to coal and can easily be cofired. [3]

## 8.2.2    Forest Residues

Forest residues are the biomass material remaining in forests from harvest operations that are left in the forest after stem wood removal, such as branches, foliage and roots. Harvesting may occur as thinning in young stands or cutting in older stands for timber or pulp that also yields tops and branches usable for biomass energy production. Harvesting operations usually remove only 25–50% of the volume, leaving the residues available as biomass for energy. Stands damaged by insects, disease or fire are additional sources of biomass. Forest residues normally have low density and fuel values that keep transport costs high, and so it is economical to reduce the biomass density in the forest itself.

Since lumber mills and other processing facilities use timber of certain quality, forest residues are left in the forest. These residues could be collected after a timber harvest and used for energy purposes. Typically, forest residues are either left in the forest or disposed of via open burning through forest management programs. The primary advantage of using forest residues for power generation is that an existing collection infrastructure is already available to harvest wood in many areas.

### 8.2.2.1 Forest Thinnings

Thinning of forest removes surplus trees to concentrate timber production plantation resulting in increased diameter growth and producing more valuable trees. Underbrush and saplings smaller than 2 inches in diameter, as well as fallen or dead trees, are also considered as forest thinnings. However, the actual business of harvesting, collecting, processing and transporting loose forest thinnings is costly and presents an economic barrier to their recovery and utilization for energy. Typically, the wood waste from forest thinnings is disposed of through controlled burning due to the expense of transporting it to a power generation facility.

### 8.2.3 Agriculture Residues

Agriculture residues are traditionally considered as "trash", or agricultural waste is increasingly being viewed as a valuable resource. Agricultural residues such as straw, rice husk, coconut shell, groundnut shell and sugar cane bagasse are an excellent alternative to using virgin wood fibre as they have many advantages. These are available in abundance and are renewable. Rice produces both straw and rice husks at the processing plant which can be conveniently and easily converted into energy. Significant quantities of biomass remain in the fields in the form of cob when maize is harvested which can be converted into energy. Sugar cane harvesting leads to harvest residues in the fields while processing produces fibrous bagasse, both of which are good sources of energy.

Rice, wheat, sugar cane, maize (corn), soybeans and groundnuts are just a few examples of crops that generate considerable amounts of residues. Following the harvest of these agricultural crops, residues such as crop stalks, leaves and cobs are left in the field. A segment of these residues could potentially be collected and combusted to produce energy. These residues constitute a major part of the total annual production of biomass residues and are an important source of energy for both domestic and industrial purposes. However, transportation of residues can be expensive and not highly energy efficient.

### 8.2.4 Energy Crops

Dedicated energy crops are another source of woody biomass for energy. These crops are fast-growing plants, trees or other herbaceous biomass which are harvested specifically for energy production. Crops such as switchgrass, hybrid poplars (cottonwoods), hybrid willows and sugar cane are being studied for their ability to serve as energy crops for fuel. These crops have the greatest potential for dedicated energy use over a wide geographic range. One of their great advantages is that they are short rotation crops; they re-grow after each harvest, allowing multiple harvests without having to re-plant. Corn and sorghum serve a dual purpose as they can be grown for fuel, with the leftover by-products being used for other purposes, including food. Similarly, sugar cane crops are used for food (sugar) and energy in the form of ethanol. Oil-bearing trees such as soya bean, corn, sunflower, soybean, castor and wild plants such as jatropha and karanj are used to produce biodiesel. Sugar cane is a major source for bioethanol. Jatropha is a shrub that grows up to 5 m in height. It can be planted even in desert and can be grown in places where no crop is grown normally. These can be used for energy production after 2–3 years.

Karanj is another wild tree which is generally used to stop soil corrosion along highways and canals. The natural growth of karanj is along coasts and river banks in India and Myanmar. It is now also grown in the Philippines, Malaysia, Australia and Seychelles. Miscanthus (commonly known as Elephant Grass) is a high-yielding energy crop that grows over 3 m tall, resembles bamboo and produces a crop every year without the need for re-planting. The rapid growth, low mineral content and high biomass yield of Miscanthus increasingly make it a favourite choice as a biofuel, outperforming maize (corn) and other alternatives. Biodiesel can be produced from non-food feedstock. This means that the feedstock does not serve to diminish food supply. It also means that feedstock used for biodiesel can be shielded from global commodity prices.

Thus, a significant and increasing fraction of agricultural land worldwide is likely to be dedicated to the production of energy crops in the near future. This is a matter of concern that cultivation of energy crops might reduce land availability for food production.

### 8.2.5 Food and Industrial Wastes

While there is an obvious need to minimize the generation of wastes in cities and to reuse and recycle them, the technologies for recovery of energy from wastes can play a vital role in mitigating the problems. Besides recovery of substantial energy, these technologies can lead to a substantial reduction in the overall waste quantities requiring final disposal, which can be better managed for safe disposal. Food processing wastes are being used throughout the world as biomass feedstocks for energy generation. These wastes include house garbage, also called municipal solid waste (MSW), trash that comes from plant or animal products, food waste from restaurants. Many food materials are processed at some stage to remove components that are inedible or not required such as peel/skin, shells, husks, cores, pips/stones, fish heads, pulp from juice and oil extraction. Through a process called co-digestion, many treatment plants are enhancing AD with organic waste. Adding fat, oil and grease (FOG), as well as food waste, to AD accelerates this process, producing more methane gas for beneficial use and reducing the amount of solid waste conveyed to landfills.

Lawn clippings and leaves are all examples of biomass trash. Animal farms are also producing wastes that are not converted into fuels at present. These wastes are contributing to environmental degradation because many municipalities have not yet used them in energy conversion. Because of the necessity to generate energy locally, biomass resources such as MSW and sewage will become important in the future. MSW can be used to produce energy by either burning MSW in waste-to-energy plants or capturing biogas.

At an MSW combustion facility, MSW is first sorted, and items that can be recycled are removed from the waste. The material that is left is sent into a combustion chamber to be burnt. The heat released from burning is used to convert water to steam. The steam is then sent to a turbine generator to produce electricity.

## 8.3    Biomass Conversion Process

Biomass can be converted into useful forms of energy using a number of different processes. Factors that influence the choice of conversion process are the type and quantity of biomass feedstock; the desired form of the energy, that is, end-use requirements; environmental standards; economic conditions; and project-specific factors. The main application of biomass is to

1) transform the chemical energy into electrical energy and
2) production of biofuel.

Currently, conversion of biomass to energy is undertaken using three main process technologies [4]:

- thermochemical
- biochemical/biological
- mechanical extraction (with esterification)

## 8.4    Thermochemical Conversion

The main processes used for the thermochemical conversion of biomass are combustion, gasification, pyrolysis and liquefaction. The thermochemical processes can convert both food and non-food biomass to fuel products via pyrolysis and gasification [5, 6]. Charcoal is one of the major commercial products of biomass pyrolysis and is the largest single biofuel produced today. Common feedstock for gasification includes agricultural crop residues, forest residues, energy crops, organic municipal wastes and animal waste.

### 8.4.1    Combustion

Biomass conversion into heat energy is still the most popular and perhaps the most efficient process using combustion. The most common application of biomass energy in developing countries is its use as a source of heat for cooking, sometimes called traditional biomass use. The burning of biomass in air, that is, combustion, is used over a wide range of outputs to convert the chemical energy stored in biomass into heat, mechanical power or electricity. Combustion of biomass produces hot gases at temperatures around 800–1000°C. It is possible to burn any type of biomass, but in practice, combustion is feasible only for biomass with a moisture content <50%, unless the biomass is pre-dried. High-moisture-content biomass is better suited to biological conversion processes. The heat is then used in a manufacturing process or to raise steam in a boiler which can drive a steam turbine to generate electricity. Net bio-energy conversion efficiencies for biomass combustion power plants range from 20% to 40%. The higher efficiencies are obtained with systems over 100 MW or when the biomass is co-combusted in coal-fired power plants.

The size of combustion system has a very wide range from a few kilowatts of thermal input such as a single gas ring for cooking to huge coal-fired combustion boilers with inputs of 3–5 GW in a single unit.

### 8.4.2 Gasification

Gasification is a thermochemical process where solid biomass is converted into gaseous fuel without leaving any solid residue. It is a well-established technology and was extensively used in World War II to power vehicles for transport. Gasification is the conversion of biomass into a combustible gas mixture by the partial oxidation of biomass at high temperatures, typically in the range of 800–900°C. The main components of this gas are CO, $H_2$, $CO_2$, $CH_4$, $H_2O$ and $N_2$. However, a variety of tars are also produced during the gasification reaction. The partial oxidation can be carried out using air, oxygen, steam or a mixture of all of these. Air gasification produces poor quality of gas in terms of heating value. Oxygen gasification provides better quality gas in terms of energy contents. The production of syngas from biomass allows the production of methanol and hydrogen, each of which may have a future a fuel for transportation. Figure 8.2 shows a schematic diagram that illustrates the biomass gasification technology.

The conversion of biomass by gasification into a fuel is most suitable for use in an IC engine.

Basic chemical reactions in the gasification process for producing syngas are

$$C + O_2 \rightarrow CO_2 \tag{8.2}$$
$$C + H_2O \rightarrow CO + H_2 \tag{8.3}$$
$$C + CO_2 \rightarrow 2CO \tag{8.4}$$
$$CO + H_2O \rightarrow CO_2 + H_2 \tag{8.5}$$
$$C + 2H_2 \rightarrow CH_4 \tag{8.6}$$

Since there is an interaction of air or oxygen and biomass in the gasifier; these are classified according to the way air or oxygen is introduced in it as downdraft, updraft, cross-draft and fluidized bed. In an updraft gasifier, air or oxygen passes through the biomass from the bottom and the combustible gases come out from the top.

The **updraft gasifier** is the simplest type of reactor for the gasification of biomass that is easy to operate and has high conversion efficiency, although it produces high levels of tar. Because it has high tar content, these are not suitable for internal combustion engine. Updraft gasifiers shown in Fig. 8.3 achieve the highest efficiency as the hot gas passes through fuel bed and leaves the gasifier at a low temperature. The sensible heat

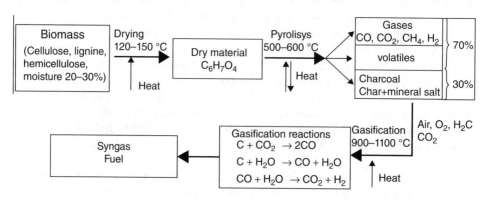

**Figure 8.2** Gasification flow diagram.

**Figure 8.3** Updraft gasifier.

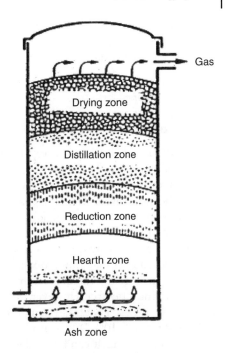

given by gas is used to pre-heat and dry fuel. The gas produced has practically no ash content.

In **downdraft gasifiers** shown in Fig. 8.4, gas is drawn from the bottom of the reactor while the hottest reaction zone is in the middle. It is suitable for a variety of biomass materials. The biomass fuel enters through the hopper and flows down, gets dried and pyrolyzed before being partially combusted by the gasifying media entering at the nozzles. In the downdraft gasifier, air contacts the pyrolyzing biomass before it contacts the char and supports a flame. The heat from the burning volatiles maintains the pyrolysis. When this phenomenon occurs within a gasifier, the limited air supply in the gasifier is

**Figure 8.4** Downdraft gasifier.

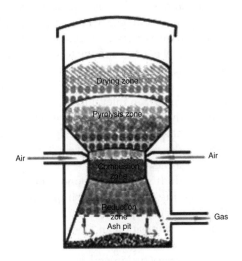

rapidly consumed, so that the flame gets richer as pyrolysis proceeds. At the end of the pyrolysis zone, the gases consist mostly of about equal parts of $CO_2$, $H_2O$, $CO$ and $H_2$. It is possible to distinguish four separate zones in this gasifier:

- Drying zone
- Pyrolysis zone
- Oxidation zone
- Reduction zone

Solid fuel is introduced at the top. In the drying zone, the drying of the fuel takes place. As the fuel moves down, it enters the pyrolysis zone. Here large molecules (e.g. cellulose, hemicellulose and lignin) break down into medium-size molecules and carbon (char) during the fuel heating. Then the products of the pyrolysis process flow downwards into the hotter zones of the gasifier. Some of these are burnt in the combustion zone, and the rest (depending on the residence time in the hot gasifier zone) will break down to even smaller molecules of hydrogen, methane, carbon monoxide, ethane, ethylene and so on.

The throat allows maximum mixing of gases in a high-temperature region, which aids tar cracking. Since the volatile matter in the fuel gets cracked within the reactor, the output gas is almost tar-free. However, the gas, as it comes out of the reactor, contains small amounts of ash and soot. They can produce as much as 20% char, but in most cases, the char content is 2–10%. While production of char reduces the quantity of energy contained in the syngas, it can be used as a fuel (charcoal) and again burnt in the gasifier. Char can also be used for soil amendment. Because char often has a high value, gasifiers are sometimes operated to produce high quantity of char with reduction in gas production. The gas from the downdraft gasifiers can be cleaned to very high purity such that it can be used in IC engines or for direct heating applications where the purity of gas is most important.

Downdraft gasifiers are of two types, single throat and double throat. Single-throat gasifiers are used for stationary applications, whereas double-throat ones are used in automotive.

The ***cross-draft-type*** gasification is one of the simplest types of gasification. These were adapted for the use of charcoal as a fuel. They have certain advantages over the updraft and downdraft gasifiers, and unlike these types, the ash bin, fire and reduction zone in cross-draft gasifiers are separated. This reactor operates on a small scale and virtually produces no tar. The reactor for this gasification is much as the updraft gasifier in that the fuel enters from the top and the thermochemical reaction occurs progressively as this fuel descends into the reactor. As shown in Fig. 8.5, the main difference is that the air enters the gasifier from the side of the reactor, instead of from the top or the bottom. Normally, an inlet nozzle is used to bring the air to the centre of the combustion zone as shown in the figure. The velocity of the air as it enters the combustion zone is considerably higher in this design, which creates a hot combustion zone. The start-up time for this reactor is relatively short, and high temperatures can be attained using this type of gasification.

Cross-draft gasifiers respond rapidly to load changes (it takes less time to start the gasifier). They are normally simpler to construct and more suitable for running engines than the other types of fixed-bed gasifiers. However, they are sensitive to changes in biomass composition and moisture content.

***Fluidized-bed gasifier*** is unique among the biomass gasifiers in one important capability: biomass fuel in any particle size range, any moisture content and any ash or grit

**Figure 8.5** Cross-draft gasifier.

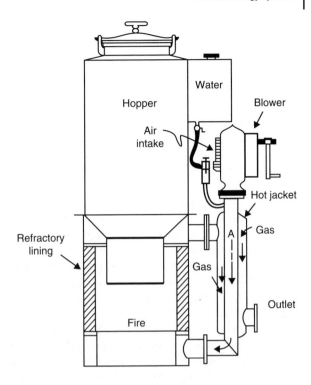

content can be gasified. The fluidizing process is the forced flow of gaseous constituents through a stacked height of solid particles. At high-enough gas velocities, the gas/solid mass exhibits fluid-like properties, thus the term fluidized-bed.

A fluidized-bed gasifier is shown in Fig. 8.6. Air, steam and mixture of oxygen and steam are commonly used fluidization media. The bed is usually composed of

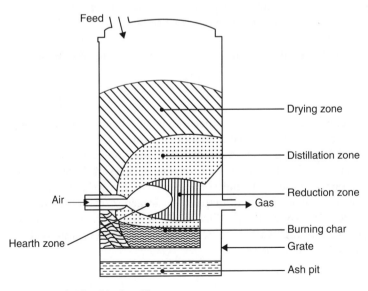

**Figure 8.6** Fluidized-bed gasifier.

sand, limestone, dolomite or alumina. There are two types of fluidized-bed gasifiers: bubbling and circulating. Bubbling-bed reactors operate at relatively lower gas velocities (less than 1 m/s) compared to circulating-bed reactors that operate at gas velocities of 3–10 m/s. Bubbling fluidized-bed gasifier consists of a horizontal air distributor with an array of bubble caps. This provides the fluidizing air to the lower furnace bed material. The bubble caps are closely spaced so that airflow is distributed uniformly over the furnace plan area. The lower furnace is filled with sand or other non-combustible material such as crushed limestone or bed material from prior operation. Airflow is forced upwards through the material, and the bed expands. The airflow through the bed is very uniform due to a high number of bubble caps.

The circulating fluidized-bed furnace has a flat-floor horizontal air distributor with bubble caps. It provides fluidizing air to the lower furnace bed material. The bubble caps are closely spaced for uniform air distribution over the furnace plan area. 50–70% of the total combustion air enters the furnace through the bubble caps with the balance injected through over-fire air (OFA) ports. Air is blown through a bed of solid particles at a sufficient velocity which drags them upwards with the gas flow and keeps these in a state of suspension. The flow of gas into the reactor must be sufficient to float the coal particles within the bed but not so high as to entrained them out of the bed. However, as the particles are gasified, they will become smaller and lighter and separated in a cyclone and then recycled to the bottom of the fluidized bed. The maximum operating point of the reactor is limited by the melting point of the bed material and is usually between 800 and 900°C. Such systems are less sensitive to fuel variations but produce larger amounts of tar and dust. They are more compact but also more complex and usually used at larger scales.

### 8.4.2.1 Applications

The main applications of biomass gasifiers are in shaft power system, direct heat application and chemical production of methanol and formic acid. The application of shaft power system is in driving farm machinery such as tractors and harvesters. These can also be used in small-scale electricity generators. It can also be used for production of hydrogen and hydrocarbons.

Direct heat applications are used in drying crops and refrigeration system. Stirling engines also use gas from these gasifiers.

### 8.4.3 Pyrolysis

Pyrolysis is a thermochemical decomposition process in which the organic material is converted into a carbon-rich solid and volatile matter by heating the biomass in the absence of air to around 500°C (Fig. 8.7). Pyrolysis is a high-temperature process in which biomass is rapidly heated in the absence of oxygen. As a result, it decomposes to generate mostly vapours and aerosols and some charcoal. The products that are formed in this process are water, charcoal, methane, hydrogen, carbon dioxide and carbon monoxide. Figure 8.7(b) shows the production of charcoal using pyrolysis. The charcoal or coke is generally of high carbon content and may contain around half the total carbon of the original organic matter.

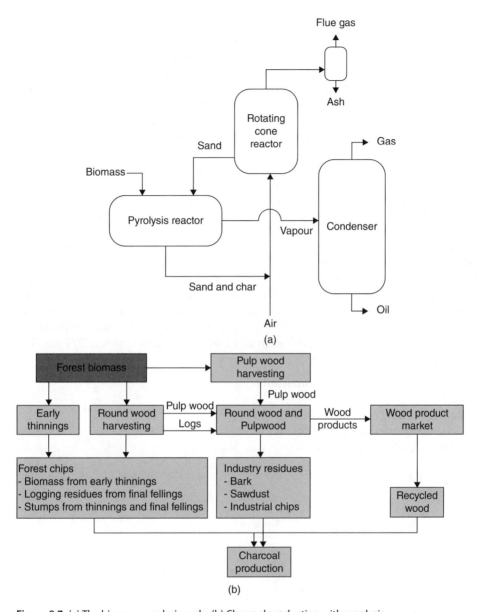

**Figure 8.7** (a) The biomass pyrolysis cycle. (b) Charcoal production with pyrolysis.

Pyrolysis is the simplest and certainly the oldest method of processing biomass to produce charcoal, a better combusting fuel. This requires relatively slow reaction at very low temperatures to maximize the solid yield. More recently, studies on the mechanisms of pyrolysis have suggested ways of substantially changing the proportions of the gas, liquid and solid products by changing the rate of heating, temperature and residence time.

Pyrolysis can be performed at relatively small scale and at remote locations which enhance the energy density of the biomass resource and reduce transport and handling costs. Temperature is the most important factor for the product distribution of pyrolysis; the best temperature for the production of the pyrolysis products is between 625 and 775 K. At lower process temperature, the product is mainly charcoal. At high temperature, the biomass will mainly produce gases, and moderate temperature is optimum for producing liquids. For highly cellulosic biomass feedstocks, the liquid fraction usually contains acids, alcohols, aldehydes, ketones, esters, heterocyclic derivatives and phenolic compounds.

Pyrolysis processes can be categorized as slow pyrolysis or fast pyrolysis. Fast pyrolysis is currently the most widely used pyrolysis system. Fast pyrolysis process is shown in Fig. 8.8 and can be described as: a process in which organic materials are rapidly heated to 450–600°C in the absence of air. Under these conditions, organic vapours, pyrolysis gases and charcoal are produced.

Fast pyrolysis is characterized by high heating rates and short vapour residence times. This generally requires a feedstock prepared as small particle sizes and a design that removes the vapours quickly from the hot solids. There are a number of different reactor configurations that can achieve this, including ablative systems, fluidized beds, stirred or moving beds and vacuum pyrolysis systems. A moderate (in pyrolysis terms) temperature around 500°C is usually used. The main product of fast pyrolysis is 60–80% bio-oil and takes only seconds for completion. In addition, it gives 20% char and 20% gas. The essential features of a fast pyrolysis are as follows:

- Dry feedstock: (less than 12% moisture)
- Small biomass particles: <3 mm
- Biomass heating time of 1–2 s
- Moderate temperatures (400–500°C)
- Vapour residence time of 1 s.

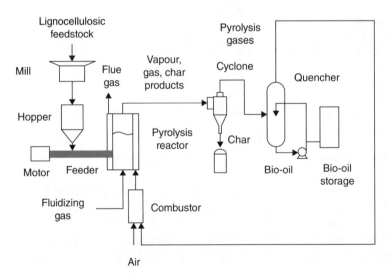

**Figure 8.8** Fast pyrolysis system. Source: Saber et al. (2016) [7]. Reproduced with permission from Elsevier.

The slow pyrolysis has been known for a long time and is characterized by longer residence time of 30 min to days and lower temperature range. Its main product is charcoal, and the equipment are available for small- as well as large-scale production.

The pyrolysis process is an endothermic reaction and consumes energy. In terms of energy demand in pyrolysis, the main factor is the water content of the starting biomass. The heat of vaporization of pure water is 2.26 kJ/g at 100°C, while the chemical energy content of wood is only about 18.6 kJ/g. If the wood has high moisture content, then the net energy yield of the pyrolysis process will be very low because the energy necessary for the pyrolysis and gasification processes comes mainly from combustion of one or more of the products of pyrolysis.

### 8.4.3.1 Torrefaction

Torrefaction is a mild pyrolysis used for the pretreatment of biomass to increase the heating value and hydrophobicity. Its potential applications are for making torrefied pellets, which can be used as a high-quality feedstock in gasification for high-quality syngas production and as a substitute for coal in thermal power plants and metallurgical processes. Torrefied biomass is more brittle, making grinding easier and less energy intensive. In this process, biomass is heated to a temperature of approximately 250–350°C in an atmosphere with low oxygen concentrations, so that all moisture is removed. Although the weight loss in this process is about 30%, the energy loss is only 10%. The main product is the solid, torrefied biomass. Inert gas is used during the process to limit the conversion of carbon into other compounds as well as of volatiles in the biomass. The most common inert gas being used is nitrogen. During the process, the biomass partly decomposes giving off various condensable and non-condensable gases. The chemical composition of the biomass material is an important factor. Because of the relatively low temperature of the torrefaction process, most critical chemical fuel components (alkali metals, chloride, sulfur, nitrogen, heavy metals and ash) remain in the fuel after torrefaction. This means only clean biomass feedstocks can be used for torrefaction.

Torrefaction of biomass produces three primary products: (i) a solid product of a dark/brown colour, (ii) condensable liquids, including moisture, acetic acids and some oxygenates; and (iii) non-condensable gases that are composed mainly of CO, $CO_2$ and small amounts of other hydrocarbons such as methane. There are two main torrefaction methods: (i) the wet process and (ii) the dry process. In wet torrefaction also known as hydrothermal pretreatment, biomass is treated with hot-compressed water in an inert atmosphere, while dry torrefaction does not use water. Temperatures of 175–225°C and pressures of up to 700 psi are needed for wet torrefaction. The solid product has 55–90% of the mass and 80–95% of the fuel value of the original biomass. For dry torrefaction, the temperature is maintained at 200–300°C. It can recover 60–80% of mass and 70–90% energy value.

The principal advantages of torrefied products are as follows:

Torrefaction results in a high-quality fuel, with characteristics compatible with coal. The increase in the calorific value is caused by the removal of moisture and some organic compounds from the original biomass. A fundamental difference with charcoal is the difference in volatile matter; in torrefaction processes, the aim is to maintain the volatile matter (and thereby energy) as much as possible in the fuel.

It does not absorb moisture. The moisture content of torrefied biomass is very low (from 1% to 3%).

The fixed carbon content of torrefied biomass is high, between 25% and 40%, while the ash content is low.

Torrefaction reduces the oxygen-to-carbon ratio through reduction in oxygen [8]. This makes a biomass better suited for gasification due to its lower oxygen/carbon ratio.

Torrefied biomass has better combustion properties as it takes less time for ignition due to less moisture and burns for a longer time due to a larger percentage of fixed carbon compared to raw biomass.

### 8.4.4 Liquefaction

Liquefaction is a thermochemical conversion process of organic material into liquid bio-crude and co-products. Depending on the process, it is usually conducted at moderate temperatures (300–400°C) and pressures (10–20 MPa) with added hydrogen or CO as a reducing agent. Unlike coal, the biomass is "wet", or at least wetter than coal, and can be processed as an aqueous slurry. Liquefaction of biomass can be hydrothermal or gaseous. In hydrothermal liquefaction, water simultaneously acts as reactant and catalyst, and this makes the process significantly different from pyrolysis. Hydrothermal liquefaction is pyrolysis in hot-compressed water of around 300°C and 10 MPa. Biomass is converted into gas, liquid and solid, similarly to common pyrolysis in gas phase.

Hydrothermal processing is divided into three separate processes, depending on the temperature of the operating conditions. At temperatures below 520 K, it is known as hydrothermal carbonization. The main product is a hydrochar which has a similar property to that of a low-rank coal. At intermediate temperature ranges between 520 and 647 K, the process is defined as hydrothermal liquefaction resulting in the production of a liquid fuel known as biocrude. Biocrude is similar to petroleum crude and can be upgraded to the whole distillate range of petroleum-derived fuel products. At higher temperatures above 647 K, gasification reactions start to dominate, and the process is defined as hydrothermal gasification, resulting in the production of a synthetic fuel gas.

## 8.5 Biochemical/Biological Conversion

Another form of conversion of biomass to energy is biological in nature and relies on microorganisms to convert biomass into fuel such as ethanol and biogas. Two main processes used are fermentation and AD. Fermentation processes from any material that contains sugar could derive ethanol. Various raw materials that can be used in the manufacture of ethanol via fermentation are classified into three main types: sugars, starches and cellulose materials. Sugars (from sugar cane, sugar beets, molasses and fruits) can be converted into ethanol directly. Starches (from corn, cassava, potatoes and root crops) must first be hydrolyzed to fermentable sugars by the action of enzymes from malt or moulds. Cellulose (from wood, agricultural residues, liquor from pulp and paper mills) must similarly be converted into sugars, generally by the action of mineral acids. Once simple sugars are formed, enzymes from microorganisms can readily be used to ferment it to ethanol. In AD, microorganisms break down biodegradable material in the absence of oxygen. One of the end products is biogas, which is combusted to generate electricity and heat or can be processed into renewable natural gas and transportation fuels.

## 8.5.1 Fermentation

Fermentation is a chemical process where glucose breaks down to form ethanol and $CO_2$. Ethanol production from sugar is the most simple and common process and makes use of yeast for conversion into ethanol which is a one-step fermentation process. Ethanol was used as fuel in combustion engines as early as the 18th century. But at present, it is used mainly as an additive to gasoline to give clean burning and good octane properties However, substantial gains made in fermentation technologies now make the production of ethanol for use as a petroleum substitute and fuel enhancer both economically competitive and environmentally beneficial.

One of the most promising fermentation technologies used recently is the "Biostil" process which uses centrifugal yeast reclamation and continuous evaporative removal of the ethanol. It is a biological conversion process through the action of certain specific yeasts or enzymes produced by microbes that act on the sugar, starch or cellulosic components and convert these into ethanol and carbon dioxide. Ethanol is separated from other components using distillation process. The mixture is heated so that the ethanol boils off and can be condensed back to liquid by cooling. Ethanol from sugar or starchy crops is termed the "first-generation" ethanol as opposed to "second-generation" ethanol that comes from cellulosic biomass. A wide variety of carbohydrates containing raw materials have been used for the production of ethanol by fermentation process. These raw materials are classified under three major categories: (i) sugar-containing crops – sugar cane, wheat, beet root, fruits, palm juice and so on; (ii) starch-containing crops – grain such as wheat, barely, rice, sweet sorghum, corn and root plants such as potato, cassava; (iii) cellulosic biomass – wood and wood waste, cedar, pine, wood, agriculture residues and so on. Recent advances in the use of cellulosic feedstock may allow the competitive production of alcohol from woody agricultural residues and trees to become economically competitive in the medium term.

Cellulosic ethanol is chemically identical to first-generation ethanol. However, it is produced from different raw materials via a more complex process (cellulose hydrolysis). In contrast to first-generation bioethanol, which is derived from sugar or starch produced by food crops (e.g. wheat, corn, sugar beet, sugar cane), cellulosic ethanol may be produced from agricultural residues (e.g. straw, corn stover), other lignocellulosic raw materials (e.g. wood chips) or energy crops (Miscanthus, switchgrass, etc.). Conversion of starch and cellulosic biomass into ethanol requires an additional step. First, sets of microbes are required to break down the starch components into sugars. In the next step, sugars are converted into ethanol via fermentation by yeast. The most common microbes for starch conversion into sugars are certain microbes that produce enzymes to break down starch into sugars. The production of second-generation biofuels is non-commercial at this time, although pilot and demonstration facilities are being developed. With the commercialization of these second-generation biofuels, there will be significant reduction in $CO_2$ production. They also do not compete with food crops, and some types can offer better engine performance.

The following chart shows the various processes to produce ethanol from sugary, starchy and lignocellulosic biomass. The ethanol concentration during fermentation is around 10–18% by volume and will have to be distilled to generate high-purity ethanol and dehydrated to obtain 100% pure ethanol.

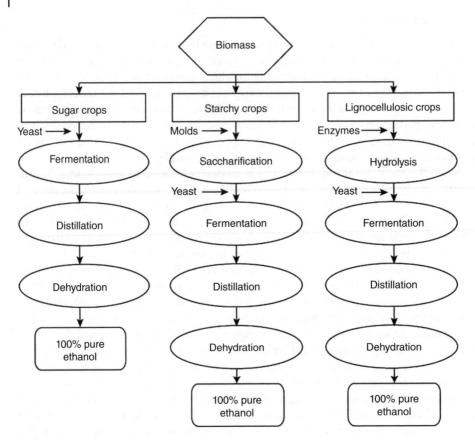

**Figure 8.9** Various processes used in fermentation.

The theoretical ethanol production from pure sugar (e.g. pure glucose) is illustrated in Eq. (8.7).

$$C_6H_{12}O_6 + \text{yeast} \rightarrow 2C_2H_5OH + 2CO_2 + \text{Heat} \qquad (8.7)$$

The density of pure ethanol is about $0.789\,\text{g/cm}^3$. Thus, 180 g of pure glucose is needed to produce 92 g of pure ethanol, while producing 88 g of carbon dioxide or 1.54 kg sugar is required to produce 1 l of ethanol. The various processes used in fermentation are shown in Fig. 8.9.

Potential ethanol yields of some sugar and starchy crops in l/kg are shown in Table 8.1.

Distillation: To increase the concentration of ethanol, distillation is required which is heat intensive usually supplied by crop residues. Bioethanol can be used to produce ethanol gas, a clean burning fuel that consists of bioethanol bound in a hydrated cellulose-thickening agent.

To produce ethanol from cellulosic biomass, it is hydrolyzed by enzymes into glucose sugar or by chemical method using sulfuric acid which is then fermented to form ethanol.

**Table 8.1** Potential ethanol yields.

| Crop | Dry matter (%) | Lignin (%) | Carbohydrates (%) | Ethanol yield (l/kg) |
|---|---|---|---|---|
| Barley | 88.7 | 2.90 | 67.10 | 0.41 |
| Corn | 86.2 | 0.60 | 73.70 | 0.46 |
| Rice | 88.6 | | 87.50 | 0.48 |
| Wheat | 89.1 | | 38.85 | 0.40 |
| Oat | 89.1 | 4.0 | 65.60 | 0.41 |
| Sugar cane | 26.0 | | 67.00 | 0.50 |
| Bagasse | 71.0 | 14.50 | 67.15 | 0.28 |

### 8.5.2 Anaerobic Digestion

AD is a process that uses bacteria to convert animal slurries, silage, food processing and other organic wastes to methane-rich biogas. An aerobic digester is a large sealed vessel from which air is excluded. Anaerobic bacteria in the absence of oxygen are used to break down the organic matter of biomass, and during the conversion, a mixture of methane and carbon dioxide gases is produced. When functioning well, the bacteria convert about 90% of the feedstock energy content into biogas (containing about 60% methane), which is a readily useable energy source for cooking and lighting. In this process, organic compounds are solubilized and hydrolyzed by the microbial action to smaller compounds and volatile fatty acids (VFAs), which are then degraded into methane and carbon dioxide. The sludge produced after the manure has passed through the digester is non-toxic and odourless. In addition, it has lost relatively little of its nitrogen or other nutrients during the digestion process, thus making a good fertilizer.

AD can occur at two main temperature ranges: mesophilic conditions, between 20 and 45°C, usually 35°C and thermophilic conditions, between 50 and 65°C, usually 55°C.

AD is a commercially proven technology and is widely used for treating high-moisture-content organic wastes, that is, +80% to 90% moisture. Biogas can be used directly in spark ignition gas engines and gas turbines and can be upgraded to higher quality, that is, natural gas quality, by the removal of $CO_2$. Used as a fuel in spark ignition gas engines (s.i.g.e.) to produce electricity only, the overall conversion efficiency from biomass to electricity is about 10–16%. As with any power generation system using an internal combustion engine as the prime mover, waste heat from the engine oil and water-cooling systems and the exhaust could be recovered using a combined heat and power system. Figure 8.10 shows an AD system also often referred to as "biogas systems". Depending on the system design, biogas can be combusted to run a generator producing electricity and heat (called a co-generation system), burned as a fuel in a boiler or furnace or cleaned and used as a natural gas.

There are three stages of AD that include stage I, hydrolysis and acidogenesis; stage II, acetogenesis; and stage III, methanogenesis. In AD, hydrolysis and acidogenesis are the essential first step where polymeric materials such as carbohydrates, proteins and lipids are hydrolyzed into smaller, water-soluble compounds such as sugars, amino acids and

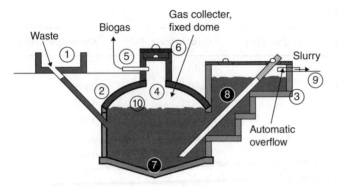

**Figure 8.10** Fixed dome.

long-chain fatty acids (LCFA) by enzymes produced by the microorganisms. In general, hydrolysis is a chemical reaction in which the breakdown of water occurs to form H+ cations and OH– anions. The hydrolytic activity is of significant importance in wastes with high organic content and may become rate limiting. Chemicals can be added during this step in order to decrease the digestion time and provide a higher methane yield. Thus, in hydrolysis and acidogenesis, sugars, amino acids and fatty acids produced by microbial degradation of biopolymers are successively metabolized by fermentation end products such as lactate, propionate, acetate and ethanol by other enzymatic activities which vary tremendously with microbial species.

In the second stage, acetogenic bacteria, also known as acid formers, produce an acidic environment in the digestive tank to convert the products from the first stage into simple organic acids, carbon dioxide and hydrogen. The principal acids produced are acetic acid, butyric acid, propionic acid and ethanol. While acidogenic bacteria further break down the organic matter, it is still too large and unusable for the ultimate goal of methane production, so the biomass must next undergo the process of acetogenesis.

In general, acetogenesis is the creation of acetate, a derivative of acetic acid, from carbon and energy sources by acetogens. These microorganisms catabolize many of the products created in acidogenesis into acetic acid, $CO_2$ and $H_2$. Acetogens break down the biomass to a point at which methanogens can utilize much of the remaining material to create methane as a biofuel.

The final step in the AD process is the conversion of methane from the final products of acetogenesis as well as from some of the intermediate products from hydrolysis and acidogenesis by the action of methane-producing microbes (methanogenesis).

There are two general pathways involving the use of (i) acetic acid and (ii) carbon dioxide, to produce methane in methanogenesis: the main mechanism to create methane in methanogenesis is the path involving acetic acid. The conversion of hydrogen into methane utilizing the carbon dioxide produced during the process is called hydrogenotrophic (Eq. (8.9)). These reactions are shown in Eqs (8.8) and (8.9), respectively, including the estimated amounts of products formed by volume (mole basis) and by weight (kg).

$$CH_3COOH \rightarrow CH_4 + CO_2$$
$$(60\,kg) \rightarrow (16\,kg) + (44\,kg)$$

(8.8)

$$4H_2 + CO_2 \rightarrow CH_4 + 2H_2O$$
$$(8\,kg) + (44\,kg) \rightarrow (16\,kg) + (36\,kg)$$

(8.9)

Throughout this entire process, large organic polymers that make up the biomass are broken down into smaller molecules by chemicals and microorganisms. Upon completion of the AD process, the biomass is converted into biogas, namely carbon dioxide and methane, as well as digestate and wastewater [9]. The composition of solid residue remaining after digestion depends on the system and original feedstock, but it contains much of nitrogen and other plant nutrients from the original feedstock and can be used as a fertilizer and soil conditioner.

### 8.5.3 Anaerobic Digestion Technologies Suitable for Dairy Manure

The increase in production and concentration of intensive livestock operations along with increased urbanization of rural regions has resulted in greater awareness and concern for the proper storage, treatment and utilization of livestock manure. In particular, the management, treatment and disposal of liquid and solid manure at dairy operations are receiving increased attention. The optimal manure management system should provide a sustainable approach designed to minimize environmental impacts and maximize resource recovery. AD under controlled conditions offers a manure treatment [10, 11] solution that not only stabilizes the wastewater but also produces a significant amount of energy in the form of biogas, controls odours, reduces pathogens, minimizes environmental impact from waste emissions and maximizes fertilizer nutrient and water recovery for reuse. To increase the methane production, manure from dairy cows can be co-digested with additional substrates such as agricultural residues and food-processing waste as shown in Figure 8.11.

The main equipment for dairy manure digestion system are digester and manure separator. Many of the problems associated with the handling and the storage of animal slurry containing a high proportion of the solid material can be overcome by mechanical separation. The products of separation are an easily handled solid and readily pumped liquid. The output of liquid and fibre, together with the dry matter content of both fractions, depends on the dry matter content of the raw slurry delivered to the separator; however, it has been found that separation provides about 25% solids and 75% liquid from raw dairy manure. Separation can be used as a first stage in a biological treatment system.

A digester is a device for optimizing the AD of biomass and/or animal manure, mainly used to recover biogas for energy production. It is a containment vessel designed to exclude air and promote the growth of methane bacteria. Commercial digester types include complete mix, continuous flow (horizontal or vertical plug flow, multiple tank and single tank) and covered lagoon. It may be designed to heat or mix the organic material. Complete-mix digesters are the most flexible of all digesters as far as the variety of wastes that can be accommodated. Wastes with 2–10% solids are pumped into the digester, and the digester contents are continuously or intermittently mixed to prevent separation. Complete-mix digesters are usually aboveground, heated, insulated round tanks. Mixing can be accomplished by gas recirculation, mechanical propellers or circulation of liquid. Advantages of anaerobic digestion processes are as follows:

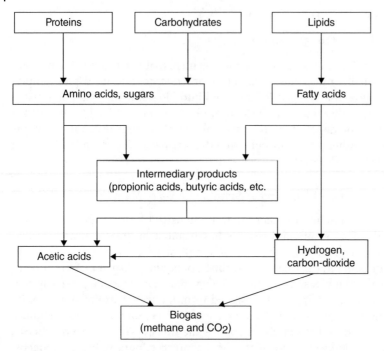

**Figure 8.11** Anaerobic digestion.

1) Renewable combustible gas is produced.
2) The digestion of solids removes objectionable odours after the conversion process.
3) A highly valued sludge is co-produced, which is considered an excellent source of nutrients and has fertilizer value.
4) There is reduction in the amount of wastes to handle at the end of the conversion process.
5) The waste resources are contained in an enclosed vessel, preventing spoilage and generation of odour.

## 8.6   Classification of Biogas Plants

Classification of biogas plants depends upon the plant design and mode of working. They are classified as (i) batch type and (ii) continuous type. Batch-type biogas plants are loaded with substrate completely. Once charged, they start supplying the gas after 8–10 days for about 50–60 days. After this time, it is emptied and recharged.

The continuous-type biogas plants are loaded with relatively small portions of substrates daily. Almost all biogas plants built now are of continuous type. These plants are further classified as (i) floating drum or constant pressure type and (ii) fixed dome type or constant volume type.

Floating-drum plants consist of an underground digester (cylindrical or dome-shaped) and a moving gas holder. The gas holder floats either directly on the fermentation slurry or in a water jacket of its own. The gas is collected in the gas drum, which rises or moves down, according to the amount of gas stored. The gas drum is prevented from tilting by a guiding frame. When biogas is produced, the drum moves up, and when it is consumed, the drum goes down. If the drum floats in a water jacket, it cannot get stuck, even in the substrate with a high solid content. With the availability of cheap fixed-dome plants, the floating-drum plants became obsolete as they involve high investment and maintenance cost along with other design weakness.

A fixed-dome-type biogas plant consists of a digester with a fixed, non-movable gas holder, which sits on top of the digester (Fig. 8.10). When gas production starts, the slurry is displaced into the compensation tank. Gas pressure increases with the volume of gas stored and the height difference between the slurry level in the digester and the slurry level in the compensation tank (Fig. 8.10). The costs of a fixed-dome biogas plant are relatively low. It is simple as no moving parts exist. There are also no rusting steel parts, and hence, a long life of the plant (20 years or more) can be expected. While the underground digester is protected from low temperatures at night and during cold seasons, sunshine and warm seasons take longer to heat up the digester. No day/night fluctuations of temperature in the digester positively influence the bacteriological processes. The construction of a fixed-dome plants is labour-intensive, thus creating local employment. Fixed-dome plants are not easy to build. They should only be built where construction can be supervised by experienced biogas technicians. Otherwise, plants may not be gas-tight (porosity and cracks).

The slurry (waste) enters from an inlet as shown in Fig. 8.10. The gas is stored in the upper part of the digester. When gas production commences, the slurry is displaced into the compensating tank. Gas pressure increases with the volume of gas stored, that is, with the height difference between the two slurry levels. If there is little gas in the gas holder, the gas pressure is low. The gas occupies about 10% of the total volume of digester.

## 8.7 Mechanical Extraction (with Esterification)

Extraction is a mechanical conversion process used to produce oil from the seeds of various biomass crops, such as oilseed rapeseed, cotton and groundnuts. Oil separation, or the extraction of oil from seeds or plant parts, is accomplished by mechanical pressing, sometimes followed by chemical extraction. Crude vegetable oils are recovered from the oil seeds by applying a mechanical pressure using screw press (expeller). Screw press can be applied in two ways: pre-pressing and full pressing. In pre-pressing, only part of the oil is recovered and cake with 18–20% oil is further treated by solvent extraction. There are various ways of extracting oils from oil seed crop. Although the most common extraction method is mechanical oil separation using an oil seed press or extrude, other methods are also used. The oil from seeds may also be obtained via solvent extraction. The most popular solvent is hexane, while a cheaper alternative is petroleum ether. The product of this process is the crude oil. The crude oil then undergoes various refining steps to remove the waxes, phosphorus (degumming), FFAs (by neutralization) and other impurities. Other oil extraction methods include enzymatic processing with the use of inert gas such as $CO_2$.

The direct use of vegetable oils as fuels has not been satisfactory in diesel engines, because of their high viscosity, low volatility, polyunsaturated compound content, high content of free fatty acids, resins and rubber and cinder deposits. One of the most popular biofuels from biomass is biodiesel, which is simply the ester of the oil [12].

The biodiesel is produced by transesterification of fatty acids in vegetable oils, using an alkaline catalyst (sodium or potassium hydroxide at 80°C), a reaction that is responsible for high yield. However, the overall biofuel production process is complex. Commercial biodiesel producers would normally begin with a refined, bleached and deodorized oil. Before oil seed extraction, the seed kernels must be cleaned and dried, or the hulls must be removed. More importantly, the biodiesel produced must pass ASTM D6751 standards before it can be blended with diesel. This standard includes more than 15 parameters to which the biodiesel product must conform before it can be used in engines.

Biodiesel can be blended with the diesel fuel easily and may be used in conventional diesel engines without engine modification. Diesel fuel may be blended with biodiesel, and the blending ratio by some engine manufacturers can be as much as 20%. This is denoted as B20, meaning 20% biodiesel with 80% petroleum diesel.

The process of making biodiesel is very simple. The chemical equation and the simple product balance for the biodiesel conversion process are shown as follows.

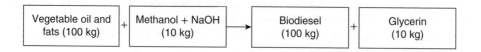

Although most commonly used oils are soybean, rapeseed and palm or sunflower, biodiesel can be produced from more than 300 vegetable species, depending on their availability in the biodiesel production area. The use of waste cooking oils has the advantage that they are recycled, avoiding a significant environmental problem.

## 8.8 Municipal Solid Waste to Energy Conversion

MSW is typically known by the public as "trash" or "garbage". Basically, MSW consists of solid waste generated by households, commercial establishments and industries. It includes product packaging, grass clippings, furniture, clothing, bottles, food scraps, newspapers, appliances and batteries. Disposal of MSW is a major problem in big cities because of large quantity of waste generated. Some of the items in MSW are recycled wherever economical and/or practical. In addition, materials not generally included as MSW but sent to landfills are construction and demolition debris, municipal wastewater treatment sludges and non-hazardous industrial wastes. In other words, MSW does "not" include wastes of other types or from other sources such as automobile bodies, municipal sludges, combustion ash and industrial process wastes that might also be disposed of in municipal waste landfills.

Residential waste sources (including waste from multi-family dwellings) were estimated to be 55–65% of total MSW generation. Commercial waste sources (including waste from schools, some industrial sites where packaging is generated and businesses) constitute between 35% and 45% of MSW generation. In India, MSW comprises 30–55% of biodegradable (organic) matter, 40–55% inert matter and 5–15% recyclables. Composition of waste varies with the size of the city, season and income group.

The composition of MSW after selective collection is shown in Table 8.2.

**Table 8.2** Selective composition of MSW in the United States.

| S. no | Waste material | Percentage weight |
|-------|----------------|-------------------|
| 1 | Paper | 38 |
| 2 | Food waste | 10.9 |
| 3 | Yard waste | 12.1 |
| 4 | Plastic and rubber, textiles, leather | 17 |
| 5 | Metal | 7.8 |
| 6 | Glass | 5.5 |
| 7 | Wood | 5.3 |
| 8 | Miscellaneous | 3.4 |

Among waste-to-energy technologies, incineration is very common. Other technologies are gasification, AD and pyrolysis [13]. Sometimes, incineration is used simply for disposal of garbage without generating energy. Incineration or thermal treatment of waste is much popular in countries such as Japan where there is scarcity of land. The energy generated by incineration is presently used in countries such as Denmark and Sweden.

The incineration plant used for treating MSW is moving grate. This grate is capable of hauling waste from combustion chamber to give way for complete and effective combustion. The grate systems include the following:

- Reciprocating grates
- Roller grates
- Reversed-feed grates. Grate incinerators usually have the following components:
- Waste feeder
- Incineration grate
- Bottom ash discharger
- Incineration air duct system
- Incineration chamber
- Auxiliary burners

In order to accomplish complete combustion of the gases, it is necessary for the gases to be at a temperature above 850°C for at least 2 s. The completion of the gases burnout is indicated by the levels of the carbon monoxide in the off gases. Usually, auxiliary firing systems are used to keep the combustion gases at the desirable temperature levels. The standard approach for the recovery of energy from the incineration of MSW is to utilize the combustion heat through a boiler to generate steam. Of the total available energy in the waste, up to 80% can be retrieved in the boiler to produce steam. The steam can be used for the generation of power via a steam turbine and/or used for heating. An energy recovery plant that produces both heat and power is commonly referred to as a CHP plant, and this is the most efficient option for utilizing recovered energy from waste via a steam boiler.

Biogas may be generated by digesting the organic fraction of MSW (OFMSW). The produced biogas may be utilized for CHP production or for transport fuel production as $CH_4$-enriched biogas. When used to produce transport fuel, some of the biogas is used in a small CHP unit to meet electricity demand on site. This generates a surplus thermal product.

## 8.9 The Production of Electricity from Wood and Other Solid Biomass

Solid biofuels used for electricity generation are mainly wood and/or wood waste from forestry and the subsequent industries, solid residues from the sugar-based bioethanol, industry residues from the pulp and paper production as well as wood pellets. Wood from short-rotation crop plays only a minor role in the electric power generation using biomass.

Direct combustion biomass power plant is shown in Fig. 8.12. It is a classic example of the steam cycle, which in turn is the most common industrial application of the basic "heat engine" or Carnot cycle. It is a mature, commercially available technology that can be applied on a wide range of scales from a few MW to 100 MW or more. There are two main components of combustion-based biomass power plant.

1) The biomass-fired boiler that produces steam
2) The steam turbine used to generate electricity

Most biopower plants use direct-fired combustion systems. They burn biomass directly to produce high-pressure steam that drives a turbine generator to make electricity. Two most common forms of boilers are stoker and fluidized bed. The fluidized-bed boiler shown in Fig. 8.12 is the most common type of boiler recommended for biomass fuel, which is burned within a hot bed of inert particles, typically sand.

The fuel–particle mix is suspended by an upward flow of combustion air within the bed. As velocities increase, the gas/solid mix exhibits fluid-like properties. Fluidized-bed boilers are categorized as either atmospheric (AFBC) or pressurized units. In AFBC, coal is crushed to a size of 1–10 mm depending on the rank of the coal, the type of fuel feed and fed into the combustion chamber. The atmospheric air, which acts as both the fluidization air and combustion air, is delivered at a pressure and flows through the bed after being pre-heated by the exhaust flue gases. The velocity of fluidizing air is in the range of 1.2–3.7 m/s. The rate at which air is blown through the bed determines the amount of fuel.

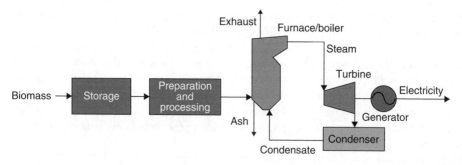

**Figure 8.12** Direct combustion steam turbine system.

Atmospheric fluidized-bed boilers are further divided into bubbling-bed and circulating-bed units; the fundamental difference between bubbling-bed and circulating-bed boilers is the fluidization velocity (higher for circulating). Circulating fluidized-bed boilers separate and capture fuel solids entrained in the high-velocity exhaust gas and return them to the bed for complete combustion. Atmospheric-pressure bubbling fluidized-bed boilers are most commonly used with biomass fuels. The type of fluid bed selected is a function of the as-specified heating value of the biomass fuel. Bubbling bed technology is generally selected for fuels with lower heating values. The circulating bed is most suitable for fuels with higher heating value.

Stoker boilers employ direct fire combustion of solid fuels with excess air, producing hot flue gases, which then produce steam in the heat exchange section of the boiler. The steam is used directly for heating purposes or passed through a steam turbine generator to produce electric power. Mechanical stokers are the traditional technology that has been used to automatically supply solid fuels to a boiler. All stokers are designed to feed fuel onto a grate where it burns with air passing up through it. Travelling grate is a moving platform for the combustion of fuel to remove the ash residue after combustion. Stoker units use mechanical means to shift and add fuel to the fire that burns on and above the grate located near the base of the boiler.

In some biomass industries, the extracted or spent steam from the power plant is also used for manufacturing processes or to heat buildings. These CHP systems greatly increase the overall energy efficiency to approximately 80%, from the standard biomass electricity-only systems with efficiencies of approximately 20%. Seasonal heating requirements will impact the CHP system.

These can be fuelled entirely by biomass or can be co-fired with a combination of biomass and coal or other solid fuels. There are three main types of turbines which are used for generation of electricity in CHP system. These are as follows:

1) Condensing steam turbine
2) Extraction steam turbine
3) Non-condensing (backpressure steam turbine)

Extraction steam turbine can be condensing or backpressure. These are multi-stage turbines.

## 8.10  Summary

In this chapter, the term biomass energy is explained and its source as process of photosynthesis is shown. The difference between fossil fuels and biomass is explained. Further, the different resources of biomass are described.

Next, production of energy from biomass using different processes is explained. Direct utilization of heat for use is also explained.

## References

1 Hall, D.O. and House, J.I. (1994) Trees and biomass energy - carbon storage and or fossil-fuel substitution. *Biomass and Bioenergy*, **6**, 11–30.

2 Bassan, N.B. (ed.) (2011) *Handbook of Bioenergy Crops. A Complete Reference to Species, Development and Applications*, Earth Scan, London.

3 Hughes, E. (2000) Biomass cofiring: economics, policy and opportunities. *Biomass and Bioenergy*, **19**, 457–465.

4 Mckendry, P. (2002) *Energy Production From Biomass (Part 2): Conversion Technologies, Biosource Technology*, Elsevier.

5 Mukunda, H.S. *et al.* (1994) Gasifiers and combustors for biomass – technology and field studies. *Energy for sustainable development*, **1** (3), 27–38.

6 Huynh, C.V. and Kong, S. (2013) Performance characteristics of a pilot-scale biomass gasifier using oxygen-enriched air and steam. *Fuel*, **103**, 987–996.

7 Saber, M. *et al.* (2016) A review of production and upgrading of Algal bio-oil. *Renewable and Sustainable Energy Reviews*, **58**, 916–930.

8 Chew, J.J. and Doshi, V. (2011) Recent advances in biomass pretreatment- torrefaction fundamentals and technology. *Renewable and Sustainable energy Reviews*, **15** (8), 4212–4222.

9 Bridgwater, A.V. (2003) Renewable fuels and chemicals by thermal processing of biomass. *Chemical Engineering Journal*, **91**, 87–102.

10 Pullen, T. (2015) *Anaerobic Digestion - Making Biogas - Making Energy*, Earth Scan, London.

11 Mata, A. *et al.* (2000) Anaerobic digestion of organic solid wastes. An overview of research achievements and perspectives. *Bioresource Technology*, **74** (1), 3–16.

12 Naik, S.N. *et al.* (2010) Production of first and second generation biofuels: a comprehensive review. *Renewable and Sustainable Energy Reviews*, **14**, 578–597.

13 Murphy, J.D. and Mckeogh, E. (2004) Technical, economic and environmental analysis of energy production from municipal solid waste. *Renewable Energy*, **29** (7), 1043–1057.

# 9

# Geothermal Energy

## 9.1 The Origin of Geothermal Energy

The geothermal energy is basically thermal energy stored in the Earth's crust. It is clean and sustainable. It is one of the less-recognized forms of renewable energy, and this is the only form of renewable energy which is not dependent on the Sun. Geothermal energy comes from the natural generation of heat primarily due to the decay of the naturally occurring radioactive isotopes of uranium, thorium and potassium within the Earth. Resources of *geothermal energy* range from the shallow ground to hot water and hot rock found a few miles beneath the Earth's surface and even deeper down to the extremely high temperatures of molten rock called magma. It is estimated that the annual energy due to the internal heat generation and flowing from the interior of the Earth to the surface is about $10^{21}$ J per year.

Geothermal energy has been used by the Romans, Japanese, Turks, Icelanders, Central Europeans and the Maori of New Zealand for bathing, cooking and space heating. Baths in the Roman Empire, the middle kingdom of the Chinese and the Turkish baths of the Ottomans were some of the early users of geothermal energy [1, 2].

Geothermal energy is present everywhere below the Earth's surface, but the most desirable resources are concentrated in the regions of active or geologically young volcanoes. Most geothermal reservoirs are deep underground with no visible sign of it available on the Earth's surface. But sometimes, geothermal energy erupts to the surface in the form of

- volcanoes
- hot springs
- geysers.

The total thermal energy contained in the Earth is estimated to be $12.6 \times 10^{12}$ EJ and that of the crust of the order of $5.4 \times 10^9$ EJ up to depths of 50 km. Stored thermal energy down to 3 km depth in the continents is estimated to be $42.67 \times 10^6$ EJ (Source EPRI report 1978). With the present global energy consumptions for all types of energy, the Earth's energy to a depth of 10 km could theoretically supply all of mankind's energy needs for 6 million years. At present, it is only possible to extract small portion of this stored energy due to limitations in drilling technology and rock permeability. Commercial utilization is concentrated on areas where drilling to depths up to 4 km can access fluids at temperatures of 180°C to more than 350°C.

*Operation and Control of Renewable Energy Systems,* First Edition. Mukhtar Ahmad.
© 2018 John Wiley & Sons Ltd. Published 2018 by John Wiley & Sons Ltd.

The first use of geothermal energy for electric power production was in Italy, with experimental work by Prince Ginori Conti between 1904 and 1905. The first commercial power plant (250 kW) was commissioned in 1913 at Larderello, Italy. Now power is generated in plants installed in Thailand, Argentina, Taiwan, Australia, Costa Rica, Austria, Guatemala, Ethiopia, with the latest installations in Germany and Papua New Guinea.

## 9.2   Types of Geothermal Resources

The Earth from crust to core is divided into three areas: core, mantle and crust, as shown in Fig. 9.1. The core which is about 1900 km below the Earth's surface is extremely hot 5000–6000°C and is made mostly of iron (80%) and nickel (20%), whose inner half (by radius) is solid and whose outer half is liquid. After that, there is a thick layer in the middle called *mantle*, which makes up most of the rest (83%) of the Earth's volume and is made mostly of rocky material, whose inner part is semi-rigid and whose outer and cooler part is plastic and, therefore, can flow (thick lava). Then there is the top layer of the Earth called *crust* where temperatures are moderate for humans to survive. A semi-fluid material called *magma* originates in the lower part of the Earth's crust and in the upper portion of the mantle. There, high temperatures and pressures cause some rocks to melt and form magma. On average, the temperature of the Earth increases about 30°C/km above the mean surface ambient temperature. The temperature at the base of the crust is about 1000°C, and assuming a conductive gradient, the temperature of the Earth at 10 km inside would be over 300°C.

The distribution of layers is not the same everywhere inside the Earth. In certain locations, anomalies exist in composition and structure of these layers. There are certain regions where hot molten rock (magma) occurs at shallow depths due to the following: (i) intrusion of molten rock (magma) from depth, bringing up great quantities of heat; (ii) high surface heat flow, due to a thin crust and high temperature gradient; (iii) ascent of groundwater that has circulated to depths of several kilometres and been heated due to the normal temperature gradient; (iv) thermal blanketing or insulation of deep rocks by thick formation of such rocks as shale whose thermal conductivity is low; and (v) anomalous heating of shallow rock by decay of radioactive elements, perhaps augmented by thermal blanketing.

Most geothermal exploration and use occur in these places where the gradient is higher, and thus, drilling is shallower and less costly.

Magma is extremely hot – between 700 and 1300°C. When magma rises to the surface of the Earth through vents or during volcanic eruptions, it becomes lava. When lava stops flowing and cools, it hardens into igneous rock. Several of the world's most advanced geothermal sites are located in extinct volcanic areas.

Geothermal energy resources are characterized by geological settings, intrinsic properties and viability for commercial utilization. Most of the geothermal sites are located near the edges of the pacific plate known as ring of fire. The diversity of both the nature of the geothermal resource and its exploitation presents a challenge in the context of resource classification. There are different ways to classify the geothermal resources. Temperature is a fundamental measure of the quality of the resource and consequently is

**Figure 9.1** Earth temperatures from crust to core.

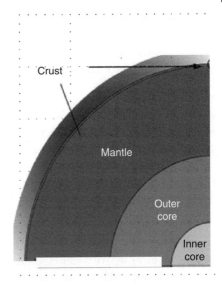

**Table 9.1** Types of geothermal resources.

| Resource type | Temperature range (°C) |
|---|---|
| Convective hydrothermal resources | |
| Vapour-dominated | 240 |
| Hot-water-dominated | 20–350 |
| Geopressured (aquifer) | 90–200 |
| Hot rock resources (conductive) | |
| Solidified (hot dry rock) | 90–650 |
| Part still molten (magma) | >600 |

the primary element of most classification systems by which geothermal energy is identified and utilized. Geothermal resources suitable for different types of uses are commonly divided into two categories, high and low enthalpy, according to their energy content. High-enthalpy resources (>150°C) are suitable for electrical generation with conventional cycles, while low-enthalpy resources (<150°C) are employed for direct heat uses and electricity generation using binary cycles [3, 4].

The geothermal resources can also be classified as *convective* (hydrothermal), *conductive* and *deep aquifers* as shown in Table 9.1 [5–7]. The hydrothermal resources include liquid- and vapour-dominated type. Conductive sources include hard rock and magma over a wide range of temperatures. Deep aquifers are circulating fluids in porous media or fractured zones. These are either geopressured or at hydrostatic pressure.

Ambient temperatures in the 5–30°C range can be used with geothermal (ground-source) heat pumps to provide both heating and cooling.

## 9.3  Hydrothermal Resources

Hydrothermal means that the transfer of heat involves water, either in liquid or in vapour state (hence the name "hydro"). Hot springs and geysers, for example, are hydrothermal features. All geothermal electricity produced today derives from the hydrothermal resource base. Hydrothermal systems occur where the Earth's heat is carried upwards by convective circulation of naturally occurring hot water or steam. Some high-temperature convective hydrothermal resources result from deep circulation of water along fractures. These systems are dominated by liquid and vapour. The presence of water facilitates not only vertical hydrothermal energy flows (convection) but also horizontal hydrothermal energy flows (through convection, advection and diffusion). These are high-enthalpy systems and found in areas of magmatic intrusions where temperatures above 1000°C can occur at less than 10 km of depth.

These geothermal reservoirs of steam or hot water occur naturally where magma comes close enough to the surface to heat the groundwater trapped in fractured or porous rocks or where water circulates at great depths along faults. The water in these resources is trapped in the underground reservoir (acquifers), and the Earth's heat is carried upwards by convective circulation of naturally occurring hot water or steam. Naturally occurring large areas of hydrothermal resources are called geothermal reservoirs, Fig. 9.2. Hydrothermal resources occur at a location, perhaps in conjunction with other hydrothermal features. They emit heat and discharge hydrothermal water. The water contains chemicals and gases. So, location, area, heat, water and chemistry become critical vital signs to monitor. Although these vital signs are observable at the surface, they also provide information about the sub-surface hydrothermal system.

Hydrothermal resources are used for different energy purposes depending on their temperature and the depth. In places where the pressure is high, the temperature may reach as high as 350°C. As the water flows through the porous medium that constitutes the aquifer, its static pressure drops and large volume of steam is produced. When this

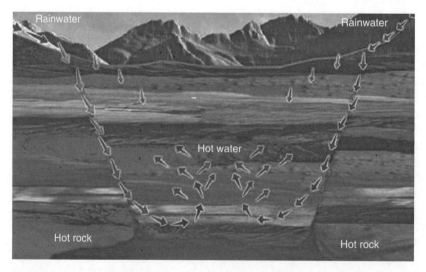

**Figure 9.2** Geothermal reservoir. *Source*: Nasruddin et al. (2016) [8]. Reproduced with permission from Elsevier.

steam escapes through cracks in the surface, it is called *fumaroles*. In areas where the water table rises near the surface, fumaroles can become hot springs or *geysers*.

Hydrothermal resources are further classified as (i) vapour-dominated, (ii) liquid-dominated and (iii) hot-water resource. Three geothermal power generation technologies are used to convert hydrothermal fluids to electricity: dry steam, flashy and binary cycle. The type of conversion technology used depends on whether the fluid is present as steam or water and its temperature.

### 9.3.1 Vapour-Dominated Systems

Vapour-dominated geothermal fields are located in the regions of recent volcanism, near the borders of tectonic plates. These reservoirs are few in number, with the largest sources located as geysers in northern California. Larderello in Italy and Matsukawa in Japan are also places where the steam is exploited to produce electric energy. In order to form a heat reservoir, the anomalous magmatic intrusion should encounter porous and permeable water-filled rock strata. In hard compact rocks, faulting may provide a channel for the magma to reach the surface. Soft or plastic rocks, when present, can flow and block the fault space, causing the magma to spread at the contact between the soft and the hard rocks. A vapour-dominated geothermal system is shown in Fig. 9.3.

In these resources, the water boils underground and generates steam at about 165°C. These systems are the best and most productive geothermal resources because the steam is largely dry and is of very high enthalpy. Currently, hydrothermal systems are the only commercially exploited geothermal systems for power generation. In a steam-dominated reservoir, the fluid distribution is controlled by the flow of steam moving up and the water moving down. If the mass fluxes of water moving down and steam moving up are roughly similar, and the vertical pressure gradient is near steam-static, the relative permeability to water must be low. The flowing steam occupies most of the fracture space, and water occupies the remaining pore space. The mass of water in the reservoir is much greater than the mass of steam. Due to the decrease of steam production, it is necessary to inject water through wells into the reservoir.

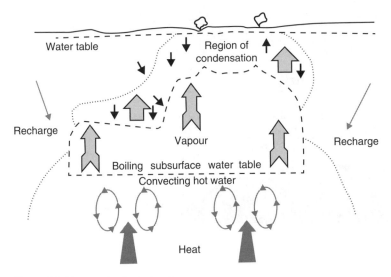

**Figure 9.3** Vapour-dominated system.

When the geothermal resource produces a saturated or superheated vapour, the steam is collected from the production wells, and appropriate measures are taken to remove any solid debris from the steam flow, as well as corrosive substances contained in the process stream. After processing, the steam is sent to a conventional steam turbine. If the steam at the wellhead is saturated, steps are taken to remove any liquid that is present or forms prior to the steam entering the turbine. Normally, a condensing turbine is used; however, in some instances, a backpressure turbine is used that exhausts steam directly to the ambient temperature.

### 9.3.2  Water-Dominated Systems

Water-dominated systems are produced by groundwater circulating to depth and ascending from buoyancy in permeable reservoirs that are at a uniform temperature over large volumes. Typically, the temperature of hot-water reservoirs varies from 60° to 100°, and they occur at depths ranging from 1500 to 3000 m. The geology of hot-water geothermal fields is quite similar to that of an ordinary groundwater system. They differ from the vapour-dominated geothermal fields in the fact that the hot-water geothermal fields are characterized by liquid water being the continuous pressure-controlling fluid phase. As the water rises in the well, its static pressure is reduced due to gravitational and frictional forces. The temperature of the water also decreases because of heat loss to the surroundings. However, because rocks have high insulating property, the decrease in temperature is not much. Figure 9.4 presents a typical hot-water field. As

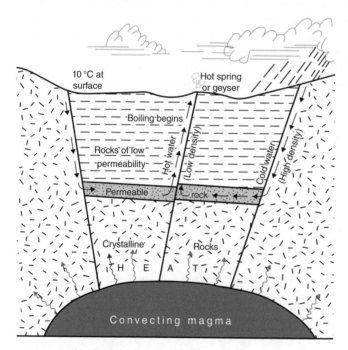

**Figure 9.4**  Hot-water geothermal system.

shown in Fig. 9.4, a hot-water geothermal field could develop in the absence of a cap rock, if the thermal gradients and the depth of the aquifer are adequate to maintain a convective circulation.

In liquid-dominated plants, geothermal plants are built upon liquid reservoirs within the Earth's surface. This liquid is sent through one or more separators in order to lower the pressure of the water, creating steam. This steam then propels a turbine generator, causing it to produce electricity. This steam is then condensed back into a liquid and placed back into the liquid reservoir it originated from. This type of geothermal plant is very common and provides a sustainable, reusable form of energy.

## 9.4  The Geopressured Resources

Geopressured resources are reservoirs of naturally high-pressured hot water. Geopressured resources are formed when there is an impermeable layer of sedimentary cap rock that traps a geothermal reservoir. In these instances, the weight of the sediment layer and the lack of permeability increase the pressure inside the reservoir. Geopressured resources typically range from 90 to 200+°C, and the increase in pressure reduces the energy required to pump the resource, making geopressured resources desirable. These sources consist of deeply buried reservoirs of hot brine under abnormally high pressure that contain dissolved methane. Thermal waters under high pressure in sand aquifers are the target for drilling, mainly as they contain dissolved methane. Geopressured geothermal systems were first identified in the deep sedimentary layers underneath the Gulf of Mexico at a depth between 6 and 8 km with pore pressures of up to 130 MPa and temperatures in the range of 150–180°C.

In geopressured systems shown in Fig. 9.5, the water temperature can range from 90 to 200°C which correspond to low saturation pressures. The pressure in the liquid is significantly higher than saturation pressure at all levels in the well.

Three forms of energy are useable in geopressured wells: (i) *thermal* from the high temperatures, (ii) *hydraulic* from the high fluid flow pressure and (iii) *chemical* from the dissolved methane in the fluids. It is possible to exploit each form singly or in combination to satisfy a variety of energy needs. Production of electrical energy from geopressured sources is discussed later.

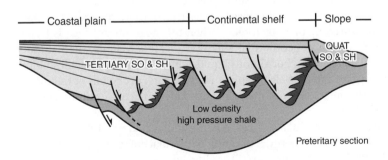

**Figure 9.5** Geopressured geothermal system.

## 9.5 Hard Rock Resources

Hot dry rock geothermal reservoirs contrast with conventional geothermal reservoirs in that the rocks are generally deep, impermeable and with little or no porosity/fluid content to transport the heat. Hence, a geothermal reservoir must be engineered by pumping cool water down an injection well and recovering hot water from a nearby production well(s), with flow between the two wells stimulated by the pressure of injected fluids.

These are further classified as (i) solidified (hard dry rock) and (ii) magma. Hot dry rock resources are much more common than hydrothermal resources and are more accessible.

### 9.5.1 Solidified (Hot Dry Rock Resources)

The hot dry rock (HDR) resource is larger and more widespread than the conventional geothermal resource at around temperatures of 200°C. These sources are accessible under a significant proportion of the world's landmass. Since there is no water which can be transported to the surface of the Earth, recovery of heat involves creating a fracture system at depth (that acts as a heat exchanger) and circulating water from an injection borehole towards a production borehole. The water is pumped into the cracks from the surface and withdrawn from another well at a distance. Thus, a number of wells are required to be drilled through hard rocks. HDR geothermal resource represented by the vast regions of hot rock at accessible depths in the Earth's crust far exceeds that of the combined total of the world's fossil energy resources. Recently, it has been recognized that very few rocks are actually completely dry, and hence, these sources are now termed as enhanced or engineered geothermal systems (EGSs).

### 9.5.2 Part Still Molten (Magma)

Another source of geothermal energy is magma, which is partially molten rock. Molten rock is the largest global geothermal resource and is found at depths below 3–10 km. Magma is the ultimate source of all high-temperature geothermal resources. Plate boundaries are the most common sites of volcanic eruptions. At several volcanic locales, magma is present within the top 5 km of the crust. Its great depth and high temperature (between 700 and 1200°C) make the resource difficult to access and harness. Thus, technology to use magma resources is not well developed.

*Magma* (molten rock) may come quite close to the surface where the crust has been thinned, faulted or fractured by plate tectonics. When this near-surface heat is transferred to water, a usable form of geothermal energy is created. Magma, the naturally occurring molten rock material is a hot viscous liquid, which retains fluidity till solidification. It may contain gases and particles of solid materials such as crystals or fragments of solid rocks. However, the mobility of magma is not much affected until the content of solid material is too large.

The heat energy available from such sources, if harvested, would constitute very large additions to the global renewable energy. Extraction of thermal energy from magma was tested during the 1980s by drilling into the still-molten core of a lava lake in Hawaii.

## 9.6 Energy Contents of Geothermal Resources

### 9.6.1 Hard Dry Rock Resources

In order to evaluate the amount of energy available in a hot dry rock resource, it is assumed that the thermal gradient $G$ is constant. Under this assumption, the temperature $T$ at a depth of $h$ will be

$$T = T_0 + Gh$$

where $T_0$ is the surface temperature.

Suppose that $h_1$ is the minimum depth needed to reach temperatures of $T_1 = 150°$ and $h_2$ is the maximum depth to which current technology allows wells to be drilled. Recall that by the definition of the specific heat of a substance, $c$, the stored thermal energy in a mass $m$ whose temperature is higher by $\Delta T$ above some reference temperature can be expressed as

$$E = mc\Delta T = mc(T - T_1) \tag{9.1}$$

where mass $m = \rho A$ and $A$ is the area of rock. Thus, the amount of stored thermal energy below a surface area $A$ between depths $H$ and $h + dh$ can be expressed as

$$dE = \rho Ac(T - T_1)dz \tag{9.2}$$

The total useful energy of the rock between depths $h_1$ and $h_2$ is given by

$$E = \int_{h_1}^{h_2} \rho Ac(T - T_1)dZ \tag{9.3}$$

$$= \rho AcG(h_2 - h_1)^2/2 \tag{9.4}$$

Since $(T - T_1) = G(h_2 - h_1)$

### Example 9.1

A hard dry rock resource has a thermal gradient of 30°C/km. The minimum useful temperature is 150°C above the surface temperature. Determine the heat content per square kilometre of HRD to a depth of 10 km, assuming density $= 2700\,kg/m^3$ and specific heat of 800 J/kg/degree.

### Solution:

$$150 = Gh_1 = 30h$$

Or

$$h_1 = 5\,km$$

Therefore, $h_2 - h_1 = 10 - 5 = 5$

From Eq. (9.4) $\dfrac{E}{A} = \rho cG(h_2 - h_1)^2/2 = 2700 \times 10^9 \times 820 \times 30 \times \dfrac{25}{2} = 1.66 \times 10^{18}\,J/km^2$

## 9.7 Exploration of Geothermal Resources

Since geothermal resources are underground, exploration methods, geological, geochemical and geophysical, have been developed to locate and assess them. The objectives of geothermal exploration are to identify and rank prospective geothermal reservoirs prior to drilling and to provide methods of characterizing reservoirs (including the properties of the fluids) that enable estimates of geothermal reservoir performance. Exploration typically begins with gathering data from existing nearby wells and other surface manifestation. Most countries have existing databases of geological and hydrological data. These have usually been gathered for other purposes but may very well be useful for guiding early geothermal surveying and exploration. Every effort should be made to collect and analyze these data prior to designing and planning an additional exploration program. Remote sensing, in particular, is now playing a more significant role in preliminary surveying for geothermal resources.

In order to make the exploration program cost-effective while reducing risk, the survey design initially focuses on simpler (cheaper) methods and becomes progressively more complex and costly as early results show promise for more detailed efforts. Geophysical methods are among the three main disciplines applied on the surface to explore geothermal resources, including geology and the chemistry of thermal fluids.

The important physical parameters in a geothermal system are

- temperature;
- porosity;
- permeability;
- chemical content of fluid (salinity); and
- (pressure).

Most of these parameters cannot be measured directly through conventional geophysical methods applied on the surface of the Earth. However, there are other parameters that can be measured which are linked with these parameters and may give important information on the geothermal system. Among these parameters are

- temperature (°C);
- electrical resistivity (m);
- magnetization.
- density;
- seismic velocity;
- seismic activity;
- thermal conductivity; and
- streaming potential.

The decision to go for extracting energy from geothermal source depends upon exploration of surface features to determine first whether or not a resource is worth the large amount of investment required to drill a geothermal well. Once surface features have been investigated, several techniques are used to help identify drill targets without having to put a drill bit into the ground. These exploration technologies not only need to better locate geothermal resources but they must be able to provide more accurate imaging of the structure of the sub-surface reservoir and provide accurate reservoir temperatures at specified depths.

## 9.8  Geophysical Methods in Geothermal Exploration

The geothermal reservoirs and their immediate surroundings have certain specific physical characteristics that are susceptible to detection and mapping by geophysical methods. The temperature within the reservoir is the most important physical characteristic of a geothermal system. Electrical and magnetic properties of the rock also provide information about the location of geothermal resources.

The most important methods used in geophysical exploration of geothermal fields are as follows:

- Thermal methods
- Electrical methods
- Magnetic measurements

### 9.8.1  Thermal Methods

The base temperature constitutes the most important physical characteristic of a geothermal system; therefore, thermal exploration techniques provide the most direct method for making a first estimate of the size and potential of a geothermal system with surface geophysical exploration. Measurements can be made in holes as shallow as a few metres, but it is preferred to conduct temperature surveys in wells that are at least 100 m deep. Temperatures measured a short distance beneath the surface of the Earth are strongly affected by cyclic changes in temperature on the surface of the Earth. Drilling is usually fairly expensive and puts practical limits to the use of the method. Furthermore, shallow wells are not always adequate to get reliable values on the thermal gradient. In addition, determination of temperature in a test hole is not an easy task. In deep test holes, which must be drilled with a circulating fluid such as mud, a considerable disturbance of the normal-temperature environment will take place during drilling. This is particularly true if the gradient is relatively high and the temperature change over the well interval is relatively large. Temperature at depth can be sensed directly in boreholes or estimated by extrapolation of heat-flow measurements in both shallow and deep holes. Heat-flow measurements combine observed temperature gradients and thermal conductivity measurements to determine the vertical heat transport in areas where conduction is the primary mechanism of heat transport.

Normally, one must wait for a period of time comparable to that involved in drilling the well before the well temperatures return to within 10% of their undisturbed state. Drilling a well to several hundred metres depth may take a few days. In order to measure temperatures accurately for thermal gradient purposes, the temperatures are recorded for periods that are several times longer than the duration of drilling. However, there are methods which provide the relationship between the temperature and the time after the drilling.

### 9.8.2  Electrical Methods

Electrical methods or resistivity methods are the most important geophysical methods in the surface exploration of geothermal areas. Various methods used for measuring electrical resistivity are based on the proposition that temperature affects the electrical properties of rocks. Direct current resistivity methods measure Earth's resistivity using a direct or low-frequency alternating current source. High temperatures and hot thermal fluid circulation observed in a geothermal system have a severe impact on the electrical properties of the geological formations encountered in the geothermal field area. A large decrease in resistivity of the rocks is observed due to saline, brine hot waters that circulate in the permeable paths of it. Rocks are electrically conductive as consequences of ionic migration in pore space water and, more rarely, electronic conduction through metallic lustre minerals. The electrical conductivity of ionic conductors increases largely with temperature. Conductivity of the host rock of the geothermal field increases due to wall rock hydrothermal alteration and hydrothermal mineral deposition in fracture zones.

The maximum increase in conductivity for most electrolytes is about seven times between 20 and 350°C. Different methods of electrical measurements and setups are as follows:

- *dc Methods*, In this method, dc or low-frequency ac current ($I$) is passed through the ground. The developed electrical potential field is measured. Two current electrodes are used for passing current into the ground, while the developed potential difference, along the potential dipole length, is measured by two other potential electrodes.
- *TEM*, where current is induced by a time-varying magnetic field from a controlled source. The monitored signal is the decaying magnetic field at the surface from the secondary magnetic field.
- *MT*, magnetotelluric (MT) and the audio-magnetotelluric (AMT) use natural electromagnetic fields for energizing the ground. Since the origin of these fields is located at a large distance from the study area, the plane-wave assumption for the EM field is valid, thus simplifying the interpretational techniques.

Raw electrical field data have to be interpreted in order to derive models of the electrical structure of the sub-surface. Interpretational schemes exist for almost all of the electrical methods. Most modern interpretation methods involve either or both forward modelling and inversion. In forward modelling, as proposed resistivity model of the sub-surface is constructed and the response of this model to a particular electrical method being used is computed for comparison with the observed data. Adjustment in the model is made till the data closely matches the observed data.

### 9.8.3 Magnetic Measurements

The largest component (80–90%) of the Earth's field is believed to originate from convection of liquid iron in the Earth's outer core, which is monitored and studied using a global network of magnetic observatories and various satellite magnetic surveys. This field in first approximation is dipolar and has a strength of approximately 50,000 nT.

Magnetic surveys are an effective method to locate a prospective geothermal reservoir. It is found that all igneous and metamorphic rocks generally have a higher magnetic susceptibility than sedimentary rocks. In the magnetic method, the intensity of the natural magnetic field is measured. This includes contribution from the Earth's core and crust, as well as any secondary magnetic field induced in magnetic geological bodies, which locally creates positive and negative magnetic field anomalies activity. Ferromagnetic materials exhibit a phenomenon characterized by a loss of nearly all magnetic susceptibility at a temperature called the Curie temperature. At this temperature, ferromagnetic rocks become paramagnetic, and there are no detectable magnetic anomalies. The Curie temperature for titanomagnetic material, which is the most common magnetic mineral in igneous rocks, is less than approximately 570°C. It has been established that an increase in titanium content of titanomagnetite causes a reduction in Curie temperature, and the titanium content generally increases in the more mafic igneous rocks. The greatest potential for the magnetic method lies in its ability to detect the depth at which the Curie temperature is reached.

## 9.9 Geochemical Techniques

The basic philosophy behind using geochemical methods in geothermal exploration is that fluids on the surface (aqueous solutions or gas mixtures) reflect the physical, chemical and thermal conditions in the geothermal reservoir inside the Earth. Chemical data on hot water and steam discharges in a virgin area serve as useful indicators of the feasibility of further exploration in the area including preliminary drilling locations. Together with structural information from geological, hydrological and geophysical methods, they can guide decision-making on sub-surface exploration by drilling.

Chemical geothermometers, which relate the fluid chemistry and reservoir temperature, are routinely used in assessing the energy potential.

Geothermometers have been classified into three groups:

1) Water or solute geothermometers
2) Steam or gas geothermometers
3) Isotope geothermometers

### 9.9.1 Water or Solute Geothermometers

The most important water geothermometers are silica (quartz and chalcedony), Na/K ratio and Na-K-Ca geothermometers. Others are based on cation ratio and any uncharged aqueous species as long as equilibrium prevails. Silica geothermometers are one of the oldest and most commonly used types. Geochemical studies have shown that quartz is an important secondary mineral phase present, and therefore, it is common to compare the silica value of the thermal waters with the quartz solubility versus temperature curve for deducing the reservoir temperature. In systems above about 180°C, the silica concentration is controlled by the equilibrium with quartz. At lower temperatures, the equilibrium with chalcedony becomes important.

#### 9.9.1.1 Na-K Geothermometer

A commonly used geothermometer where geothermal waters are known to come from high-temperature environments (>180°C, up to about 200°C) is the atomic ratio of sodium to potassium (Na/K). The ratio decreases with an increase in temperature. The cation concentrations ($Na+$, $K+$) are controlled by temperature-dependent equilibrium reactions with feldspar and mica. The main advantage of this thermometer is that it is less affected by dilution and steam separation than other geothrmometers, provided there is little $Na+$ and $K+$ in the diluting water compared to the reservoir water. Na/K geothermometer fails at temperatures lower than 100–120°C and gives high temperatures for solutions with high calcium contents.

### 9.9.1.2 Na-K-Ca Geothermometer

The temperatures obtained from Na-K geothermometers are generally higher than the actual values, for high calcium contents. To overcome this, Na-K-Ca thermometers are used.

A temperature equation for a geothermometer is a temperature equation for a specific equilibrium constant referring to a specific mineral–solution reaction.

## 9.9.2 Gas Thermometers

The gas geothermometers are useful for predicting sub-surface temperatures in high-temperature geothermal systems. They are applicable to systems in basaltic to acidic rocks and in sediments with similar composition, but should be used with reservation for systems located in rocks which differ much in composition from the basaltic to acidic ones. The geothermometry results may be used to obtain information on steam condensation in up flow zones or phase separation at elevated pressures.

## 9.9.3 Isotopes

Chemical elements with the same atomic number (protons) and different atomic mass are defined as isotopes. Isotopes have identical chemical behaviour but different physical properties. Hydrogen has three isotopes, 1H, 2H, 3H, respectively, and oxygen also has three isotopes, 16O, 17O and 18O. Isotope geothermometers involving oxygen- or hydrogen-exchange reactions with water. Several isotope-exchange reactions can be used as sub-surface temperature indicators. The exchange of 18O between dissolved sulfate and water is the most useful. An isotopic fractionation occurs when steam separates from hot water. The isotopic compositions of both the steam and water in a well sample may be determined from the total discharge, whose steam and water fraction is known.

Different parts of the Earth are composed of a variety of elements in varying amounts. The Earth's crust contains a variety of noble gases, one of those being helium. Natural helium occurs as two isotopes, helium-4 (4He) and helium-3 (3He). Typically, helium-4 is more abundant in the Earth's crust, whereas helium-3 is more abundant in the mantle below. Thus, the helium-3/helium-4 ratio of the gas found in groundwater can provide an indication of the extent to which the water has interacted with volcanic rocks derived from the mantle.

Almost all the geothermometers are empirical and involve many assumptions. Different temperatures can be given by using different ones with the same set of geochemical data. A particular geothermometer may be suitable to one place but not another. This deficiency prevents geothermometers from more effective application and sometimes causes considerable problems Geochemists will gather and interpret data points from multiple geothermometers in order to make the most reliable sub-surface temperature estimates.

## 9.9.4 Drilling

Drilling represents about 30–50% of the cost of a hydrothermal geothermal electricity project. The drilling of the first production well at a geothermal resource is therefore considered to be an exploration-phase activity. A prospect is usually not considered as being in the drilling phase until after at least one production well has been drilled successfully. The drilling of the first "wildcat" well has a success rate probability of approximately 25%. The success rate of further wells approaches 60–80% in construction stages.

Drilling for geothermal energy is quite similar to rotary drilling for oil and gas. The main differences are due to the high temperatures associated with geothermal wells, which affect the circulation system and the cementing procedures as well as the design of the drill string and casing. The established deep drilling technique is the rotary drilling whereby a string of drill pipe is hung from a derrick and turned by a diesel engine. The top section of the pipe, called the "Kelly", is square in cross section to allow it to be rotated by the action of a rotary table through which it passes. The bit is a tri-cone roller bit that applies very concentrated loads on the rock face, causing it to crack and spall. The rock fragments or chips must be removed for the bit to proceed. Broken rock chips are removed by scraping or hydraulic cleaning. The drilling fluid, or "mud", is a critical element in the removal of chips.

Rotary bits drill the formation using primarily two techniques:

1) rock removal by exceeding its shear strength and
2) removal by exceeding the compressive strength.

First, a small-diameter "temperature gradient hole" is drilled (some only 200 ft deep, some over 4000 ft deep) with a truck-mounted rig to determine the temperatures and underground rock types. Either rock fragments or long cores of rock are brought up from deep down the hole, and temperatures are measured at depth. Depending on the temperature gradient, the drilling is decided.

Typically, wells are deviated from vertical to about 30–50° inclination from a "kick-off point" at depths between 200 and 2000 m. Several wells can be drilled from the same pad, heading in different directions to access larger resource volumes, targeting permeable structures and minimizing the surface impact.

While rotary cone bits have been sufficient to drill production wells to date, other conventional drilling technologies as well as potential "breakthrough" technologies could potentially be implemented into geothermal drilling to increase returns. The diamond PDC drill bit commonly employed in oil and gas operations with success is viewed as inherently more efficient than rotary cone bits.

Recently, a company called Hyper Sciences has suggested to harness geothermal energy with a new kind of drilling technology, using projectiles fired into the ground. It completely eliminates the use drilling bit which by slowly grinding away at rocks perpetually wears them out. Instead, it just shoots holes in the ground with bullets which can bore a hole 10 times faster than traditional drill.

## 9.10 Utilization of Geothermal Resource

Geothermal resources are suitable for many different types of uses but are commonly divided into two categories, *high* and *low enthalpy*, according to their energy content [6, 7, 9]. Utilization of geothermal fluid depends heavily on its thermodynamic characteristics and chemistry. High-enthalpy resources, such as dry steam and hot fluids that are found in volcanic regions and island chains, are utilized to generate electric power with conventional cycles. Low-enthalpy resources (<150°C) are employed for direct heat uses and electricity generation using a binary fluid cycle. In direct heat use, the geothermal sources are used for space heating/cooling, water heating for domestic and industrial use and in hybrid steam power plants.

### 9.10.1 Electricity Generation from Geothermal Resources

The present-day annual electrical energy produced from geothermal resources is about 57,000 GWh, which is less than 0.4% of the total worldwide electricity generation which is very modest. Power from high-enthalpy geothermal sources is generated in many places, especially those that are located on plate boundaries or tectonically active regions. A vapour-dominated (dry steam) resource can be used directly, whereas a hot-water resource needs to be flashed by reducing the pressure to produce steam. In places where low-temperature resource is available (below 150°C), a secondary low-boiling-point fluid (hydrocarbon) is required to generate the vapour, in a binary or organic Rankine cycle plant.

Presently, geothermal power plants in operation are essentially of three types: *dry steam, flash steam and binary cycle*. A combination of flash and binary technology, known as the flash/binary combined cycle, has also been used. It takes advantage of both technologies.

In general, geothermal power plants have lower efficiency when compared with fossil-fuelled plants primarily due to (i) low temperature of the steam, which is usually much below 250°C; and (ii) presence of non-condensable gases such as $CO_2$, $H_2S$, $NH_3$ and so on. in the steam, which have to be removed from the condensers of power plants. However, the capacity factor of geothermal power plants can be quite large as compared to other renewable sources such as wind and solar which are intermittent in nature. The global average was 73% in 2005 and can be as high as 96% in some cases.

The total generation from geothermal sources is growing by 3% annually because of a growing number of plants and improvements in their capacity factors, development of binary cycle power plants and improvements in drilling and extraction technology.

The total installed capacity from worldwide geothermal power plants is given in Table 9.2.

### 9.10.2 Dry Steam Power Plants

Dry steam plants were the first type of geothermal power plants that were developed for commercial use. These plants have been operating for more than hundred years longer than any other geothermal conversion technology. In a dry steam plant, the steam produced in dry steam fields runs the turbines and the generator. The availability of dry steam eliminates the need to burn fossil fuels to generate steam, eliminating the need to transport and store fuels. These plants emit only excess steam and very minor amounts of gases.

These power plants are most suitable for vapour-dominated resources where steam production is not contaminated with liquid. The reservoirs produce superheated steam at 180–225° and 4–8 MPa. This steam is piped directly from underground wells to the power plant where it is directed into a turbine/unit. Conventional dry steam turbines require fluids of at least 150°C and are available with either non-condensing (backpressure) or condensing exhausts. The non-condensing steam turbine uses high-pressure steam for the rotation of blades. This steam then leaves the turbine at the atmospheric

**Table 9.2** Total worldwide installed capacity from 1950 up to the end of 2010.

| Year | Installed capacity (MW) | Energy produced (GWh) |
|---|---|---|
| 1950 | 200 | |
| 1955 | 270 | |
| 1960 | 386 | |
| 1965 | 520 | |
| 1970 | 720 | |
| 1975 | 1,180 | |
| 1980 | 2,110 | |
| 1985 | 4,764 | |
| 1990 | 5,834 | |
| 1995 | 6,883 | 38,035 |
| 2000 | 7,972 | 48,261 |
| 2005 | 8,933 | 55,709 |
| 2010 (projected) | 10,715 | 67,246 |

*Source*: World Geothermal congress (2010) [10]. Reproduced with permission from Ruggero Bertani.

pressure or lower pressure. Therefore, this turbine is also known as the backpressure steam turbine. This low-pressure steam is used for processing, and no steam is used for condensation.

In the backpressure system, steam is passed through the turbine and vented to atmosphere. This cycle consumes twice more steam per produced kilowatt-hour (kWh), at identical turbine inlet pressure, compared to a condensing cycle.

Condensing units are more complex in design, requiring more ancillary equipment and space. The condensing turbine contains two outlets. The first outlet extracts the steam with intermediate pressure for the feeding of the heating process while the second outlet extracts the remaining steam with low-pressure steam for the condensation. The condensed water then goes back to the boiler for the regeneration of the electricity.

The steam after passing through the turbine rotates its shaft which in turn drives the generator producing electricity. A dry steam power plant is shown in Fig. 9.6.

### 9.10.3 Single-Flash Steam Power Plant

Flash steam power plants are the most common form of geothermal power plants. Flash power plants typically require resource temperatures in the range of 170–260°C. A single-flash power plant is shown in Fig. 9.7. When the geothermal wells produce a mixture of steam and liquid, the single-flash plant is a relatively simple way to convert the geothermal energy into electricity. First, the mixture is separated into distinct steam and liquid phases with a minimum loss of pressure. This is done in a cylindrical cyclonic pressure vessel, usually oriented with its axis vertical, where the two phases disengage owing to their inherently large density difference. The fluid at temperatures greater than 182°C is pumped under high pressure into a tank at the surface held at a much

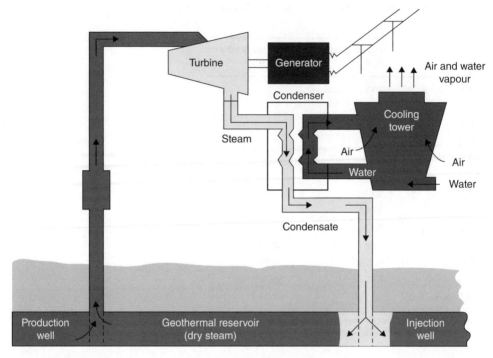

**Figure 9.6** Dry steam power plant. *Source*: Nasruddin et al. (2016) [8]. Reproduced with permission from Elsevier.

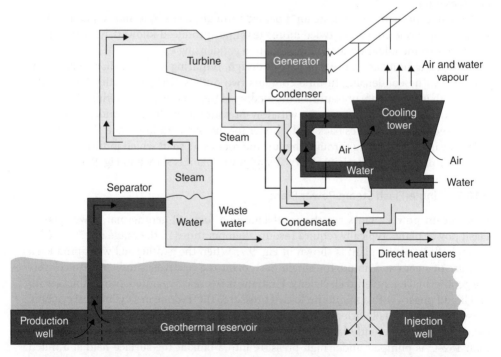

**Figure 9.7** Single-flash steam. *Source*: Nasruddin et al. (2016) [8]. Reproduced with permission from Elsevier.

lower pressure, causing some of the fluid to rapidly vaporize or "flash". The vapour then drives a turbine, which drives a generator. The liquid if not separated from the steam can cause scaling and/or erosion of piping and turbine components. The cooled water is returned to the reservoir to be heated by geothermal rocks again. Normally, a conventional condensing steam turbine is used in single-flash system, but lower steam pressures and temperatures are common. The plant therefore requires more steam per kWh than in dry steam power plant. Moreover, the bulk of the water produced may contain as unflashed hot brine which is then reinjected unless it can be used for direct heating.

The single-flash steam plant is the mainstay of the geothermal power industry. Single-flash plants account for about 32% of all geothermal plants. The turbines used must be made of corrosion-resistant materials owing to the presence of gases such as hydrogen sulfide that can corrode the ordinary steel.

### 9.10.4 Double-Flash Power Plant

The schematic diagram for a double-flash plant is shown in Fig. 9.8. The double-flash steam plant is an improvement on the single-flash design, and it can produce 15–25% more power output for similar geothermal fluid. It is a simple extension of the single-flash cycle which makes use of the energy remaining in the separated brine. By directing this brine to a low-pressure separator, additional steam at a lower pressure than the primary steam can be generated which can increase the total power generated by more than 50%. This additional power generation is limited by the low flash separation pressure, which is generally maintained above atmospheric. The plant is more complex, more costly and requires more maintenance. The turbine used may be a dual-admission, single-flow machine as shown in Fig. 9.8, where the low-pressure steam is admitted to the steam path at an appropriate stage so as to merge smoothly with the partially expanded high-pressure steam. Other designs are possible; for example, two separate turbines could be used, one for the high-pressure steam and one for the low-pressure steam. Because of the additional flash process, and the hotter resource, the waste brine becomes more highly concentrated in silica than in a single-flash plant. Thus, silica scaling problem is more serious in this case.

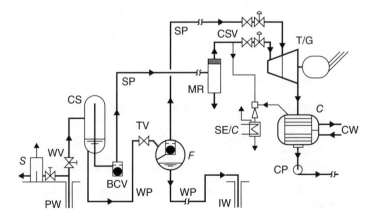

**Figure 9.8** Double-flash steam. *Source*: Dipippo (2008) [6]. Reproduced with permission from Elsevier.

### 9.10.5  Binary Cycle Power Plant

Among geothermal energy resources, the medium- and low-temperature water-dominated systems, with temperatures between 110 and 160°C, are the most abundant. Binary cycle geothermal power plants are similar to conventional fossil or nuclear plants in that the working fluid undergoes an actual closed cycle. Binary power plants (or Organic Rankine Cycle units, ORC) are the best energy conversion systems to exploit them, both from a technical and an environmental point of view. Binary cycle geothermal power generation plants differ from dry steam and flash steam systems in that the water or steam from the geothermal reservoir never comes in contact with the turbine/generator units. In these plants, the heat is recovered from the geothermal fluid, via a heat exchanger, to vaporize a low-boiling-point organic fluid and drive an organic vapour turbine. The heat-depleted geothermal brine is pumped back into the source reservoir, thus securing sustainable resource exploitation. Because the geothermal water never flashes in air-cooled binary plants, the total water can be injected back into the system through a closed loop. This serves the dual purpose of reducing already low emissions to near zero and maintaining the reservoir pressure, thereby extending the project lifetime.

A simple binary cycle power plant is shown in Fig. 9.9. In the binary process, the geothermal fluid, which can be either hot water, steam or a mixture of the two, heats another liquid such as isopentane or isobutane (known as the "working fluid") that boils at a lower temperature than water. The two liquids are kept completely separate through the use of a heat exchanger used to transfer heat energy from the geothermal water to the working fluid. When heated, the working fluid vaporizes into gas, and (as steam) the force of the expanding gas turns the turbines that power the generator.

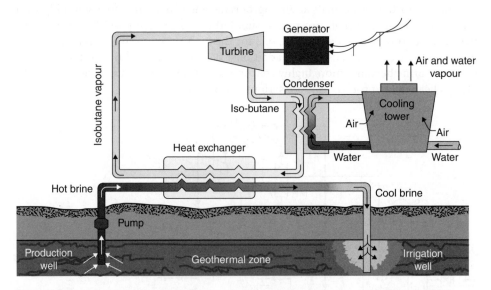

**Figure 9.9** Binary cycle. Image credit: Idaho national laboratory, DOE. *Source*: Dipippo (2008) [6]. Reproduced with permission from Elsevier.

## 9.11    Enhanced Geothermal Systems

At places where no natural geothermal resources in the form of steam or hot water exist, the heat of the rock can be used by creating artificial permeability for fluids extracting that heat. Known as "hot dry rock" technology (HDR), this method is under development since the 1970s. In an EGS, fluid is injected into the sub-surface under carefully controlled conditions, which cause pre-existing fractures to re-open, creating permeability. The principle of EGS is simple: in the deep sub-surface where temperatures are high enough for power generation (150–200°C), an extended fracture network is created and/or enlarged to act as a new pathway. Water from the deep wells and/or cold water from the surface is transported through this deep reservoir using injection and production wells and recovered as steam/hot water. Injection and production wells as well as further surface installations complete the circulation system. The extracted heat can be used for district heating and/or for power generation.

EGS plants, once operational, can be expected to have great environmental benefits ($CO_2$ emission is zero).

### 9.11.1    Combined or Hybrid Plants

Because of relative simplicity and reliability of single-flash plants, they are often the first type of plants installed at a newly developed field. However, their utilization efficiency is lower than that of a double-flash plant. Since single-flash plants have a significant amount of waste liquid from their separators that is still fairly hot, typically 150–170°C, the possibility of using it to generate more power can be an option instead of directly reinjecting. In many places, combined single- and double-flash plants have been built to increase the efficiency. The single- and double-flash combined system shown in Fig. 9.10 consists of two single-flash units, Units 1 and 2, and a third unit, Unit 3. Unit 3 appears to be simply another single-flash unit, but the power plant as a whole is an integrated single- and double-flash facility since the original geofluid experiences two stages of flashing.

Combined flash and binary types, where a binary plant is used as a bottoming cycle with a flash steam plant, increase the overall thermal efficiency, improve load-following capability and efficiently cover a wide resource temperature range. A combination of flash and binary technologies, known as the flash/binary combined cycle, has been used effectively to take advantage of the benefits of both technologies. In this type of plant, the flashed steam is first converted to electricity with a steam turbine, and the low-pressure steam exiting the backpressure turbine is condensed in a binary system. This allows for the effective use of air-cooling towers with flash applications and takes advantage of the binary process. The flash/binary system has a higher efficiency where the well fields produce high-pressure steam.

### 9.11.2    Combined Heat and Power (CHP) Plants

In many places, it is common to combine both power generation and direct heat usage in a single geothermal plant. By capturing some of the waste heat in the leftover brine before it is reinjected, the overall utilization efficiency of the resource is enhanced.

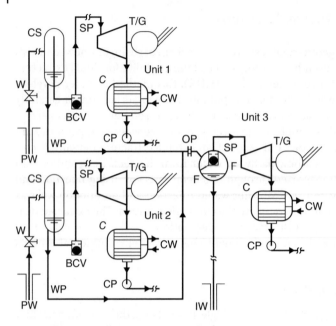

**Figure 9.10** Combined single and double flash.

CHP essentially takes the "waste" heat produced by geothermal electric plants and uses co-generation plants, or combined or cascaded heat and power plants (CHP) produce both electricity and hot water for direct use. Relatively small industries and communities of a few thousand people provide sufficient markets for CHP applications.

In this case, each unit operates with common controls, fluid collection and reinjection systems. The plant requires close monitoring of the injection water temperature in combined cycle systems, as declines could occur that lead to scaling. Iceland has three geothermal co-generation plants with a combined capacity of 580 MW in operation.

## 9.12 Direct Use of Geothermal Energy

Direct use of geothermal energy is one of the oldest, most versatile and also the most common form of utilization of geothermal energy [11, 12]. Direct use of geothermal resources is primarily for direct heating and cooling. Direct or non-electric utilization of geothermal energy refers to the immediate use of the heat energy rather than to its conversion to some other form such as electrical energy. The primary forms of direct use include swimming, bathing, space heating and cooling including district heating, agriculture (mainly greenhouse heating, crop drying and some animal husbandry), aquaculture (mainly fish pond and raceway heating), industrial processes and heat pumps (for both heating and cooling). In general, the geothermal fluid temperatures required for direct heat use are lower than those for economic electric power generation.

According to the present estimate, district heating represents 88% of the installed capacity and 89% of the annual energy use. Geothermal (ground-source) heat pumps (GHPs) are used for both heating and cooling. Direct use of geothermal resources normally involves temperatures below 150°C.

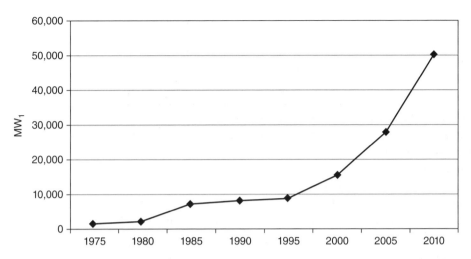

**Figure 9.11** Worldwide growth of installed geothermal direct-use capacity.

For space heating, two basic types of systems are used: open or closed loop. Open-loop (single-pipe) systems directly utilize the geothermal water extracted from a well to circulate through radiators. Closed-loop (double-pipe) systems use heat exchangers to transfer heat from the geothermal water to a closed loop that circulates heated freshwater through the radiators. This system is commonly used because of the chemical composition of the geothermal water. In both cases, the spent geothermal water is disposed of into injection wells and a conventional backup boiler may be provided to meet peak demand. It is also used for greenhouse and covered ground heating as well as for agriculture crop drying.

Worldwide growth of installed geothermal direct-use capacity is shown in Figs 9.11 and 9.12. The main growth in the direct-use sector has during the last decade been the

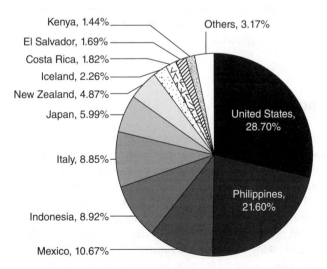

**Figure 9.12** Breakdown of geothermal energy production in different countries. *Source*: U.S. Energy Information Administration [13], Monthly Energy Review, March, 2012.

**Table 9.3** Summary of the various categories of direct-use worldwide for the period 1995–2015.

| | Capacity (MW) | | | | |
| --- | --- | --- | --- | --- | --- |
| | 2015 | 2010 | 2005 | 2000 | 1995 |
| Geothermal heat pumps | 49,898 | 33,134 | 15,384 | 5,275 | 1,854 |
| Space heating | 7,556 | 5,394 | 4,366 | 3,263 | 2,579 |
| Greenhouse heating | 1,830 | 1,544 | 1,404 | 1,246 | 1,085 |
| Aqua pond heating | 695 | 653 | 616 | 605 | 1,097 |
| Industrial uses | 610 | 533 | 484 | 474 | 544 |
| Bathing and swimming | 9,140 | 6,700 | 5,401 | 3,957 | 1,085 |
| Agriculture drying | 161 | 125 | 157 | 74 | 67 |

*Source*: World geothermal congress (2015) [14]. Reproduced with permission from John W. Lund.

use of geothermal (ground-source) heat pumps. This is due, in part, to the ability of geothermal heat pumps to utilize groundwater or ground-coupled temperatures anywhere in the world. The annual energy use for these units grew 2.29 times at a compound annual rate of 18.0%. The installed capacity grew 2.15 times at a compound annual rate of 16.6%. This is due to better reporting and to the ability of geothermal heat pumps to utilize groundwater or ground-coupled temperatures anywhere in the world. Geothermal (ground-source) heat pumps have the largest energy use and installed capacity, accounting for 68.3% and 47.2% of the worldwide capacity and use. The installed capacity is 33,134 MW, and the annual energy use of 200,149 TJ/year, with a capacity factor of 0.19 (in the heating mode). Almost all of the installations occur in North American, Europe and China, increasing from 26 countries in 2000 to 33 countries in 2005, to the present 43 countries. The equivalent number of installed 12 kW units (typical of the US and Western European homes) is approximately 2.76 million, over double the number of units report for 2005 and four times the number for 2000. Summary of the various categories of direct-use worldwide for the period 1995–2015 is shown in Table 9.3.

## 9.13 Environmental Impact

Certain environmental impacts associated with the development of geothermal sites and the operation of plants are inevitable. However, under normal conditions, they are generally confined to the immediate vicinity of the plant and are of lesser impact than those of other electric power generation technologies, particularly those using carbon-based fossil fuels and nuclear fuels. There have now been more than one hundred years of experience in developing geothermal fields and in building, operating, upgrading and even decommissioning geothermal plants of various types. In the earliest days, drilling of wells could be a hazardous undertaking and the behaviour of geothermal reservoirs was mysterious. Early developers and operators learned by doing so, and eventually a scientific understanding of the nature of the resource evolved. Along with this came the technology of how best to exploit geothermal energy and how to deal with the potential environmental impacts.

## 9.14  Summary

Geothermal energy has the potential to provide long-term, secure base-load energy and greenhouse gas (GHG) emission reductions. Accessible geothermal energy from the Earth's interior supplies heat for direct use and to generate electric energy. History of geothermal energy and types of geothermal resources are described in this chapter.

Until recently, the main interest in geothermal resources has been on the availability of steam or high-temperature hot water in porous rock formations and use of extracted steam to generate electricity. The future use of geothermal energy from advanced technologies such as the exploitation of hot dry rock/hot wet rock systems, magma bodies and geopressured reservoirs is briefly discussed. While the viability of hot dry rock technology has been proven, research and development are still necessary for the other two sources. Direct use of geothermal energy is also discussed briefly.

**Example 9.2**
Calculate the useful heat contents per square kilo metre of dry rock granite to a depth of 8.5 km. Take the geothermal temperature gradient at 28 K/km. The minimum useful temperature is 135 K above the surface temperature.

$$\rho_r = 2700\,\text{kg/m}^3, c_r = 820\,\text{J/kg K}$$

Temperature at 8.5 km $= 28 \times 8.5 = 238$ K
Useful temperature occurs at $135/28 = 4.82$ km.
Useful energy will be available from granite for $8.5 - 4.82 = 3.68$ km.
Temperature difference $= 238 - 135 = 103$ K

$$\text{Energy/km}^2 = \rho_r \times c_r \times \text{useful depth} \times \text{Temperature difference}/2$$
$$= 2700 \times 820 \times 3.68 \times 10^9 \times 103/2$$
$$= 4.19 \times 10^{17}\,\text{J/km}^2$$

## References

1 Armstead, H.C.H. (1983) *Geothermal Energy*, E. & F. N. Spon, London.
2 Fridleifsson, I.B. (2001) Geothermal energy for the benefit of the people. *Renewable and Sustainable Energy Reviews*, **5**, 299–312.
3 Stober, I. and Bucher, K. (2013) *Geothermal Energy: From Theoretical Models to Exploration and Development*, Springer.
4 Gupta, H. and Roy, S. (2007) *Geothermal Energy-An Alternative Resource for 21st Century*, 1st edn, Springer.
5 Lund, J.W. (2007) *Characteristics, Development and Utilization of Geothermal Resources*, GHC Bulletin.
6 Dipippo, R. (2008) *Geothermal Power Plants: Principles, Applications, Case Studies and Environmental Impact*, 2nd edn, Elsevier.

7 Bertani, R. (2007) World Geothermal generation in 2007, Proceedings European Geothermal Congress, Unterhaching, Germany, April-May, 2007

8 Nasruddin *et al.* (2016) Potential of geothermal energy for electricity generation in Indonesia-a review. *Renewable and Sustainable Energy Reviews*, **53** (2016), 733–740.

9 Ackerman, E., Digging for Geothermal Energy with Hypersonic Projectiles, IEEE Spectrum.

10 Bertani, R. (2010) Geothermal power generation in the world 2005-2010 update report Proceedings World geothermal congress Bali Indonesia.

11 Dickson, M.H. and Fanelli, M. (2003) *Geothermal Energy Utilization and Technology*, Unesco Publishing Renewable energy series.

12 Bloomquist, R.G. (2002) Direct use geothermal resources. *IEEE Power Engineering Society Summer Meeting*, **1**, 15–16.

13 U.S. Energy Information Administration (2012), Monthly Energy Review, March, 2012

14 Lund, J.W. and Boyd, T.L. (2015) Direct Utilization of Geothermal Energy 2015 Worldwide Review, Proceedings World Geothermal Congress 2015 Melbourne, Australia.

# 10

# Ocean Energy

## 10.1 Energy from Ocean

Oceans, covering over 70% of the Earth's surface, are the world's largest collector and retainer of the Sun's vast energy. This vast energy from the oceans is available in the form of *tides*, *ocean thermal* (due to temperature difference), *ocean currents* and *waves* [1–3]. Tidal energy is the oldest form of renewable energy, which was used in the water mills by the Romans when they occupied England. Tidal mills were also built in Spain, France, United Kingdom and China during medieval period, around AD 1100. However, it was not until the 20th century that it was investigated as a potential source of electricity. The first ever tidal power plant was opened in 1966 in La Rance, France. It has reliably generated power for more than 15 h/day and has delivered 500–600 GWh annually since 1967. However, this type of plant was not erected in other areas owing to the high cost of power plant components and difficulty in locating sites in which such power plants could be built. Now in order to find new renewable energy sources, there is renewed interest in these technologies. Significant growth has occurred in the number of devices developed for ocean energy conversion since 2003. Many countries are now involved in finding new technology and in research and development for application of ocean energy.

Tidal energy is clean renewable green energy and is non-polluting, reliable and predictable. In 1687, Sir Isaac Newton explained that ocean tides result from the gravitational attraction of the Sun and the Moon on the oceans of the Earth. Newton's law of universal gravitation states that the gravitational attraction between two bodies is directly proportional to the product of their masses and inversely proportional to the square of the distance between the bodies. The tidal generating forces vary inversely as the cube of the distance of the tide generating object. Thus, Moon plays a more significant role in the creation of tides compared to the Sun because of its proximity to the Earth although the mass of the Sun is much larger. The gravitational attraction between the Earth and the Moon is the strongest on the side of the Earth that is facing the Moon, because it is closer. This attraction causes the water on this "near side" of the Earth to be pulled towards the Moon. On the other hand, the Earth is also being pulled towards the Moon (and away from the water on the far side). As the Earth rotates, these two bulges travel at the same rate as the Earth's rotation. The Moon rotates around the Earth with respect to the Sun approximately 29.5 days (lunar month) in the same direction that the Earth rotates every 24 h. Ocean levels fluctuate daily as the Sun, Moon and Earth interact. As the Moon travels around the Earth and as they, together, travel around the Sun, the combined gravitational forces cause the world's oceans to rise and fall. Since

*Operation and Control of Renewable Energy Systems,* First Edition. Mukhtar Ahmad.
© 2018 John Wiley & Sons Ltd. Published 2018 by John Wiley & Sons Ltd.

the Earth is rotating at the same time when this phenomenon takes place, there are two high tides and two low tides each day. However, these periods do not happen at the same time each day. This is because the Moon takes slightly longer than 24 h to line up again exactly with the same point on the Earth – about 50 min more. Therefore, the timing of high tides is staggered throughout the course of a month, with each tide commencing approximately 24 h and 50 min later than the one before it. The ocean is constantly moving from high tide to low tide and then back to high tide. When the sSun and Moon are aligned, there are exceptionally strong gravitational forces, causing very high and very low tides which are called *spring tides*, occurring at full and new moon, respectively, as shown in Fig. 10.1(a) and (b). The spring in spring tide is not related with the season.

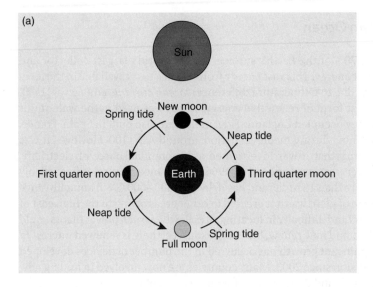

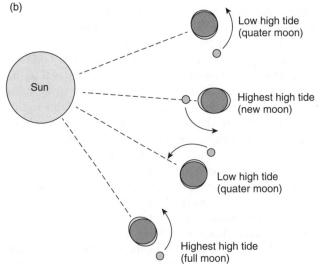

**Figure 10.1** (a) Spring and neap tides. (b) High and low tides.

Spring tides are very strong because the Moon and the Sun both are responsible for it. When the Sun and Moon are not aligned and are perpendicular to one another (with respect to the Earth), the gravitational forces cancel each other out, and the tides are of minimum range. These are called *neap tides*. Neap tides also occur twice per month during the first and third quarter cycles of the Moon. Typically, the spring tide range is about twice of the neap tide range.

For any particular location, the height and fluctuation in time of the tide depend on the location of the Sun and the Moon and on the shape of the beach, coastline, coastline depth and prevailing ocean currents. The tidal bulge of the Moon follows along the path on the Earth's surface which intersects with the orbital plane of the Moon. This plane is tilted about 23° with respect to the equatorial plane of the Earth. The result is that near the equator, the difference between high tide and low tide is very small, compared to other latitudes. The Sun also plays a major role, affecting the size and position of the two tidal bulges. The solar tidal bulges are about half the size of those caused by the Moon. Similarly to the Moon, gravitational attraction to the Sun creates one bulge towards the Sun and one away from it. Unlike the Moon, solar tides do not vary on a daily basis. If there was no Moon, the daily tidal period would be exactly 24 h. High tide would be at noon and midnight, and low tide at 6 PM and 6 AM every day. When the Earth is closest to the Sun (perihelion), which occurs around January 2 of each calendar year, the tidal ranges are enhanced. When the Earth is furthest from the Sun (aphelion), around July 2, the tidal ranges are reduced.

The energy from the oceans can be harvested using thermal energy, from the temperature difference of the warm surface waters and the cool deeper waters, as well as potential and kinetic energy (or mechanical energy) from the tides, waves and currents. The energy from the marine and tidal currents can be harnessed by building a tidal barrage across an estuary, or a bay in high-tide areas, or by extracting energy from free-flowing water.

The ocean is the world's largest solar collector. The thermal energy of the oceans can be extracted because in tropical seas, there are temperature differences of about 20–25° between the warm, solar-absorbing near-surface water and the cooler 500–1000 m "deep" water. Heat engines can operate from this temperature difference across such a large heat storage.

The energy in the ocean waves is a form of concentrated solar energy that is transferred through complex wind–wave interactions. When wind blows across the sea surface, it transfers the energy to the waves. They are powerful source of energy. The energy output is measured by wave speed, wave height, wavelength and water density. The stronger the waves, the more capable it is to produce power. Methods of capturing wave energy are being developed. These methods described later include placing devices on or just below the surface of the water and anchoring the devices to the ocean floor.

## 10.2 Harnessing the Tidal Energy

The technology that is used to produce electricity using the difference between the low and high tides is very similar to the one used for the generation of electricity in traditional hydroelectric power plants. The stronger the tide, either in water level height or in tidal current velocities, the greater the potential for tidal electricity generation.

Tidal power plants are similar to a conventional run of river hydro plants, one of the differences being lower head, which requires large axial-flow turbines with low speed. For a tidal barrage to be feasible, the difference between high and low tides must be at least 16 ft. There are only a few places where this tide change occurs around the Earth. Some power plants are already operating using this idea. La Rance Station in France, the world's first and still largest tidal barrage, has a rated capacity of 260 MW and has been operating since 1966. However, tidal barrages have several environmental drawbacks, including changes to marine and shoreline ecosystems, most notably fish populations. Tidal power can be classified into two main types with a theoretical third variant currently being subject to development. These are tidal stream generators, tidal barrage generators and dynamical tidal power.

### 10.2.1 Tidal Barrage Power

The tidal barrage system is a dam that traps water at high tides and then releases it through defined channels that carry it through a turbine. A tidal barrage is usually made of reinforced concrete and spans an estuary, bay, river or other ocean inlet. The electricity is generated by water flowing into and out of gates and turbines installed along a dam or barrage built across a tidal bay or estuary. When high tides arrive, the doors are shut to keep the water from withdrawing over to the sea as it ordinarily would. After that, the water is released through particular channels that direct it through turbines.

The level of water in the large oceans of the Earth rises and falls according to predictable patterns. The main periods *of* these tides are diurnal at about 24 h 50 min and semi-diurnal at about 12 h 25 min. The change in height between successive high and low tides is in the range between about 0.5 m in general and about 10 m at particular sites near continental land masses. The movement of the water produces tidal currents, which may reach speeds of 5 m/s in coastal and inter-island channels. In contrast to other clean sources, such as wind, solar, geothermal, tidal energy can be predicted for centuries ahead from the point of view of time and magnitude [4].

In fact, people used the phenomenon of tides and tidal currents long before the Christian era. The earliest navigators, for example, needed to know periodical tide fluctuations as well as where and when they could use or would be confronted with a strong tidal current. However, the serious study and design of industrial-size tidal power plants for exploiting tidal energy only began in the 20th century with the rapid growth of the electric industry. Overall, the tidal current energy technology is not as widely researched as wave energy. However, the status of developments has advanced significantly in recent years.

However, this energy source, similarly to wind and solar energy, is distributed over large areas, which presents a difficult problem for collecting it. Four large-scale working tidal power plants are the La Rance Plant (France, 1967), the Kislaya Guba Plant (Russia, 1968), the Annapolis Plant (Canada, 1984) and the Jiangxia Plant (China, 1985). All existing tidal power plants are constructed using the design of conventional river hydropower stations.

### 10.2.2 Tidal Barrage Technologies

Power generation using tidal barrage technology is shown in Fig. 10.2. As shown in Figure 10.2, the Tidal barrage consist of a large, dam-like structure built across the

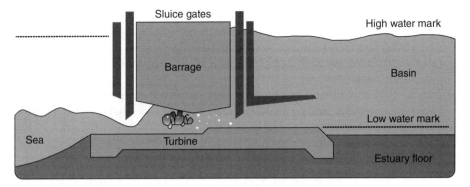

**Figure 10.2** Tidal barrage technology.

mouth of a bay or an estuary in an area with a large tidal range. The common model for tidal power facilities involves erecting a tidal dam, or barrage, with a sluice across a narrow bay or estuary. As the dam includes a sluice that is opened to allow the tide to flow into the basin, the sluice is then closed, and as the sea level drops, traditional hydropower technologies can be used to generate electricity from the elevated water in the basin. The disadvantage of tidal barrages is that it is a slow generation process because it takes 12 h for the tide to move from low to high. Then it takes another 6 h for the tide to retreat back to low again. In addition, building dams and barrages across tidal estuaries is an expensive process; therefore, the best tidal sites are those that exist where a tidal bay has a narrower opening, thus reducing the length of the dam required.

### 10.2.3 Tidal Stream Power

There is interest in another form of technologies that capture the tidally driven coastal currents or tidal stream. Tidal currents are generated not only by tides but also by wind, temperature and salinity differences. Most ocean currents are driven by wind and solar heating of surface waters near the equator, while some currents result from density and salinity variations of the water column. Ocean currents are relatively constant and flow in one direction, in contrast to tidal currents along the shore.

While ocean currents move slowly relative to typical wind speeds, they carry a great deal of energy because of the density of water. Water is more than 832 times denser than air. So for the same surface area, water moving 20 km/h exerts the same amount of force at a constant 180 km/h wind. Because of this physical property, ocean currents contain an enormous amount of energy that can be captured and converted to a usable form.

Another advantage about tidal current flow is their near 98% predictability, accuracy for decades. Tidal charts are accurate to within minutes, for years ahead. Tidal current is independent of prevailing weather conditions such as wind, fog, rain and clouds that affect other renewable generation forecasts.

Extraction of energy from tidal or marine current uses a different approach from tidal barrage technology [5]. Instead of using a dam structure, the devices are placed directly "in-stream" and generate energy from the flow of water. Tidal stream power is usually accomplished through the use of an underwater turbine similar in function that resembles a small-scale wind turbine. Horizontal turbine generators called "tidal turbines" or "marine current turbines" are placed on the ocean floor, and the stream currents flow across the turbine blades, powering a generator much as how wind turns the blades of wind power turbines. In fact, in some tidal stream generation areas, the seabed looks just as underwater wind farm with arrays of tidal stream generators covering large areas. Since water is 830 times heavier than air, it therefore can generate power at lower speeds than the wind turbines.

There are a number of different technologies for extracting energy from marine currents, including horizontal- and vertical-axis turbines, as well as others such as Venturis and oscillating foils. The energy available at a site is proportional to the cube of the current velocity at the site and to the cross-sectional area. This means that, in general, the power that can be generated by a turbine is roughly proportional to its area and that achieving high-power outputs is dependent on having high flow velocities. For this reason, tidal current systems are best suited to areas where narrow channels or other features generate high velocity and are governed by the following equation. The power $P$ is given by

$$P = \rho A V^3 \tag{10.1}$$

However, a marine energy converter or turbine can only harness a fraction of this power due to losses and (10.4) is modified as follows [6–8].

$$P = \frac{1}{2} C_p \rho A V^3 \tag{10.2}$$

$C_p$ is known as the power coefficient and is essentially the percentage of power that can be extracted from the fluid stream and takes into account losses due to Betz's law and those assigned to the internal mechanisms within the converter or turbine. For marine turbines, $C_p$ is estimated to be in the range of 0.35–0.5.

### 10.2.4 Dynamic Tidal Power Generation

To losses and (10.1) is modified as follows dynamic tidal power or DTP is the most complicated, least well understood tidal power scheme which is still under development [5]. Basically, dynamic tidal power makes use of the fact that ocean tides do not operate strictly perpendicular to the shore but flow in parallel to the shore as well. This feature of tides would allow a type of barrage to be built perpendicular to the shore to harvest energy from the tides as they flow parallel to the shore. The concept is based on the fact that tides flowing in one direction will tend to build up behind a barrier, as the barrage, and so create a height difference that can be used to generate power. When the tide shifts direction, something that happens every 12 h, the height difference shifts sides.

The system only works if the barrage is at least 30 km long and only if the turbines work in BOTH directions. So far, the system is 100% theory and has never been put to test.

## 10.3 Energy of Tides

The seawater can be trapped at high tide in an estuarine basin behind a dam or barrier to produce tidal *range* power. Now this water of density $\rho$ is allowed to run out through turbines at low tide. The energy of the tide wave contains two components, namely potential and kinetic. The potential energy is the work done in lifting the mass of water above the ocean surface. This energy can be calculated as follows:

If the basin has surface area $A$ that remains covered in water at low tide, then the trapped water, having a mass $\rho AH$ at a centre of gravity $H/2$ above the low tide level, is all assumed to run out at low tide ($H$ is height of tide). The potential maximum energy available per tide if all the water falls through $H/2$ is therefore (neglecting small changes in density from the seawater) given by

$$\text{Energy per tide } E = \rho AHg\frac{H}{2} = \frac{1}{2}\rho gAH^2 \tag{10.3}$$

Taking an average $(g\rho) = 10.15\,\text{kN/m}^3$ for seawater, one can obtain for a tide cycle per square metre of ocean surface,

$$E = 1.4H^2\,\text{Wh} \tag{10.4}$$

The tidal power due to potential energy is directly proportional to the area of basin and the tide amplitude. However, in actual turbine, the energy from the turbine can be obtained till the water head reaches a value $h$. Hence, the power available is given by

$$E = 1.4\,(H - h)^2\,\text{watt hour per cycle per square metre}$$

The kinetic energy KE of the water mass M is its capacity to do work by virtue of its velocity V. It is given by

$$\text{KE} = \frac{1}{2}MV^2 \tag{10.5}$$

The total tide energy equals the sum of its potential and kinetic energy component. Knowledge of the potential energy of the tide is important for designing conventional tidal power plants using water dams for creating artificial upstream water heads. Such power plants exploit the potential energy of vertical rise and fall of the water.

In contrast, the kinetic energy of the tide has to be known in order to design floating or other types of tidal power plants which harness energy from tidal currents or horizontal water flows induced by tides. They do not involve installation of water dams. The substantial capital costs associated with construction and concerns over adverse environmental impacts make the technology somewhat unappealing in contrast to tidal current technologies.

## Example 10.1

For a tidal range at a particular place of 10 m, and the surface tidal energy harnessing plant of 9 km², if the specific gravity of water is 1025.18 kg/m³, determine the total energy potential per day of the plant.

**Solution:**

$$\text{Mass of the water} = \text{volume of water} \times \text{Specific gravity}$$

$$= \text{area} \times \text{tidal range} \times \text{mass dendsity} = 9 \times 10^6 \times 10 \times 1025.8 = 92 \times 10^9\,\text{kg}$$

$$\text{Potential energy content of the water in the basin at high tide} = \frac{1}{2} \times \text{Area}$$

$$\times \text{density} \times g \times \text{tidal range squared}$$

$$= \frac{1}{2} \times 9 \times 10^6 \times 1025.8 \times 9.81 \times 100 = 4.5 \times 10^{12}\,\text{J approx}$$

As there are two tides per day, the total energy potential per day = Energy for a single high tide $\times 2 = 9 \times 10^{12}\,\text{J}$.

## 10.4 Turbine Technologies

The method of extracting power from the water in the tidal barrage reservoir is very similar to that employed in conventional hydroelectric power. To create the necessary pressure for power generation, water is allowed to flow into or out of the reservoir through open gates before the barrage is closed/sealed. As the tidal cycle progresses, a difference in water level, or head, is created between the inside and outside of the reservoir. The turbines that are used in tidal barrages are either unidirectional or bidirectional and include bulb turbines, straflo or rim turbines and tubular turbines.

The harnessing of the energy in a tidal flow requires technologies that convert the kinetic energy of moving water into useable energy which can then drive a generator. It is, therefore, proposed by many for using a technology which has been successfully utilized to harness the wind, which is also a moving fluid.

Marine current turbines are relatively a new technology requiring development with specifications that would allow long-term operation at low maintenance whilst submerged in seawater. Three main categories of turbines can be distinguished. These are as follows:

- Horizontal axis
- Vertical axis
- Reciprocating hydrofoils

### 10.4.1 Horizontal-Axis Turbines

Horizontal-axis turbines are the most commonly used turbines meant for extraction of power from marine currents. These are much similar in design to those used for wind power. The principle of operation is identical with that of wind turbines, and the only difference is the massively increased energy density of water and the possibility of cavitation, coupled with the destructive wear and tear of a submerged structure possibility. Because of these reasons, marine turbines are stockier and stubbier than wind turbines. In addition, the blades are smaller and turn more slowly than wind turbines.

Horizontal-axis turbines utilize lift generated by blades to turn a rotor. A horizontal-axis turbine for extracting power from marine currents is shown in Fig. 10.3.

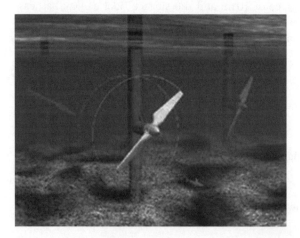

**Figure 10.3** Horizontal-axis turbines.

**Figure 10.4** Ducted.

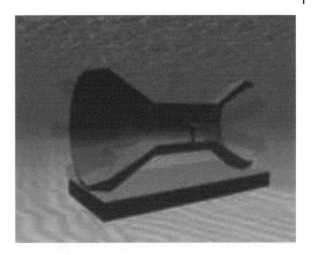

Rotor blade is the key component that extracts energy from the tide, and therefore, an efficient blade design is critical to the success of the HATT. The blades of the turbine are positioned in parallel (horizontal)) to the direction of the flow of water. The turbine rotor is driven by the tidal current. The motion of the rotor converts mechanical energy into rotational energy. Gearbox transmits shaft's rotational energy to the generator which converts this energy into electricity. Shrouds/ducts/diffusers are used to improve the efficiency of the horizontal-axis marine turbine. Shrouds/ducts/diffusers deploy diffuser and hydrofoil principles to accelerate water streamline. Power generated is proportional to the cube of steam velocity.

While the low speed of rotation of the turbines can make the use of a gearbox attractive, the difficulty of accessing devices for maintenance, especially those fixed on the seabed, makes it problematic. Variable-speed generators are used because of varying speed of tidal flows in many designs, which require frequency conversion in order to be connected to the power grid.

The horizontal-axis turbines are further classified into two categories: ducted and non-ducted. In ducted turbines, blades are enclosed in a duct. Due to the duct, the ocean current is concentrated and streamlined so that the flow and power output from the turbines increase. The ducted turbine is shown in Fig. 10.4.

### 10.4.2   Vertical-Axis Turbines

Vertical-axis turbines are not commonly used in the wind power industry now. However, several ocean power companies are developing designs for them. About 12% of tidal current turbines are vertical-axis turbines. Vertical-axis turbines that operate in marine currents are based on the same principles as the land-based Darrieus turbine. The Darrieus turbine is a cross-flow machine, whose axis of rotation meets the flow of the working fluid at right angles. The cross-flow turbines allow the use of a vertically orientated rotor so that torque can be transmitted directly to the water surface without the need of complex transmission systems or an underwater nacelle. In water, the tip speed ratio is relatively low compared to the air equivalent. In spite of its low potential efficiency, it has several advantages over horizontal-axis machines. It can accept flow from any horizontal direction, the swept area can be wider than its depth so for a given

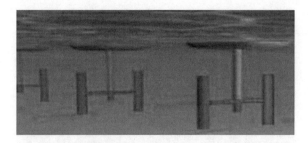

**Figure 10.5** Vertical axis.

generating capacity, the system can be located in shallower water and the generator and gearbox can be placed above the water surface.

There are several different designs in use, as fixed-pitch blades or variable-pitch blades (either controlled or freely moving). In adjustable-pitch blades, thrust is controlled by changing the pitch of the blades. The shaft often has a constant rotational speed. These can also be ducted or non-ducted. A vertical turbine is shown in Fig. 10.5.

### 10.4.3 Reciprocating Hydrofoils

Reciprocating devices have blades called hydrofoils – shaped as airplane wings – that move up and down as the tidal stream flows on either side of the blade. The up and down movement of the hydrofoils is subsequently converted into rotation to drive a rotating shaft or connected to pistons to support a hydraulic system for power generation. The advantage of reciprocating devices is that the length of the blade is not constrained by water depth.

## 10.5 Support Structure

All tidal current technologies require a support structure to keep the turbines in place so that they can withstand the harsh conditions of the sea. The choice of the foundation depends on the position of tidal current turbine in the water, depth of the water and structure of the seabed. Three types of support structures are commonly used. These are as follows:

### 10.5.1 Gravity Structures

These structures consist of large mass of concrete and steel connecting the turbine to the seabed. Gravity base structures (GBSs) are usually configured to have a cellular arrangement, similar to a "tray", so that they have sufficient buoyancy to allow them to be towed by vessel to the installation location and then filled with a ballast material (i.e. water, grout or iron ore) in order to be sunk into place on the seabed. Once in place, the ballasted GBS will be designed to have sufficient weight to resist overturning and sliding forces.

### 10.5.2 Piled Structures

Piles have been used for many years to support offshore wind turbines and in oil and gas installations throughout the world. The same technology can be used for the benefit of the marine renewable energy industry. In a piled structure, one or more beams are drilled or pinned into the seabed. The use of piles as a support structure can be generally classified as either a single monopile or a group of piles. The type of piled foundation, and the design, is dependent on the environmental and geotechnical conditions at the proposed location.

### 10.5.3 Floating Foundations

The floating foundations are connected to the seabed through either rigid or flexible wires or chains.

Another technical aspect for tidal current technologies is their deployment in the form of farms or arrays. Individual generator units are limited in capacity, so multi-row arrays of tidal turbines are built to capture the full potential of tidal currents. However, turbines have an impact on the current flows, so the configuration in which they are placed is a critical factor to determine their potential yield and output.

## 10.6 Wave Energy

The energy received by the Earth from the Sun is partly converted into wind. Waves are caused by the wind blowing over the surface of the ocean. The size of waves can vary considerably depending on the wind speed, sometimes reaching a height of more than 20 m. Most energetic waves are created between 30° and 60° latitude. In addition, *trade winds* occur at ±30° of the equator. As air warmed at the equator rises towards the poles, the rotation of the Earth causes it to deflect and flow back towards the equator. In the northern hemisphere, the winds blow from the east to the west, while they blow in the opposite direction in the southern hemisphere. These winds tend to be much stronger over open water than they are across land, which has made them ideal for sailors.

Wave energy technology is currently the most advance ocean energy technology compared to tidal current, based upon global development. Wave power conversion process is the least intrusive of all the renewable energy technologies. It does not create any waste, does not have any $CO_2$ emissions or other gas pollutants, there is no noise pollution, no visual impact and it does not threaten marine life. Using waves as a source of renewable energy offers significant advantages over other methods of energy generation. Some of the advantages are as follows:

Ocean waves are very powerful forces. Average power flow intensity of $2-3\,kW/m^2$ is available at a vertical plane perpendicular to the direction of wave propagation just below the water surface.

1) Waves can travel large distances with little energy loss.
2) Wave power devices can generate power up to 90% of the time, compared to 20–30% for wind and solar power devices.
3) Wave energy technology can be located near-shore as well as offshore.
4) Wave energy converters can also be designed for operation in specific water depth conditions: deep water, intermediate water or shallow water.

### 10.6.1 Wave Energy and Power

The amount of energy carried by a wave is related to the amplitude of the wave. A high-energy wave is characterized by a high amplitude; a low energy wave is characterized by a low amplitude.

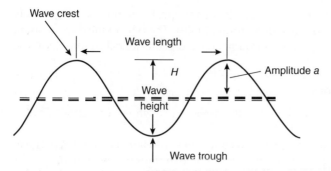

**Figure 10.6** Surface wave motion.

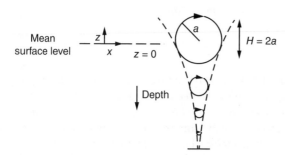

**Figure 10.7** Particle motion at various depths in water waves.

The water particles excited by the wind have circular trajectories in each location of the ocean with the largest diameter (the amplitude equals half the crest to trough height) at the surface and decreasing exponentially with depth. The conjugation of this circular motion is responsible for the wave formation and respective propagation. The characteristic of an ideal wave is shown in Fig. 10.6. The particle motion remains circular if the seabed depth $H > 0.5\lambda$, where $\lambda$ is wavelength. The amplitude becomes negligible at the sea bottom as shown in Fig. 10.7. Under these conditions (Fig. 10.7), a water particle whose mean position below the surface is $z$ moves in a circle of radius given by

$$r = ae^{kz} \tag{10.6}$$

here $k$ is the wave number, $2\pi/\lambda$ and $z$ is the mean depth below the surface (a negative quantity). The distance between the two consecutive *crests*, or two consecutive *troughs*, defines the *wavelength* $\lambda$. Wave *height* $H = 2a$ (crest to trough) is proportional to wind speed and duration. The wave *period* $T$ (crest to crest) is the time in seconds needed for the wave to travel the wavelength $\lambda$ and is proportional to sea depth. The frequency $f = 1/T$ indicates the number of waves that appear in a given position. Consequently, the wave speed is $c = \lambda/T = \lambda f$. Longer-period waves have longer wavelength and contain more power.

Ocean waves transport mechanical energy. The energy transported by a wave is directly proportional to the square of the amplitude of the wave.

The power across each metre of wave front associated with a wave of wavelength $\lambda$ and height $H$ is given by [7–9]

$$P = \frac{1}{2}\rho g H^2 \lambda \, \text{W/m} \tag{10.7}$$

For irregular waves of height $H$ (m) and period $T$ (s), an equation for power per unit of wave front can be derived as

$$P \cong 0.42\, H^2 T \, \text{kW/m} \tag{10.8}$$

Therefore, the power in the wave increases directly as the square of the wave amplitude and directly as the period. Excluding waves created by major storms, the largest waves are about 15 m high and have a period of about 15 s. According to Eq. (10.8), such waves carry about 1700 kW of potential power across each metre of wave front.

## 10.7 Wave Energy Converters

The energy in sea waves can be converted into mechanical energy by using proper wave power mechanisms known as *wave energy converters, wave devices or wave machines*. There is a wide range of wave energy technologies. Each technology uses different solutions and devices to capture energy from waves. There are three main categories: oscillating water columns (OWCs) that use trapped air pockets in a water column to drive a turbine; oscillating body converters that are floating or submerged devices using the wave motion (up/down, forwards/backwards, side to side) to generate electricity; and overtopping converters that use reservoirs to create a head and subsequently drive turbines.

### 10.7.1 Oscillating Water Column

An OWC is a wave energy converting technology that can be installed onshore preferably on rocky shores; near shore in up to 10 m of water; or offshore in 40–80 m deep water OWC-type wave energy devices are employed for converting fluid power into rotary-mechanical power. In its simplest form, this consists of a partially immersed vertical tube. An OWC comprises two key elements: a collector chamber, which takes power from the waves and transfers it to the air within the chamber, and a power take-off (PTO) system, which converts the pneumatic power into electricity or some other usable form. The most commonly used form of PTO in this application is the Wells turbine/induction generator combination. An OWC using a Wells turbine to convert wave energy to electrical energy is shown in Fig. 10.8. The motions of ocean/sea waves push an air pocket up and down behind a breakwater. Then the air passes through an air turbine. When a wave passes on to a partially submerged cavity open under the water (Fig. 10.8), it cause the column to act as a piston, moving up and down and thereby forcing the air out of the chamber and back into it. This continuous movement generates a reversing stream of high-velocity air, which is channelled through rotor blades driving an air turbine–generator group to produce electricity. When the waves rise, the air is forced up the chamber, increasing in speed. As the waves fall, air is sucked back into the chamber. A high-speed bidirectional airflow is therefore produced and is converted into unidirectional rotational motion using a Wells turbine. The doubly fed induction machine is the favoured machine for the OWC.

The problem with this pneumatic system is that the rushing air can be very noisy, unless a silencer is fitted to the turbine. But the noise is not a huge problem anyway, as the waves make lot of noise themselves.

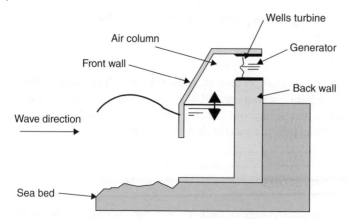

**Figure 10.8** An oscillating water-column wave energy converter.

Other forms of self-rectifying turbines, which are the impulse machines of Dresser-Rand and Oceanlinx, are also under development. The manufacturers claim that these machines offer a higher conversion efficiency compared to a Wells turbine, but there is no publically available information to demonstrate in real sea.

The main advantages of OWC devices are as follows:

- There are few moving parts, and there is no moving part inside water.
- It can be used for offshore, near-shore and shore-line locations.
- The air velocity can be increased by reducing the cross-sectional area of the air channel, so that high-speed turbines can be used.

### 10.7.2 Oscillating Body

Oscillating-body converters are either floating (usually) or submerged (sometimes fixed to the bottom). They exploit the more powerful wave regimes that normally occur in deep waters where the depth is greater than 40 m. In general, they are more complex than OWCs, particularly with regard to their PTO systems. The simplest oscillating-body device is the heaving buoy reacting against a fixed frame of reference. The rise and fall of the waves move the buoy-like structure, creating mechanical energy which is converted into electricity and transmitted to the shore by means of a secure, undersea transmission line.

In fact, the many different concepts and ways to transform the oscillating movement into electricity have given rise to various PTO systems, for example hydraulic generators with linear hydraulic actuators, linear electric generators and piston pumps. The advantages of oscillating-body converters include their size and versatility since most of them are floating devices.

### 10.7.3 Overtopping Converters (or Terminators)

Overtopping devices consist of a floating or bottom-fixed water reservoir structure. These devices use reflector arms and/or sloped surfaces to drive the waves to a reservoir of stored seawater. The potential energy, due to the height of the collected water above the sea surface, is transformed into electricity using conventional low-head hydro turbines.

As the waves hit the structure, they flow up a ramp and over the top (hence the name "overtopping"), into a raised water reservoir on the device in order to fill it. Wave energy is converted into potential energy by lifting the water up onto a higher level. As shown in Fig. 10.10, the captured potential energy of the trapped water in the reservoir is extracted using gravity as the water returns to the sea via a low-head Kaplan turbine generator located at the bottom of the wave capture device. An advantage for overtopping devices is that the low-head turbine technology used in hydropower can be applied here. These devices are often large installations and can be placed in shore line as well as offshore.

### 10.7.4 Point Absorbers and Attenuators

A large group of wave energy converters are based on directly converting the waves into an oscillating mechanical motion. The point absorber is a type of wave energy device that could potentially provide a large amount of power in a relatively small device, compared to other technologies. While there are several different designs and strategies for deploying these types of devices, they all work in essentially the same manner. Point absorbers are relatively small compared to wave length and may be bottom-mounted or floating structures. If these are floating structures, they absorb energy from all directions at or near the water surface. A point-absorber wave energy conversion system is shown in Fig. 10.9. Point-absorber devices can be designed to work at near-shore and offshore sites and at most sea states. A small (1–5 m diameter) and light (<5 tonnes) buoy such as the one used in the sea-based concept has high absorption for small and high waves. As waves become longer, absorption starts to drop. Heavy (100 s of tonnes) and big (10–25 m) point absorbers such as Wave bob are more suited to long waves (>7 s). A point absorber can be designed for short- or long-wave periods, and by using active control, a single point absorber can be designed to match most sea states. The diameter

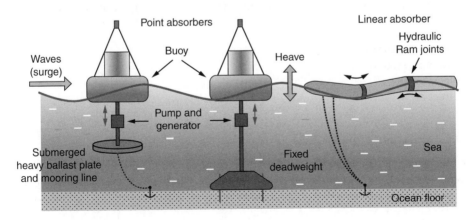

**Figure 10.9** Point-absorber wave energy converter.

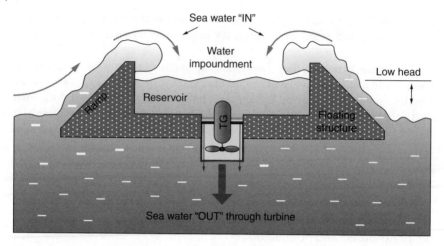

**Figure 10.10** Overtopping device for wave energy conversion. *Source*: Polinder and Scuotto (2005) [10]. Reproduced with permission from IEEE.

of a point absorber should be less than 1/6 of the wavelength; otherwise, it will start counteracting itself. For systems without control, absorption is usually 10–30%. Active control has shown absorption of 40–5% for specific wave climates (Fig. 10.10).

## 10.8   Power Takeoff Systems

The PTO system converts the captured mechanical energy of the wave into electrical energy. In both wave and tidal systems, a mechanical interface can be employed to convert the slow rotational speed or reciprocating motion into high-speed rotational motion for connection to a conventional rotary electrical generator (Fig. 10.11). Direct drive is also an option but is not in use. The PTO for wave energy can be classified as follows:

1) Air turbines for OWC
2) Hydraulic systems
3) Water turbines
4) Direct drive

### 10.8.1   Air Turbines for OWC

The energy conversion from the oscillating air column can be achieved by using Wells turbine which was introduced by Dr A. A. Wells in 1976 because of its ability to rotate in the same direction, irrespective of the airflow direction. Furthermore, this turbine is one

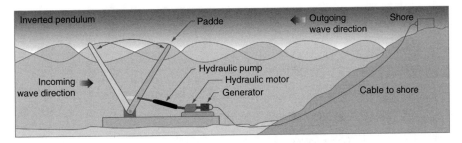

**Figure 10.11** Inverted pendulum.

of the simplest and probably the most economical turbines for wave energy conversion. However, according to a previous studies, Wells turbine has inherent disadvantages: low efficiency about 60–65% at high flow coefficient and poor starting characteristics in comparison with conventional turbines because of a severe stall. It is possible to use pitch control of the turbine blades to increase efficiency.

Recently, in order to develop a high-performance self-rectifying air turbine for wave energy conversion, an impulse turbine for bidirectional flow has been proposed. Its rotor is basically identical to the rotor of a conventional single-stage steam turbine of axial-flow impulse type. Since the turbine is required to be self-rectifying, instead of a single row of guide vanes (as in the conventional de Laval turbine), there are two rows, placed symmetrically on both sides of the rotor. These two rows of guide vanes are as the mirror image of each other with respect to a plane through the rotor disc.

### 10.8.2 Hydraulic Systems

Another method of converting the low-speed oscillating motion of the primary WEC interface is to employ a hydraulic system. Axial-flow reaction turbines are used to convert the head (typically 3–4 m at full size) created between the reservoir of an overtopping device and the mean sea level. Figure 10.12 shows a basic hydraulic PTO system with energy storage for a WEC. Components of a basic compression unit include a hydraulic cylinder, two pressure accumulators and a valve-control system. The hydraulic cylinder is used to transfer the motion of the floating body into a double-acting piston in the cylinder connected to it. Two chambers are connected respectively to a high-pressure accumulator and to a low-pressure accumulator. The motion of the cylinder results in varied pressure difference that results in a torque on the generator/motor. The flow may be controlled by adjustable inlet guide vanes. In some cases, the blades of the runner can also be adjusted (Kaplan turbines) which greatly improves efficiency over a wide range of flows; however, this can be costly and is not normally employed in the small turbines.

Hydraulic systems are efficient energy conversion systems consisting of two energy conversion systems. Accumulators are included in the circuit to provide energy storage and to maintain constant flow to the hydraulic motor. In addition, the low-pressure accumulator provides a small boost pressure to reduce the risk of cavitation. These

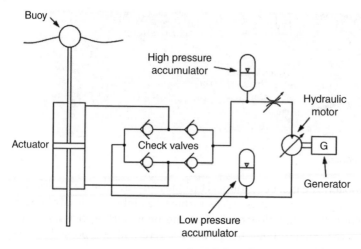

**Figure 10.12** Typical hydraulic PTO with energy storage.

hydraulic turbines may reach peak efficiencies of about 90%. Their efficiency is in general quite sensitive to the head-to-rotational-speed ratio, which makes the use of variable-speed electrical generators highly advantageous, especially in the case of Pelton turbines.

### 10.8.3 Water Turbines

Water turbines used in run of river power plants have been proposed for so-called over-topping devices, which are essentially low-head hydro systems. The principle of operation of such devices is that water during high tide is stored and has potential energy which is used by the water turbines to convert it into mechanical energy.

Waves crash over the device into a floating reservoir, which in turn feeds a water turbine coupled to conventional rotary generators.

### 10.8.4 Direct Drive

Wave power devices traditionally use conventional rotary electrical machines for power conversion. However, hydraulic systems or air turbines are required to convert the low reciprocating motion of the wave device to rotation at 1500 rpm. The concept of a direct drive system is introduced, in which a reciprocating electrical machine is driven at the same speed as the device. In a direct drive system, there is no mechanical interface coupling the device to the electrical generator. Hence, it has the potential to provide a simpler system requiring fewer moving parts, lower maintenance requirements and higher efficiency.

The direct drive system employs a linear generator which consists of a magnetic translator that is driven to reciprocate synchronously with the motion of a directly coupled buoy/floating body. A simplified structure of direct-drive conversion system is shown in Fig. 10.13.

Since the generator and prime mover are connected without any gear, the velocity of the generator is equal to that of the prime mover, being of the order of 0.5–2 m/s. In order to generate sufficient amount of power, the machine will have to exert large forces. Direct drive machines are physically very large and heavy. The concept of direct drive has been demonstrated within the Archimedes Wave Swing (AWS) device. Figure 10.14

**Figure 10.13** Direct drive.

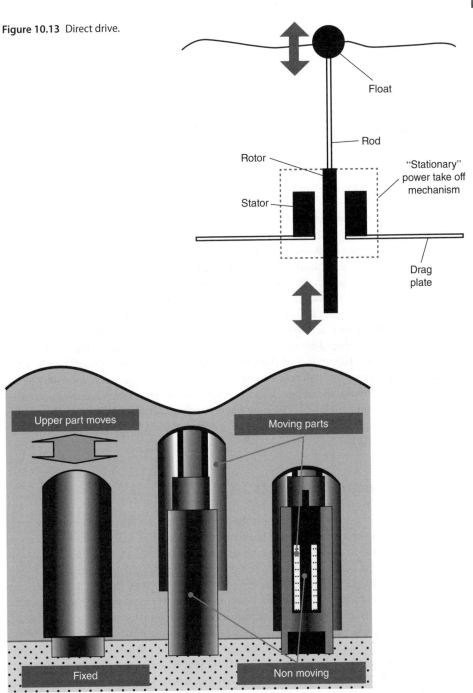

**Figure 10.14** Archimedes Wave Swing. *Source*: Polinder and Scuotto (2005) [10]. Reproduced with permission from IEEE.

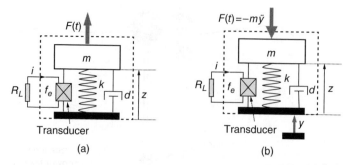

**Figure 10.15** (a) Direct-force and (b) inertial-force generators.

shows the principle of operation of the device. The device is seabed-mounted. It consists of an air-filled chamber with a floater that moves down with the crest of a wave and up with a floater that moves down with the crest of a wave and up with the trough of a wave. It is a submerged offshore device activated by the fluctuations of static pressure caused by the surface waves, thus exploiting only their potential energy. Basically, the AWS is an air-filled cylindrical steel chamber whose lid, called the floater, is a heaving body while the bottom part is fixed. The active force that moves the floater is supplied by the pressure difference acting on its top. When the wave crest is above the AWS, the chamber volume is reduced by the high water pressure. When the trough is above, the floater heaves under the action of the chamber pressure. The air within the chamber behaves as a spring whose stiffness can be adjusted by pumping water in or out of the chamber (chamber volume changes). In the AWS, direct-drive energy conversion takes place via a permanent magnet (PM) linear synchronous generator. This type of energy conversion does not allow energy storage. As a consequence, the output power exhibits poor quality (Fig. 10.15).

## 10.9 Piezoelectric Generators

Sea wave energy harvesting with piezoelectric materials can be classified into three main categories: the vortex caused by the bluff body fixed on the seabed; the vibration of heaving-based buoy on the sea surface; and the vibration of the foam cantilever substrate attached by polymer patches to absorb the small wave motion near the seabed.

The vast majority of piezoelectric energy harvesting devices use a cantilever beam structure [11]. Basically, the harvester beam is located on a vibrating host structure, and the dynamic strain induced in the piezoceramic layer(s) generates an alternating voltage output across the electrodes covering the piezoceramic layer(s). This voltage can be extracted through electrical equipment connected to the electrodes. A common piezoelectric generator is illustrated in Fig. 10.16, where a cantilever beam, with a mass attached to its tip, will oscillate and develop bending stresses along the beam. It is clearly seen that the transverse motions of the water particles have no effect on the bending of the cantilever, and only the longitudinal waves are responsible for producing energy.

The electrical-circuit analogue representing such systems is shown in Fig. 10.17. It is composed of a mechanical section, an electrical section and a transformer. In the mechanical part of the circuit, the inductor, the resistor and the equivalent

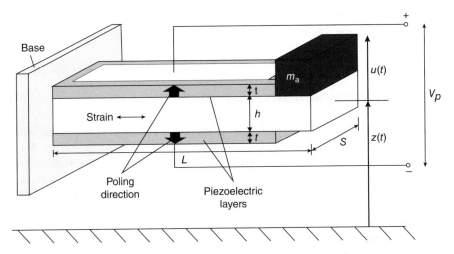

**Figure 10.16** Common piezoelectric-based power generator.

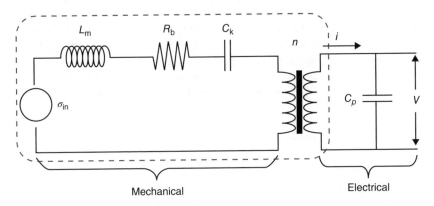

**Figure 10.17** Electrical circuit diagram.

capacitor represent the equivalent mass of the system, the mechanical damping and the mechanical stiffness, respectively. The stress generator represents the stresses induced by input vibrations and is analogous to a voltage source. In the electrical section, the capacitor represents the actual electrical capacitance of the piezoelectric bender.

A cantilever beam can have many different modes of vibration, each with a different resonant frequency. The first mode of vibration has the lowest resonant frequency and typically provides the most deflection and therefore electrical energy. A lower resonant frequency is also closer in frequency to physical vibration sources, and generally more power is produced at lower frequencies. Therefore, energy harvesters are generally designed to operate in the first resonant mode [12, 13].

### 10.9.1 Power Extraction Systems

The piezoelectric generator, behaving as a resonator, acts as a sinusoidal voltage source. When it is implemented in the ocean, the generated AC current has to be processed and rectified for proper usage, before storing the energy in a battery or supplying it directly to a load. The controller regulates the output voltage depending on the need of the specific application. The external circuit connected to the piezoelectric harvester has a great influence on the energy flow in the harvester [13].

## 10.10 Ocean Thermal Energy Conversion

Ocean thermal energy conversion (OTEC) generates electricity by harnessing the temperature difference between the warm surface of tropical oceans and the colder deep waters. The surface of ocean retains the solar radiation in tropical regions, resulting in average year-round surface temperatures of about 28 °C. The warm water layer of the sea extends to depths of about 100–200 m. There is a temperature difference of 10–25 °C, between this warm surface and cold water in about 1000 m depth. The equatorial and tropical regions have higher difference in temperature than other regions. Heat engines can generate power based on thermodynamic principle of heat transfer from a temperature difference of about 20°C. The disadvantage of OTEC is that it requires large pipelines and heat exchangers to produce relatively modest amounts of electricity, making it very expensive. In order to handle such a large amount of water flow, the heat exchangers of an OTEC plant will have to be very large. Early estimates of the total heat exchanger surface area ranged from 7 to 9 m²/kW of electrical capacity. These high fixed costs dominate the economics of OTEC to the extent that it cannot compete with conventional power systems, except in limited niche markets.

From thermodynamics, it is known that the maximum efficiency of heat engines (Carnot cycle) is given by –

$$\text{Thermal efficiency (Carnot cycle)} = 1 - \frac{T_c}{T_H}$$

where $T_c$ and $T_H$ are low and high temperatures measured in kelvin.

For example, if the warm surface layer temperature is 298 K (25 °C) and the cold temperature is 278 K (5 °C), then

$$\eta = 1 - \frac{278}{298} = \frac{20}{298} \ m = 6.7\%$$

In practice, the thermal efficiency of a practical system may be only 2–4% when energy required in pumping cold water and losses are included. Thus, large amount of water is required to produce sufficient electricity.

### 10.10.1 Technology for OTEC

OTEC technology is not new. In 1881, Jacques Arsene d'Arsonval, a French physicist, proposed for the first time utilization of thermal energy of the ocean. His student Georges Claude, in 1930, built the first OTEC plant in Cuba. The system produced 22 kW of electricity with a low-pressure turbine. In 1935, Claude constructed another plant aboard a 10,000-ton cargo vessel moored off the coast of Brazil. In 1956, French scientists designed another 3-mW OTEC plant for Abidjan, Ivory Coast, West Africa. The plant was not completed, because it was too expensive. OTEC power systems operate as cyclic heat engines based on Rankine cycle. As shown in Section 1.5.5, the complete conversion of thermal energy into electricity is not possible. A portion of the heat extracted from the warm seawater must be rejected to a colder thermal sink. The thermal sink employed by OTEC systems is seawater drawn from the ocean depths by means of a submerged pipeline.

There are three kinds of OTEC systems: closed-cycle, open-cycle and hybrid [14].

### 10.10.1.1 Closed-Cycle

Closed-cycle systems use fluids with low boiling point such as ammonia, freon or propane which are evaporated using warm surface water. The vapour is then used to rotate a turbine to generate electricity. Warm surface seawater is pumped through a heat exchanger, where the low-boiling-point fluid is vaporized. The expanding vapour turns the turbo generator. The vapour drives a generator that produces electricity; the working fluid vapour is then condensed by the cold water taken from deep ocean and pumped back in a closed system. Figure 10.18 shows a simplified schematic diagram of a closed-cycle OTEC system. The principal components are the heat exchangers, turbo generator and seawater supply system. These components account for most of the local power consumption and a significant fraction of the capital cost. In this system, heat transfer from warm surface seawater occurs in the evaporator, producing a saturated vapour from the working fluid. Electricity is generated when this gas expands to lower pressure through the turbine. Latent heat is transferred from the vapour to the cold seawater in the condenser, and the resulting liquid is pressurized with a pump to repeat the cycle. It is best suited to areas near the equator, where the intense solar radiation produces sufficiently warm surface water.

Most suitable working fluids for closed-cycle OTEC applications are ammonia and various refrigerants. Their primary disadvantage is the environmental hazard posed by leakage; ammonia is toxic in moderate concentrations, and many refrigerants have been banned. The Kalina cycle is a variant of the OTEC closed cycle. Whereas simple closed-cycle OTEC systems use a pure working fluid, the Kalina cycle proposes to employ a mixture of ammonia and water with varying proportions at different points in the system (Fig. 10.19).

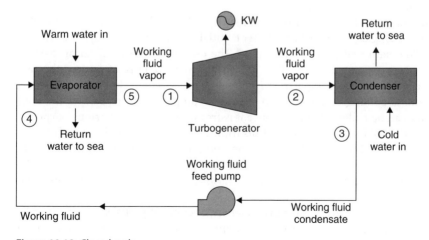

**Figure 10.18** Closed cycle.

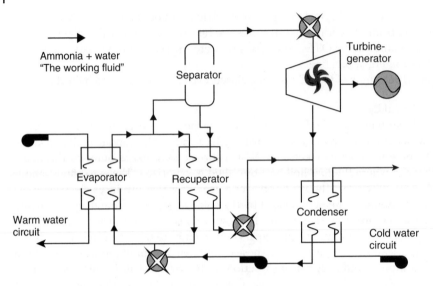

**Figure 10.19** Kalina cycle.

#### 10.10.1.2 Open-Cycle

Open-cycle systems use the tropical oceans' warm surface water to make electricity.

The open cycle consists of the following steps: (i) warm seawater is placed in a low-pressure container for flash evaporation; (ii) expansion of the vapour is used to drive a turbine to generate power; (iii) heat transfer to the cold seawater thermal sink resulting in condensation of the working fluid; and (iv) compression of the non-condensable gases (air released from the seawater streams at the low operating pressure) to pressures required to discharge them from the system.

The steam, which has left its salt behind in the low-pressure container, is almost pure, freshwater. Only less than 0.5% of the mass of warm seawater entering the evaporator is converted into steam. Thus, an open-cycle OTEC plant can provide a large quantity of desalinated water.

The name "open cycle" comes from the fact that the working fluid (steam) is discharged after a single pass and has different initial and final thermodynamic states; hence, the flow path and process are "open". The essential features of an open-cycle OTEC system are presented in Fig. 10.20. Warmer surface water is introduced through a valve in a low-pressure compartment and flash evaporated. The vapour drives a generator and is condensed by the cold seawater pumped up from below. The condensed water can be collected and, because it is freshwater, used for various purposes. Additionally, the cold seawater pumped up from below, after being used to facilitate condensation, can be introduced in an air-conditioning system, Furthermore, the cold water can potentially be used for aquaculture purposes, as the seawater from the deeper regions close to the seabed contains various nutrients, such as nitrogen and phosphates.

The entire system, from evaporator to condenser, operates at partial vacuum, typically at 1–3% of atmospheric pressure. Vacuum compressor is used for initial evacuation of the system and removal of non-condensable gases during operation. Effluent from the condenser must be discharged to the environment. Liquids are pressurized to ambient

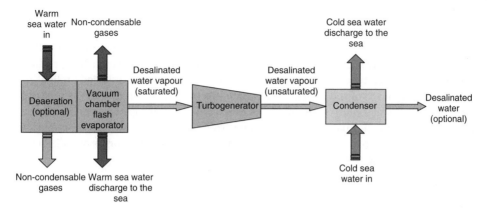

**Figure 10.20** Open cycle.

levels at the point of release by means of a pump or, if the elevation of the condenser is suitably high, can be compressed hydrostatically.

### 10.10.1.3 Hybrid Systems

Hybrid cycles combine the potable water production capabilities of open-cycle OTEC with the potential for large electricity generation capacities offered by the closed cycle. Several hybrid-cycle variants have been proposed. A hybrid system is shown in Fig. 10.21. In a hybrid system, warm seawater enters a vacuum chamber, where it is flash-evaporated into steam, similarly to the open-cycle evaporation process. This

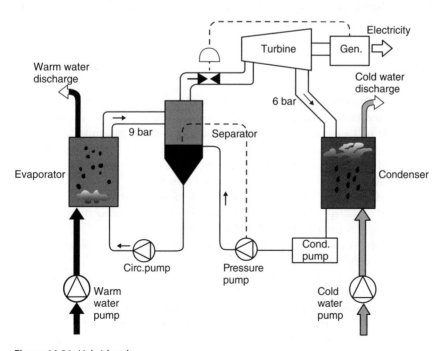

**Figure 10.21** Hybrid cycle.

produces freshwater. The steam vaporizes a low-boiling-point fluid, usually ammonia (in a closed-cycle loop) that drives a turbine to produce electricity. Subsequently, the warm seawater discharged from the closed-cycled OTEC is flash-evaporated similarly to an open-cycle OTEC system and cooled with the cold water discharge. This produces freshwater. All three types of OTEC can be land-based, sea-based or based on floating.

Another advantage of the hybrid cycle related to freshwater production is that condensation occurs at significantly higher pressures than in an open-cycle OTEC condenser, due to the elimination of the turbine from the steam flow path. This may, in turn, yield some savings in the amount of power consumed to compress and discharge the non-condensable gases from the system.

## 10.11  Summary

In this chapter, it is shown that the energy potential in tides is very high. The ocean can produce two types of energy: thermal energy from the Sun's heat and mechanical energy from the tides and waves. A barrage (dam) is typically used to convert tidal energy into electricity by forcing the water through turbines, activating a generator. For wave energy conversion, there are three basic systems: channel systems that funnel the waves into reservoirs; float systems that drive hydraulic pumps; and OWC systems. All these systems are described in this chapter.

In addition, ocean thermal energy which is used for many applications, including electricity generation, is also discussed in this chapter. There are three types of electricity conversion systems: closed-cycle, open-cycle and hybrid. Closed-cycle systems use the ocean's warm surface water to vaporize a working fluid, which has a low boiling point, such as ammonia; all these systems are discussed.

## References

1 Seymor, R.J. (1992) *Ocean Energy Recovery: The state of art*, ASCE Publications.
2 Charlier, R.H. and Finkl, C.W. (2009) *Ocean Energy Tide and Tidal power*, Springer.
3 Hardisty, J. (2014) *Wave and Tidal Research Review*, 2nd edn, Hull University.
4 Gorlov, A. M. (1979) Some new conceptions in the approach to harnessing tidal energy, Proc. Miami Int. Natinal conf. Alt energy sources.
5 Dunlap, R.A. (2015) *Sustainable Energy*, Cengage Learning.
6 Charlier, R.H. (1982) *Tidal Energy*, Van Nostrand Reinhold, New York.
7 Benelghali, S. *et al.* (2010) *Marine Tidal Current Electric Power Generation Technology: State of the Art and Current Status*, HAL archives.
8 Peppas, L. (2008) *Ocean, Tidal, and Wave Energy: Power from the Sea (Energy Revolution)*, Prime Publishing.
9 Chris, G. and Patrick, C. (2005) The power potential of tidal currents in channels. Proc of the Royal Society.
10 Polinder, H. and Scuotto, M. (2005) Wave energy converters and their impact on power systems, International conference on future power systems, pp. 9
11 Jbaily, A. and Yeung, R.W. (2015) Piezoelectric devices for ocean energy: a brief survey. *Journal of Ocean Engineering and Marine Energy*, **1**, 101.

12 Wang, D.A. and Ko, H.H. (2010) Piezoelectric energy harvesting from flow induced vibration. *J Micromech Microeng*, **20** (2), 025019.

13 Shu, Y.C. and Lien, I.C. (2006) Analysis of power output for piezoelectric energy harvesting systems. *Smart Materials and Structures*, **16**, 1499.

14 Kempener, R. and Neumann, F. (2014) Ocean thermal energy Conversion Technology brief IRENA Ocean energy Technology brief 1.

# 11

# Fuel Cells

## 11.1 Fuel Cell Technologies

A fuel cell is an electrochemical device that converts chemical energy (of a fuel) directly into electrical energy. Since the chemical energy of the fuel is directly converted to electricity, a fuel cell can operate at much higher efficiencies than internal combustion engines, extracting more electricity from the same amount of fuel. Fuel cells are capable of converting 40% of the available fuel to electricity. This can be raised to 80% with heat recovery. The fuel cell itself has no moving parts, offering a quiet and reliable source of power.

The fuel cell was invented by William R Grove in 1839. It was then called *gaseous voltaic battery*. In 1882, Lord Rayleigh developed a new form of gas battery which was an attempt to improve the efficiency of the cell [1]. A fuel cell is similar to a battery because both batteries and fuel cells convert chemical potential energy into electrical energy. In this process, heat energy is also produced as a by-product. However, the basic difference between fuel cell and batteries is that in batteries, the chemical energy is stored in the substances located inside them. When this energy has been converted to electrical energy, the battery must be discarded or recharged appropriately. In a fuel cell, the chemical energy is provided by a fuel and an oxidant stored outside the cell in which the chemical reactions take place. As long as the cell is supplied with the fuel (hydrogen) to the anode and source of oxygen usually air as oxidant, electrical power can be generated.

In the fuel cell, there is no combustion involved; instead, oxidation of the hydrogen takes place electrochemically in a very efficient way. During this process, electricity, water and heat are produced. Fuel cells are power-generating devices having a wide range of applications including stationary power generation (MW), portable power generation (kW) and transportation (kW). Because of this, they may be used in almost every application where local electricity generation is needed. For example, these cells find applications in automobiles, buses, utility vehicles, scooters, bicycles and submarines. Fuel cells offer tremendous flexibility in terms of distributed power generation, at a level of individual homes, buildings or a community. As a backup power generator, fuel cells offer several advantages over either internal combustion engine generators (noise, fuel, reliability, maintenance) or batteries (weight, lifetime, maintenance). Small fuel cells are attractive for portable power applications, either as replacement for batteries (in various electronic devices and gadgets) or as portable generators.

Every fuel cell has two electrodes, one positive and one negative, called, respectively, the anode and cathode. It also has an electrolyte, which carries electrically charged

*Operation and Control of Renewable Energy Systems,* First Edition. Mukhtar Ahmad.
© 2018 John Wiley & Sons Ltd. Published 2018 by John Wiley & Sons Ltd.

particles from one electrode to the other, and a catalyst, which speeds the reactions at the electrodes. The reactions that produce electricity take place at the electrodes.

Most common electrolyte in fuel cell is hydrogen, as it is the simplest element. An atom of hydrogen consists of only one proton and one electron. It is also available in large quantity. But despite its simplicity and abundance, hydrogen does not occur naturally as a gas on the Earth – it is always combined with other elements. For example, water is a combination of hydrogen and oxygen.

The most significant advantages of fuel cells are low emission of greenhouse gases, quiet operation and high power density. The energy density of a typical fuel cell is 200 Wh/l, which is nearly 10 times that of a battery [1–4]. A fuel cell responds rapidly to sudden increase or decrease in power demands, because the electricity is generated by a chemical reaction.

## 11.2 Types of Fuel Cells

Fuel cells are available in various power ranges: from very small devices producing only a few watts of electricity [5] to large power plants producing many megawatts. However, all fuel cells are designed using two electrodes separated by a solid or liquid electrolyte that carries electrically charged particles between them. A catalyst is often used to speed up the reactions at the electrodes. Fuel cell types are generally classified according to the nature of the electrolyte they use. This classification determines the kind of electrochemical reactions that take place in the cell, the kind of catalysts required, the temperature range in which the cell operates, the fuel required and other factors. These characteristics, in turn, affect the applications for which these cells are most suitable.

Fuel cells can be classified into six different categories based on the type of electrolyte used. These are as follows:

- Proton exchange membrane fuel cell (PEMFC)
- Solid oxide fuel cell (SOFC)
- Molten carbonate fuel cell (MCFC)
- Phosphoric acid fuel cell (PAFC)
- Alkaline fuel cell (AFC)
- Direct methanol fuel cell (DMFC).

## 11.3 Proton Exchange Membrane (PEM) Fuel Cell

Among all kinds of fuel cells, PEMFCs are being rapidly developed as the primary power source in movable power supplies and distributed generation (DG), because of their high energy density, low working temperature, quick start-up and simplicity. A PEM fuel cell is an electrochemical cell that is fed hydrogen, which is oxidized at the anode, and oxygen that is reduced at the cathode. The protons released during the oxidation of hydrogen are conducted through the proton exchange membrane at the cathode. PEM fuel cells use a solid polymer as an electrolyte and porous carbon electrodes containing a platinum or platinum alloy catalyst. The electrodes must be porous because the reactant gases are fed from the back and must reach the interface between the electrodes and the membrane, where the electrochemical reactions take place on the catalyst surface. The large amount of platinum in original PEM fuel cells is one of the reasons why fuel cells were not commercialized earlier. New research has been directed in reducing the use of platinum in electrodes.

PEMFC is the only fuel cell which is being considered for powering passenger cars. It is also being developed for stationary and portable power generation. PEM fuel cell gets its name as it uses a proton conducting polymer membrane as electrolyte. They need only hydrogen, oxygen from the air and water to operate. The polymer membrane has unique property as it is impermeable to gases but conducts protons. It acts as the electrolyte and is squeezed between the two porous, electrically conductive electrodes. These electrodes are typically made out of carbon cloth or carbon fibre paper. A layer with catalyst particles usually of platinum supported on carbon acts as the interface between the porous electrode and the polymer membrane [6–8]. The polymer electrolyte membrane is a solid, organic polymer, usually poly(perfluorosulfonic) acid. The most frequently used membrane in PEM is made of Nafion produced by DuPont. Other types of membranes being researched are polymer–zeolite nanocomposite proton exchange membrane, sulfonated polyphosphazene-based membrane and phosphoric-acid-doped poly(bisbenzoxazole) high-temperature ion-conducting membrane. The Nafion layer is essentially a carbon chain which has a fluorine atom layer attached to it. The membrane must be hydrated for this operation. The electron and proton then meet at the cathode where, in the presence of oxygen, water is formed. Since high temperatures are not necessary to hydrate the membrane, the PEM can be run at very low temperatures, typically at $80\,^{\circ}C$ or lower.

All electrochemical reactions in a fuel cell consist of two separate reactions: an oxidation at the anode and a reduction at the cathode. Normally, these two reactions would occur very slowly at the low operating temperature of the PEM fuel cell. In order to speed up the reaction, each of the electrodes is coated on one side with a catalyst layer (CL). It is usually made of platinum powder very thinly coated onto carbon paper or cloth. The catalyst is rough and porous, so the maximum surface area of the platinum can be exposed to the hydrogen or oxygen. The platinum-coated side of the catalyst faces the PEM. Platinum-group metals are critical to catalyzing reactions in the fuel cell, but they are very expensive.

### 11.3.1 Water Management

A PEM fuel cell is shown in Fig. 11.1. The anode and the cathode (the electrodes) consist of a heterogeneous, porous layer that serves as electron conductor, ion conductor and gas transport region. The faces of the electrodes in contact with the membrane (generally referred to as the active layers) contain, in addition to carbon, polymer electrolyte and a platinum-based catalyst. When hydrogen is supplied at the positive electrode (anode), it breaks into electron and proton due to chemical reaction. The proton travels to negative electrode (cathode) through the conductive membrane. The electrons travel through electrically conducting electrodes, through current collectors and through outside circuit and reach other side of the membrane.

In this system, oxygen may be either pure oxygen or oxygen from the air. The catalytic reaction occurs at the three-phase contact with the catalyst. Here, the catalyst is shown as a platinum site, but any variety of catalysts, usually noble metals, may be employed

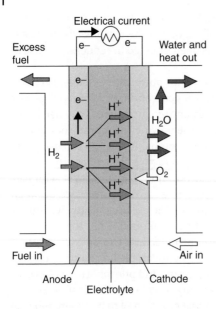

**Figure 11.1** PEMC fuel cell.

under different conditions. As the gases react, protons generated at the anode cross the membrane to the cathode as electrons flow through the external circuit to drive reduction. The membrane provides ionic conduction and prohibits direct reaction of the hydrogen and oxygen at the platinum catalyst.

Water management is of vital importance to achieve maximum performance and durability from PEMFCs. On the one hand, to maintain good proton conductivity, the relative humidity of inlet gases is typically held at a large value to ensure that the membrane remains fully hydrated. On the other hand, the pores of the CL and the gas diffusion layer (GDL) are frequently flooded by excessive liquid water. In addition, severe drying conditions can lead to irreversible membrane degradations within about 100 s.

During operation, due to the effect of electro-osmosis, water molecules move from the anode to the cathode resulting in membrane dehydration on the anode side of the membrane and flooding on the cathode side (additional water is produced at the cathode by the reaction

$$O_2 + 4H + 4e^- \rightarrow 2H_2O \tag{11.1}$$

Since the chemical reaction in the cell produces water, it is important that the electrolyte must not be flooded. Flooding of the electrodes causes a decrease in surface area in which the separation of hydrogen or the formation of water takes place. As mentioned earlier, the membrane needs to be hydrated; thus, a balance must be achieved between water removal and hydration of the membrane. The water accumulated in the cathode is usually removed out of the porous electrode via evaporation, water-vapour diffusion and capillary transport through GDL into the flow channels of the flow field and then exhausted out of the system, or else excess water will block the flow.

In removing the water from the cell, care must be taken not to remove too much water from the cathode; otherwise, the membrane and the anode will dry out. Since the electrode is very thin (50 μm), it is possible for the water to leak back to the anode, which would be the ideal situation if the exact amount were to migrate. Several complications arise in this process. The first is that the water naturally moves towards the cathode, about 1–2.5 molecules per proton. This "electro-osmotic drag" is problematic at high current densities because all the water can be removed from the anode, thus causing an abrupt loss in fuel pressure since no water will be present to transfer new protons (this is a form of a mass transportation loss). A further problem in water management is the possibility that the air at high temperature can dry out the water. Studies have shown that at temperatures over 60 °C, the air dries out the cathode. To solve this problem, it is necessary to add water to the system to keep everything hydrated, without flooding. Although these problems have been solved, it is still quite necessary to understand these issues since design of a cell is critically based on water management.

Removal of excess water is also possible by using air which is supplied to the cell to provide oxygen to the cathode. In order not to remove too much water from the cathode, which can dry the membrane and the anode out, it is necessary to have the correct airflow.

### 11.3.2 Fuel Requirement

For the full commercialization of PEMFC system, a stable supply of high-purity hydrogen is essential. Since hydrogen does not occur naturally as a gas on the Earth, it has to be generated from primary energy sources. PEMFCs function best with high-purity hydrogen gas as the fuel source, but pure hydrogen is unlikely to be the fuel source in the future due to technical and economic considerations in production and storage, especially in applications such as transportation and stationary power generation. Therefore, hydrogen from reformed fuels such as natural gas, gasoline or alcohols is likely to be the fuel that is supplied to the fuel cell. These gas streams will contain small amounts of carbon monoxide (CO) which poisons the platinum anode catalyst. Thus, PEMFCs require that the concentration of carbon monoxide present in the fuel gas should not be more than 10 ppm. In addition, whether carbon dioxide is present in the fuel cell is another important consideration. Presence of $CO_2$ in the fuel gases deteriorates the performance of the fuel cell. Hydrogen can also be produced by reforming of bioethanol, which would then also be a source of renewable hydrogen.

Hydrogen can be produced by the electrolysis of water: splitting of water into its component elements. This process takes place in an electrolyzer, which can be described as a "reverse" fuel cell: instead of combining hydrogen and oxygen electrochemically to produce electricity and water as a fuel cell does, an electrolyzer uses an electrical current and water to produce hydrogen and oxygen.

Electrolysis of water requires source of the electrical current. If grid electricity is used which is normally produced using fossil fuels, the advantage of no pollution is lost. However, if the electricity is obtained from renewable energy such as wind or solar power, the hydrogen can be produced in a completely carbon-free way. Electrolyzers exist, and many commercial versions of various capacities are available on the market. In general, the demand for purity of hydrogen decreases with high operating temperature. High-temperature PEMFCs are now being developed and researched.

### 11.3.3 Reforming Technologies

Since PEMFC requires hydrogen for its operation, there is an increasing interest in converting current hydrocarbon-based marine fuels such as natural gas, diesel and gasoline into hydrogen-rich gases acceptable to the PEM fuel cells. Reforming can take place either on a very large scale at source or locally at the point of use by small reformers integrated with the fuel cell. Generally, there are two different kinds of reforming for production of hydrogen: external reforming, which is carried out before the fuel reaches the fuel cell itself, and internal reforming, which takes place within the fuel cell stack. External reforming could be carried out at a refinery or chemical plant and the hydrogen delivered by pipeline to filling stations, thereby reducing system cost.

Steam reforming and partial oxidation are methods of hydrogen production used on a large scale industrially, most notably in the production of ammonia and can be used for the production of hydrogen. Fuel is mixed with steam in the presence of a base metal catalyst to produce hydrogen and carbon monoxide. This method is the most well developed and cost-effective for generating hydrogen and is also the most efficient, giving conversion rates of 70–80% on a large scale.

The principal process of converting hydrocarbons into hydrogen by steam reforming involves the following reactions:

$$CH_4 + H_2O \rightarrow CO + 3H_2$$
$$CO + H_2O \rightarrow CO_2 + H_2$$

In order to obtain a good utilization of the feed for hydrogen production, it is necessary to operate the steam reformer with an outlet temperature around 800–950 °C. Heat has to be supplied to the process to achieve this outlet temperature.

### 11.3.3.1 Partial Oxidation

In partial oxidation, the methane and other hydrocarbons in natural gas react with a limited amount of oxygen (typically from air) that is not enough to completely oxidize the hydrocarbons to carbon dioxide and water. With less than the required amount of oxygen available, the reaction produces primarily hydrogen and carbon monoxide (and nitrogen, if the reaction is carried out with air rather than pure oxygen) and a relatively small amount of carbon dioxide and other compounds. Subsequently, in a water–gas shift reaction, the carbon monoxide reacts with water to form carbon dioxide and more hydrogen. Partial oxidation is an exothermic process – it gives off heat. The process is, typically, much faster than steam reforming and requires a smaller reactor vessel. However, this process initially produces less hydrogen per unit of the input fuel than is obtained by steam reforming of the same fuel.

### 11.3.4 Hydrogen Storage

Hydrogen is the lightest chemical element and offers the best energy-to-weight ratio of any fuel. The major drawback of using hydrogen is that it has the lowest storage density of all fuels. However, it is possible to store large quantities of hydrogen in its pure form by compressing it to very high pressure and storing it in containers which are designed and certified to withstand the pressures involved. In this way, it can be either stored as a gas or cooled to below its critical point and stored as a liquid.

Hydrogen can also be stored in solid form, in chemical combination with other elements (there are a number of metals which can "absorb" many times their own weight in hydrogen). The hydrogen is released from these compounds by heating or the addition of water.

### 11.3.5 Catalysts for PEM Fuel Cell

The CLs play an important role in the performance of PEMFC. In order to improve the performance of the cell, the catalyst must fulfil the following requirements. It must provide high intrinsic activities for electrochemical oxidation of a fuel at the anode and for reduction of dioxygen at the cathode. It should also have good electrical conductivity, be durable and less expensive.

When pure hydrogen is used as the fuel, almost all of the performance losses occur at the cathode. Because of the slower kinetics of oxygen reduction, platinum-based electrocatalysts remain the only practical catalyst material, since they combine both activity and stability in the fuel cell environment. At the anode, losses in cell potential are usually small (less than 50 mV at 2.0 A/cm$^2$) due to the extremely facile nature of the hydrogen oxidation reaction.

However, platinum is very expensive and not available in large quantity on the Earth. Now a group of researchers has shown how to get the same kind of reactivity using nickel as catalyst that is a thousand times less expensive than platinum. Electrodes made using the new catalyst would be about 20% cheaper than those made of platinum.

## 11.4 Solid Oxide Fuel Cell

SOFCs are high-temperature fuel cells and are operated at temperatures between 600 and 1000 °C. In addition to generating electricity, the high operating temperature also enables waste heat to be used for heating purposes as well as hot water production in homes and for industrial processes. Since these cells operate at high temperatures, they can tolerate relatively impure fuels. These characteristics enable the use of a variety of fuels in these cells. High-temperature SOFCs offer very low levels of NO$_x$ and SO$_x$ emissions, clean and pollution-free technology to electrochemically generate electricity at high efficiencies.

SOFCs' relatively simple design combined with the significant time required to reach operating temperature and quick response to changes in electricity demand make them suitable for large to very large stationary power applications. In addition, pressurized SOFCs can be successfully used as replacements for combustors in gas turbines; such hybrid SOFC–gas turbine power systems are expected to reach efficiencies over 70%.

The SOFC was first considered following the discovery of solid oxide electrolytes in 1899 by Nernst [9]. He reported that the conductivity of pure metal oxides increased very slowly with temperature and remained relatively low, whereas mixtures of metal oxides can possess dramatically higher conductivities.

The SOFC is characterized by having a solid ceramic electrolyte (hence the alternative name, ceramic fuel cell), which is a metallic oxide. The principle of operation of a SOFC shown in Fig. 11.2 can be described as follows. SOFC essentially consists of two porous electrodes separated by a dense, oxygen ion-conducting electrolyte through which the oxide ions migrate from the cathode side to the anode side where they react with the fuel (H, CO, etc.) to generate the voltage. It is a two-phase system because the fuel and oxidants are fed as gases, and the fuel cell reactions occur at the solid electrode/gas interfaces. After the reaction, it again produces steam and CO$_2$ which are gases.

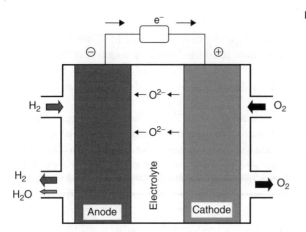

**Figure 11.2** Solid oxide fuel cell.

## 11.4.1 Electrolytes for SOFC

A major challenge in reducing the operating temperature of SOFCs is the development of solid electrolyte materials with sufficient conductivity to maintain low ohmic losses during operation. Three SOFC electrolytes, yttria-stabilized zirconia (YSZ), rare-earth-doped ceria and rare earth-doped bismuth oxide have been widely investigated as electrolytes for a fuel cell. Of these materials, YSZ has been most successfully employed with a strontium-doped lanthanum manganite (La1 2 $x$Sr$x$MnO$_3$) cathode and a mixed nickel/YSZ cermet anode and doped lanthanum chromite (LaCrO$_3$) as the interconnect. For optimum cell performance, the YSZ electrolyte must be free of porosity so as not to allow gases to permeate from one side of the electrolyte to the other.

Ceria-based materials have found potential application as electrolyte materials for the intermediate-temperature (500–750 °C) SOFC. ITSOFC is favourable compared to high-temperature SOFC. Several advantages are obtained with the reduction in operating temperature; they are as follows: less prone to thermal and mechanical stress, wide range of material selection, short start-up time, easy maintenance, better thermal management, much more economical.

In general, SOFC consists of two main sections: a reformer part, where the reforming reaction takes place, and a fuel cell part, where the electrochemical reactions occur. The external feed such as methane (in natural gas) is converted to hydrogen and carbon monoxide at the reformer section. In external reforming, the endothermic steam reforming reaction and the fuel cell reactions are operated separately.

Direct reforming of the fuel on the anode offers the simplest and most cost-effective solution and, in principle, provides the greatest system efficiency with least loss of energy. In direct reforming, the anode must perform the following tasks: firstly, as a reforming catalyst, catalyzing the conversion of hydrocarbons to hydrogen and CO; secondly as an electrocatalyst responsible for the electrochemical oxidation of H$_2$ and CO to water and CO$_2$, respectively; and finally, as an electrically conducting electrode. High efficiency results from utilizing the heat from the exothermic electrochemical reaction to reform the hydrocarbon fuel, this being a strongly endothermic reaction. However, one of the major problems with direct reforming is that it gives rise to a sharp cooling effect at the cell inlet, creating non-uniform distribution of temperature. Being endothermic, the direct internal reformation process can have the beneficial effect of reducing the airflow requirement for cell cooling different units.

## 11.5 Molten Carbonate Fuel Cell

Another high-temperature fuel cell is MCFC which operates at 600–700 °C where the alkali carbonates form a highly conductive molten salt, with carbonate ions providing ionic conduction. MCFC technology operating in a pressurized mode has the potential of a fuel-to-electricity efficiency in the range of 55–60% with $NO_x$ emissions of less than 1 ppm.

Internal reforming is possible for this cell. Internal reforming uses the heat produced by the exothermic hydrogen oxidation for the endothermic steam reforming reaction, which simplifies thermal management of the stack.

A MCFC is shown in Fig. 11.3. The defining characteristic of a MCFC is the material used for the electrolyte. The electrolyte is usually mixture of molten carbonate salts (lithium carbonate + potassium carbonate/sodium carbonate) retained in a ceramic matrix ($LiAlO_2$). In the early days of the MCFC, the electrodes were made of precious metals. As the technology developed, nickel was found to be adequate both as a metal for the anode and as nickel oxide (NiO) for the cathode. At the high operating temperatures in MCFCs, Ni (anode) and nickel oxide (cathode) are adequate to promote reaction. Noble metals as catalysts are not required for operation [10]. At the anode, hydrogen reacts with the ions to produce water, carbon dioxide and ions. The electrons travel through an external circuit to provide electrical power before they return to the cathode. At the cathode, oxygen from the air and carbon dioxide recycled from the anode react with the electrons to form $CO_3^-$ ions that replenish the electrolyte and transfer current through the fuel cell.

**Figure 11.3** Molten carbonate fuel cell.

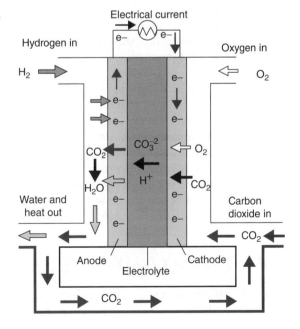

The electrochemical reactions occurring in MCFCs are as follows:

$$\text{Anode} \quad H_2 + CO_3^{2-} \rightarrow H_2O + CO_2 + 2e^- \tag{11.2}$$

$$\text{Cathode} \quad \frac{1}{2}O_2 + CO_2 + 2e^- \rightarrow CO_3^{2-} \tag{11.3}$$

with the overall reaction as

$$H_2 + \frac{1}{2}O_2 + CO_2 \rightarrow H_2O + CO_2 \tag{11.4}$$

One of the advantages of the high operating temperature of the MCFC is that the overall thermal efficiency is high, with a potential of 50–60% conversion of the fuel (natural gas) LHV to electricity without recovery and conversion of the exhaust heat. IF the exhaust heat from the MCFC which is at relatively high temperatures (650 °C) may be recovered for the generation of steam which further increases the efficiency. Efficiencies >60% may potentially be achieved with the incorporation of a bottoming cycle. On the other hand, the higher operating temperature places severe demands on the corrosion stability and life of cell components.

Unlike all other fuel cell types, the MCFC relies on a balance in capillary pressures within the pores of the anode and cathode to establish the interfacial electrode/electrolyte boundaries. The thickness of the matrix also plays an important part in the overall ohmic loss associated with the electrolyte.

The major obstacle to the commercial development of MCFCs is the dissolution of cathode. The dissolution of cathode can be described from the equation

$$NiO + CO_2 \leftrightarrow Ni^{2+} + CO_3^{2-} \tag{11.5}$$

Lithium as a cobalt oxide, $LiCoO_2$, has been selected as candidate material for MCFC cathodes because its solubility is small and the rate of dissolution into the melt is slower than that for nickel oxide. On the other hand, the electrical conductivity of $LiCoO_2$ is lower than that of nickel oxide. Thus, nickel oxide has been coated with stable $LiCoO_2$ in carbonate by a PVA-assisted sol–gel method to give a $LiCoO_2$-coated NiO (LC-NiO) cathode. The dissolution of the cathode material has also been minimized by changes in operating conditions and also by altering composition of the melt. Li/Na carbonate is emerging as a preferred electrolyte for MCFCs, although further optimization is likely.

MCFCs demand such high operating temperatures that most applications for this kind of cell are limited to large, stationary power plants. There are two main causes of performance degradation of cell; loss of electrolyte and dissolution of the cathode. Initially, electrolyte is consumed by corrosion reactions with metal hardware in the fuel cell. Over longer time, and at a more constant rate, electrolyte is primarily lost by vaporization into the fuel gas. A careful electrolyte management with improved pore structure to maintain electrolyte within the electrodes has reduced the electrolyte loss due to vaporization. Electrolyte loss is the main cause of performance degradation in atmospheric systems, and NiO dissolution is the main cause in pressurized systems.

## 11.6 Phosphoric Acid Fuel Cell

Fuel cells which use phosphoric acid solution as the electrolyte are called PAFCs. PAFCs were the first fuel cells to be commercialized. Developed in the mid-1960s and field-tested since the 1970s, they have improved significantly in stability, performance and cost. Such characteristics have made the PAFC as most suitable for stationary application [11].

The number of these cells exceeds any other fuel cell technology and is being used to power many commercial premises. The operating temperature of this cell ranges from 150 to 220 °C. The PAFC operates at greater than 40–45% efficiency in generating electricity. If co-generation is also employed, the overall efficiency is approximately 85%. Furthermore, the waste heat is produced at temperatures capable of heating hot water or generating steam at atmospheric pressure.

A PAFC is shown in Fig. 11.4. The principle of operation is similar to that of PEMFC. The PAFC consists of a pair of porous electrodes made of a finely dispersed platinum catalyst on carbon and a silicon carbide structure that holds the phosphoric acid electrolyte. In PAFCs, protons move through the electrolyte to the cathode to combine with oxygen and electrons, producing water and heat. The charge carrier in this type of fuel cell is the hydrogen ion ($H^+$, proton). At the anode, pure hydrogen or reformate fuel gases with the main component being hydrogen is supplied, and air is supplied at the cathode; the resulting electrochemical reaction yields an electric power output. At the fuel electrode, hydrogen reacts at the electrode surface to become hydrogen ions and electrons, and the hydrogen ions migrate towards the cathode within the electrolyte.

The phosphoric acid in aqueous solution dissociates into phosphate ions and hydrogen ions given by Eq. (11.6); the hydrogen ions ($H^+$) act as the charge carrier.

$$H_3PO_4 \rightarrow H^+ + H_2PO_4^- \tag{11.6}$$

Phosphoric acid has the advantages of good thermal, chemical and electrochemical stability, good capillary properties and low vapour pressures, although it is a poor ionic conductor. Among the common acids, it has the lowest volatility, and this property allows PAFCs to operate even at a temperature of 200 °C (473 K) for several thousand hours.

**Figure 11.4** PAFC fuel cell.

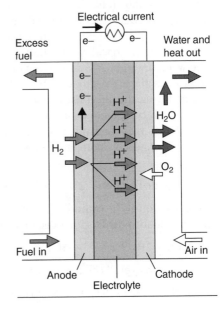

It also has an extremely wide change of vapour pressure across the face of the cell and as a function of current density, without showing a tendency to dry out or become too dilute. It exhibits high tolerance for reformed hydrocarbons, removal of CO is possible by a shift reaction and $CO_2$ is rejected naturally. The major limitation of this acid is that oxygen reduction is very slow even at high temperatures and pressures.

The voltage obtained from a single fuel cell is from 0.6 to 1.0 V; thus, more single cells are connected in series for achieving a reasonable operating voltage. Together with the electrical connection, the bipolar plates are usually machined so that they can act as gas channels. Since heat is generated in the course of the electrochemical reaction of hydrogen with oxygen, cooling plates are therefore inserted at regular intervals between fuel cells. Cooling water is passed through them to maintain a cell operating temperature of about 200 °C (473 K).

The PAFC system is the most advanced fuel cell system for terrestrial and stationary power generation applications.

## 11.7  Alkaline Fuel Cell

AFCs are also known as Bacon fuel cells after their British inventor Bacon [12]. Bacon was inspired by the idea of William Grove – that the electrolysis of water could be reversible. Along with PAFCs, AFCs were one of the earliest FCs developed and have been used by NASA. Starting with applications in space, the alkaline cell provided high-energy conversion efficiency with no moving parts and high reliability. In 1939, Bacon built a fuel cell with economically viable catalysts and practical high-pressure equipment and seals that had potassium hydroxide (27%) as electrolyte, a pure asbestos cloth as diaphragm and nickel gauze electrodes. They consume hydrogen and pure oxygen, to produce water, heat and electricity. These products make AFCs ideal for use in space where carbon dioxide production can pose a major threat.

An AFC is shown in Fig. 11.5. It consists of two porous electrodes with liquid KOH electrolyte between them. The hydrogen fuel is supplied to the anode electrode, while oxygen from air is supplied to the cathode. The working temperature ranges from 293 to 363 K. The electrical voltage between the anode and the cathode of a single fuel cell is between 0.9 and 0.5 V depending on the load and the electrochemical reactions taking place at these electrodes. The hydrogen is usually compressed, and the oxygen is obtained from the air. It uses an aqueous solution of potassium hydroxide as the electrolyte, with typical concentrations of about 30%. One advantage of using an alkaline electrolyte is that electrode material other than platinum can be used with less risk of corrosion, especially compared to acid electrolytes. The overall chemical reactions are given by

$$\text{Anode reaction} \quad 2H_2 + 4OH^- \rightarrow 4H_2O + 4e^- \tag{11.7}$$

The reaction at the cathode occurs when the electrons pass around an external circuit and react to form hydroxide ions, $OH^-$, shown as follows:

$$O_2 + 2H_2O + 4e^- \rightarrow 4OH^- \tag{11.8}$$

resulting in overall cell reaction as

$$2H_2 + O_2 \rightarrow 2H_2O^+ + \text{electric energy} + \text{heat} \tag{11.9}$$

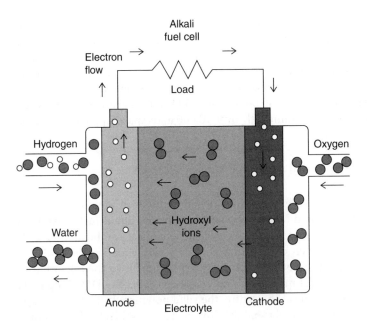

**Figure 11.5** Alkali fuel cell.

By-product water and heat have to be removed. This is usually achieved by re-circulating the electrolyte and using it as the coolant liquid, while water is removed by evaporation.

AFCs were used in the Apollo space shuttle which took the first men to the moon. Due to the success of the AFC in the space shuttle, they were tested in many different applications including agricultural tractors, power cars and provided power to offshore navigation equipment and boats. The space programme remained an important research on AFCs, and since then, AFCs have improved.

Alkaline electrode fuel cells operate at a wide range of temperatures and pressures, and their application is very limited. As a result, there are different types of electrodes used. Some of the various types of electrodes are explained as follows.

Carbon-supported catalysts are commonly used in the current production of electrodes. The electrodes consist of a double-layer structure: an active electro CL and a hydrophobic layer. The active layer consists of an organic mixture (carbon black, catalyst and PTFE) which is ground and then rolled at room temperature to cross-link the powder to form a self-supporting sheet. The use of non-platinum electrodes greatly reduces the cost of producing the electrodes. The major drawback of AFCs is their exquisite sensitivity to carbon dioxide. Even trace amounts of carbon dioxide can affect the cell's operation substantially by converting the potassium hydroxide electrolyte into potassium carbonate. Potassium carbonate is a solid that blocks pores in the cathode. This reduces the ionic conductivity of fuel cell and diminishes the speed with which the reaction can proceed. The problem of carbon dioxide is present when carbon-supported catalysts are used in the electrode. Several methods have been proposed to reduce the carbon dioxide concentrations: soda lime or ethanol amine scrubbers, physical absorption or removal of $CO_2$ by diffusion of the gas through a membrane. Guzlow in 1996 used an anode based on granules of Raney nickel mixed with PTFE which does not use carbon-supported catalysts to try and solve this problem. Apparently, this type of electrode does not react to the $CO_2$, making it highly favourable for use in this type of application. There is currently some effort in research to produce an AFC in which the potassium hydroxide is replaced at various intervals to maintain operation.

The lifetime of an AFC can, in general, be well over 5000 h for inexpensive terrestrial AFCs and has been shown to be significantly over 10,000 h for space-application AFCs.

## 11.8 Direct Methanol Fuel Cell

Significant efforts during the last few years have been focused on the direct electro-chemical oxidation of alcohol and hydrocarbon fuels. Organic liquid fuels have high energy density, and the enf associated with their electrochemical combustion to $CO_2$ is comparable to that of hydrogen combustion to water. A schematic of a DMFC is shown in Fig. 11.6. The operation of the DMFC system is similar to the operation of the PEM in terms of the physical manufacturing of the cell [13]. The DMFC consists of a proton-conducting membrane at the centre, the solid polymer electrolyte which is used to provide ionic conductivity and avoid the flow of electrons. Electrodes generally consist of expensive noble metal catalyst used to achieve sufficient rate of reaction at a low temperature. Normally, platinum is used as cathode and platinum–ruthenium alloy at anode for DMFC. In DMFC, to obtain useful current densities, the electrodes should have high surface area with respect to geometric area.

The major difference with PEMFC is in the fuel cell supply. Here the fuel is a mixture of water and methanol; it reacts electrochemically (methanol being oxidized) at the anode to produce carbon dioxide, protons and electrons as shown in Eq. 11.10.

$$\text{Anode reaction} \quad CH_3OH + H_2O \rightarrow CO_2 + 6H^+ + 6e^- \tag{11.10}$$

$$\text{And cathode reaction} \quad 1.5O_2 + 6H^+ + 6e^- \rightarrow 3H_2O \tag{11.11}$$

**Figure 11.6** Direct methanol fuel cell.

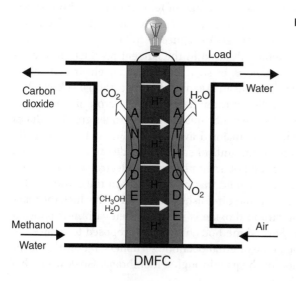

DMFC

And the overall cell reaction is given by

$$CH_3OH + 1.5O_2 + H_2O \rightarrow CO_2 + 3H_2O \tag{11.12}$$

The cell terminal voltage is 1.15 V.

Methanol and water are supplied and converted to carbon dioxide, protons and electrons at the anode. Methanol can be fed as aqueous solution or in the form of vapour to the anode. The produced electrons from the anode reaction are subsequently transferred via the external circuit (protons and electrons reduce oxygen (from air) to form water at the cathode. This overall reaction occurs at 100% coulombic efficiency with the best electrocatalysts: Pt–Ru for anode and Pt for cathode.

In early 1990s, DMFCs were designed with proton-conducting membranes as the electrolyte using Nafion. With a proton exchange membrane electrolyte (Nafion), the structure of fuel cell is vey similar to that of PEMFC. The structure and composition of the oxygen electrode in DMFC are practically the same as that in PEMFC. The methanol electrode requires a noble metal electrocatalyst loading five to ten times more than for a hydrogen electrode because of the poisoning of the electrode by CO gas. The Teflon contents in the substrate and diffusion layers are lower in the methanol electrode than in hydrogen electrode in order to facilitate the transfer of the liquid fuel methanol to the active layer of the electrode.

One of the most promising applications of DMFCs presently concerns with the field of portable power sources. In this regard, increasing interest is devoted towards the miniaturization of these fuel cell devices in order to replace the current Li-ion batteries. Theoretically, methanol has a superior specific energy density (6000 Wh/kg) in comparison with the best rechargeable battery.

In principle, methanol should be oxidized spontaneously when the anode potential is above 0.046 V, with respect to the reversible hydrogen electrode (RHE). Similarly, oxygen should be reduced spontaneously when the cathode potential is below 1.23 V. In reality, and in common with all fuel cell types, poor electrode kinetics (kinetic losses) cause the electrode reactions to deviate from their ideal thermodynamic values so as to incur a practical reduction of the extremely high theoretical efficiency possible from the cell.

Methanol crossover through the polymer membrane is known to be one of the most challenging problems affecting the performance of DMFCs. The overall efficiency of a methanol fuel cell is determined by both voltage and faradaic efficiency for the consumption of methanol. The faradaic efficiency is influenced mainly by methanol crossover through the membrane. The methanol crossover is usually measured indirectly by determining the amount of $CO_2$ produced at the cathode by the oxidation of methanol on the Pt surface. This $CO_2$ can be monitored on-line by using an IR detector. A more accurate method consists of a chromatographic analysis of samples of the cathode outlet stream. The crossover of methanol is influenced by both membrane characteristics and temperature, as well as by the operating current density. In general, an increase in temperature causes an increase in the diffusion coefficient of methanol and determines a swelling of the polymer membrane. Both effects contribute to an increase in methanol crossover rate.

The crossover rate of methanol at the open-circuit voltage of DMFC at 60 °C is about 100 mA/cm². This also reduces the open-circuit voltage of the cell as compared to hydrogen fuel cell. The crossover rate decreases with increase in current density and increases with increased methanol concentration. The crossover rate can also be reduced by using alternate membranes, membrane composites or operating at higher temperatures. Presently, research is ongoing in all these directions.

### 11.8.1 CO Removal

Another problem faced by PEMFC is poisoning of the anode by carbon monoxide which is present in reformed fuel cells. After reforming and water–gas shift, the CO in the reformer gas is usually reduced to 1–2%. A PEMFC requires a low level of CO to less than 10 ppm. This gas is more strongly adsorbed on the surface of the catalyst than hydrogen, effectively blocking the sites where hydrogen oxidation would normally occur. There are a number of ways to clean the raw reformer gas of CO. One method to prevent this from happening is to remove the CO by oxidizing it to carbon dioxide, which does not adsorb on the catalyst, before it can reach and react with the anode surface. This can be accomplished by using the reconfigured anode developed by Los Almos National laboratory. Alternatively, ultra-pure hydrogen without any contaminant can be produced using Pd-membrane technology.

## 11.9 Fuel Cell Stacks

Since the voltage of a single cell is a very small typical value, the voltage is about 0.7 V when drawing a useful current. It is therefore common to connect a number of fuel cells in series to form a stack. Normally, approximately 50 or more cells are connected to produce a usable voltage. A fuel cell stack consists of a number of single cells stacked up so that the cathode of one cell is electrically connected to the anode of the adjacent cell. Since these cells are connected in series, exactly the same current passes through each of the cells. A better way of forming a stack is to make the connections with bipolar plates made from materials with good conductivity such as graphite or stainless steel. These plates make connections all over one cathode to the anode of the next cell, hence bipolar. The conditions that they must fulfil are that there has to be a good electric connection between the electrodes and that the different gases must be separated.

The bipolar configuration is the best for larger fuel cells since the current is conducted through relatively thin conductive plates; thus, it travels a very short distance through a large area. This causes minimum electroresistive losses, even with a relatively poor electrical conductor such as graphite (or graphite polymer mixtures). The bipolar plate allows the cells to be connected in series and allows gas to be supplied to the anode and cathode.

The choice of materials for producing bipolar plates in commercial fuel cell stacks is dictated not only by the performance considerations but also by cost. However, for small cells, it is possible to connect the edge of one electrode to the opposing electrode of the adjacent cell by some kind of connector. This is applicable only to very small active area cells because current is conducted in the plane of very thin electrodes, thus travelling relatively long distance through a very small cross-sectional area.

Power conditioning includes controlling current (amperes), voltage, frequency and other characteristics of the electrical current to meet the needs of the application. Fuel cells produce electricity in the form of direct current (dc). If the fuel cell is used to power equipment that uses ac, the dc will have to be converted to alternating current.

The key aspects of a fuel cell stack design are as follows:

- Maintainance of uniform distribution of reactants to each cell
- Maintainance of uniform distribution of reactants inside each cell
- Maintenance of required temperature in each cell
- Minimum resistive losses
- No leak of reactant gases (internal between the cells or external)
- Mechanical sturdiness (internal pressure including thermal expansion, external forces)

In order to maintain the desired temperature inside the cells, the heat generated as a by-product of the electrochemical reactions must be taken away from the cells and from the stack. Different heat management schemes may be applied as described as follows.

### 11.9.1  Cooling with Separate Airflow

Forced convection airflow is a convenient way to bring the waste heat out of the stack. This will result in a more compact stack structure and increase the cooling capability. However, very high cathode airflow velocity or a very large gas channel is necessary for removal of waste heat. When the power of the fuel cell is high, a more effective cooling approach must be applied. Essentially, air cooling method is simple and needs fewer accessories compared to liquid cooling methods, but as the output power increases, it becomes harder to maintain a uniform temperature distribution within the stack by air cooling method and the parasitic losses associated with the cooling fan also increase.

Considering the fact that increasing the cathode air supply for cooling can cause dry out of the membrane, it is reasonable to have separate channels for cooling air to flow through. Thus, separate cooling plates are provided through which air is blown in a separate air cooling system. The advantage of this structure is that it can extract more heat from the stack without affecting the cathode airflow. Air-cooled designs using a bipolar-plate-integrated coolant flow are common for low-temperature PEMFC stacks in the range of 100 W–2 kW. For PEMFC stacks with power output greater than 5 kW, cooling with air may not be sufficient or not as advantageous as the liquid cooling.

### 11.9.2  Liquid Cooling

Units below 2 kW can be air cooled, and cells between 2 and 10 kW need a careful choice regarding whether air or water cooling should be used. For cells larger than 10 kW, liquid cooling is generally necessary. The liquid coolant is usually deionized water, which has the advantage of very high heat capacity, or an antifreeze coolant, for example, mixture of ethylene glycol and water, for operation under sub-zero conditions. Similarly to the cooling with a separate airflow, liquid coolant flows in the cooling channels, which are usually integrated in the bipolar plates. It is possible to place more than one cell between every two liquid cooling layers to reduce the number of cooling layers. However, it was found that the stack performance would decrease if the number of cells between the two successive cooling layers was increased.

Liquid cooling requires a more complex design: the temperature and pressure of the cooling water must be monitored, and the flow of cooling water must be supplied by a water pump. Stack cooling in DMFCs is relatively simpler since increasing circulation of dilute methanol solution at the anode could remove more waste heat from the stack.

## 11.10  Fuel Cell Applications

Fuel cells can generate power from a fraction of a watt to hundreds of kilowatts. Because of this, they may be used in almost every application where local electricity generation is required. Fuel cell applications may be classified as being either mobile or stationary applications. The mobile applications primarily include transportation systems and portable electronic equipment while stationary applications primarily include combined heat and power (CHP) systems for both residential and commercial needs. Applications such as automobiles, buses, utility vehicles, scooters, bicycles, submarines have been already demonstrated. Small fuel cells are attractive for portable power applications, either as replacement for batteries (in various electronic devices and gadgets) or as portable power generators.

As a backup power generator, fuel cells offer several advantages over either internal combustion engine generators (noise, fuel, reliability, maintenance) or batteries (weight, lifetime, maintenance). Fuel cell and fuel cell system design are not necessarily the same for each of these applications. On the contrary, each application, besides power output, has its own specific requirements, such as efficiency, water balance, heat utilization, quick start-up, long dormancy, size, weight, fuel supply.

### 11.10.1  Application in Automobile Industry

For almost a century, electric vehicles have been dependent on lead–acid batteries. The problem with battery is the limit in energy storage: it limits a practical vehicle to a range of about 100 km and a maximum speed of 100 km/h. However, steady advances in fuel-cell technology, new opportunities for hydrogen production and a growing commitment to building hydrogen infrastructure have led many major automakers to think about producing next generation of hydrogen fuel-cell vehicles. All major car manufacturers have demonstrated prototype fuel cell vehicles and have plans for production and commercialization. About 60 million new cars are sold worldwide each year. Several automotive industry leaders have speculated that fuel cell vehicles could account for 20–25% of new car sales within the next 20–25 years, a potential market of 12–15 million vehicles each year. FCVs are currently more expensive than conventional vehicles and hybrids. However, costs have decreased significantly, and it is expected that by 2020, the cost may become competitive.

Buses for city and regional transport are considered to be the most likely types of vehicles for an early market introduction of fuel cell technology. Buses require more power than passenger automobiles, typically about 250 kW or more. They operate in a more demanding operating regime with frequent starts and stops. Nevertheless, the average fuel economy of a bus fuel cell system is roughly 15% better than that of a diesel engine. Buses are almost always operated in a fleet and refueled in a central facility. This makes refueling with hydrogen much easier.

### 11.10.2  Stationary Power Applications

The stationary application of fuel cells includes CHP, uninterruptible power systems (UPS) and primary power units. Stationary fuel cells can be used as a primary power source. It is often used to power houses that are not connected to the grid or to provide supplemental power. In hybrid power systems, fuel cells can be connected to photovoltaics, batteries, capacitors or wind turbines, providing primary or secondary power. Fuel cells can also be used as a backup or energy power generator providing power when grid supply is not available. Stationary fuel-cell power systems are suitable for DG, allowing the utility companies to increase their installed capacity following the increase in demand more closely, rather than anticipating the demand in huge increments by adding gigantic power plants.

The system design for stationary power application and automotive power applications is similar in principle. The main differences are in the choice of fuel, power conditioning and heat rejection. There are also some differences in requirements for automotive and stationary fuel cell systems. For example, size and weight requirements are very important in automotive applications, but not so significant in stationary applications. The acceptable noise level is lower for stationary applications, especially if the unit is to be installed indoors. The fuel cell itself of course does not generate any noise, but it may be coming from air and fluid handling devices. Automobile systems are expected to have a very short start-up time (fraction of a minute), while the start-up of a stationary system is not time-limited, unless operated as a backup or emergency power generator. Both automotive and stationary systems are expected to survive and operate in extreme ambient conditions, although some stationary units may be designed for indoor installation only. And finally, the automotive systems for passenger vehicles are expected to have a lifetime of 3000–5000 operational hours, systems for buses and trucks somewhat longer, but the stationary fuel-cell power systems are expected to operate for 40,000–800,000 h (5–10 years).

To the end users, the fuel cells offer reliability, energy independence, "green" power and, ultimately, lower cost of energy. Stationary fuel cells may be used in different applications, for example:

### 11.10.3 Portable Applications

There is a growing demand to supply power for portable electronic equipment both in the consumer market and for the military. The portable market is often regarded as secondary after transport and stationary power, but is nevertheless significant in terms of volume of research and the potential size of the market. Portable fuel cells are lightweight, long-lasting, portable power sources that prolong the amount of time a device can be used without recharging. In comparison, secondary (rechargeable) batteries have battery charger systems that consist of ac chargers that require an outlet to be charged or dc chargers that will recharge batteries from other batteries. Rechargeable batteries are not practical for certain portable and military electronic applications because they can be heavy and not meet the power requirements. Some portable fuel cell applications include laptops, cellular phones, power tools, military equipment, battery chargers, unattended sensors and unmanned aerial and underwater vehicles. Fuel cells offer greater energy density coupled with simple and rapid recharging to give extended operation.

Although much smaller, the military market may well be leading the race to portable fuel cell devices and is a significant source of funding in this area.

## 11.11   Modelling of Fuel Cell

A simple model of PEM FC is developed here [14]. In order to simplify the modelling and reduce the computation time, the following assumptions are drawn. The gases are ideal and uniformly distributed inside the anode and cathode. The stack is fed with humidified hydrogen and air because the use of humidified fuel and air improves the efficiency of the FC.

The mathematical model equations that describe the operation of the FC consist of the voltage–current characteristics and a relationship for the consumption of the reactants as a function of the current drawn from the fuel cell. The temperature is constant and uniform for each experiment. The gas channels along the electrodes have a fixed volume with small lengths, so that it is necessary only to define a single pressure value in their interior.

### 11.11.1   Steady-State Model

*V-I* characteristics of fuel cells can be computed based on physical foundations of fuel cells. In order to model a fuel cell stack, some parameters are required to fit the model. Figure 11.7 shows the steady-state fuel cell characteristic. Typical characteristics of FC are normally given in the form of polarization curve, which is a plot of cell voltage versus cell current density.

To determine the voltage–current relationship of the cell, the cell voltage has to be defined as the difference between an ideal Nerrnst voltage and a number of voltage losses. The main losses are categorized as activation, ohmic and concentration losses.

The steady-state fuel cell voltage, *VFC*, is then calculated as

$$V_{cell} = E_{nernst} - V_{act} - V_{ohm} - V_{con} \tag{11.14}$$

$E_{nernst}$ voltage is related to the change in Gibbs free energy and to pressure effect. The activation loss is caused by the slowness of the reaction at the surface of the electrodes. The ohmic loss is due to internal resistance of the fuel cell. The concentration losses are

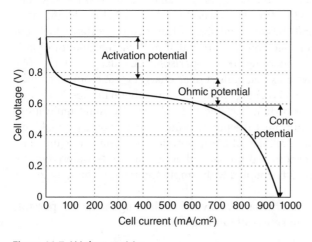

**Figure 11.7** *V-I* characterisitc.

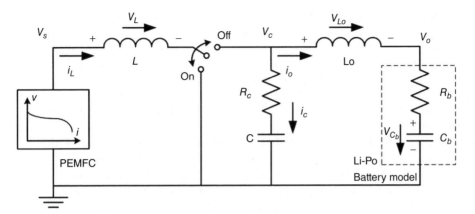

**Figure 11.8** Equivalent circuit.

generated by the depletion of the reactants at the surface of the electrodes as the fuel is consumed.

Depending on the amount of current drawn, the fuel cell generates the output voltage according to (1). The electric power delivered by the system equals the product of the stack voltage $V_{cell}$ and the current drawn $I$:

$$P = V_{cell} I$$

The *V-I* characteristic shown in Fig. 11.7 can be modelled using the equivalent circuit shown in Fig. 11.8.

## 11.12 Summary

A fuel cell combines hydrogen with oxygen (from air) in a chemical reaction, producing water, electricity and heat. Fuel cells do not "burn" the fuel; instead, the conversion takes place electrochemically without combustion. Fuelled with pure hydrogen, they produce zero emissions of pollutant and greenhouse gases at the location of the power plant. In this chapter, different types of fuel cells are presented along with their construction and working.

Fuel cells are connected to form stacks to generate sufficient power. Since lot of heat is produced in the cell during chemical reaction, various methods of cooling are described.

Next, various applications of fuel cells in automobile and stationary applications are discussed. Finally, a simple model of fuel cell for simulation is provided.

## References

1  Gregor, H. (ed.) (2013) *Fuel Cell Technology Hand book*, CRC Press.
2  O'Hayre, R.P. *et al.* (2005) *Fuel Cell Fundamentals*, John Wiley & Sons.
3  Srinivasan, S. (2006) *Fuel Cells from Fundamentals to Applications*, Springer.
4  Tim, Z. (ed.) (2007) *Advances in Fuel Cells*, Elsevier.

5 Revankar, S.T. and Majumdar, P. (2014) *Fuel Cells: Principles, Design, and Analysis*, CRC Press.

6 Frano, B. (2013) *PEM Fuel Cells Theory and Practice*, 2nd edn, Academic Press.

7 Christoph, H. and Christina, R. (2012) *Polymer Electrolyte Membrane and Direct Methanol Fuel Cell Technology*, vol. 1, Woodhead Publishing.

8 Hu, L. *et al.* (2010) *Proton Exchange Membrane Fuel Cells*, CRC Press.

9 Singhal, S.C. (2000) *Advances in Solid Oxide Fuel Cell Technology Solid State Ionics*, Elsevier.

10 Wilimeski, G. and Wolf, T.L. (1983) Molten carbonate fuel cell performance model. *Journal of the Electrochemical Science and Technology*, **130** (1), 48–55.

11 Blomen, L.J.M.J. and Mugerwa, M.N. (1993) *Fuel Cell Systems*, Plenum Press.

12 Hiroko, S. (2009) Alkaline fuel cellsUnesco EOLSS sample chapters, in *Energy Carriers and Conversion Systems* (ed. T. Ohta), EOLSS.

13 Hamnett, A. (1997) Mechanism and electrocatalysis in the direct methanol fuel cell. *Catalysis Today*, **38** (4), 445–457.

14 Ziogou, C. *et al.* (2010) Modeling and experimental validation of a PEM fuel cell system. 20th European Symposium on Computer Aided Process Engineering – Elsevier.

# 12

# Small Hydropower Plant

## 12.1 Hydropower

Earlier when commercial electric power generation using hydropower was not available, it was used for irrigation and operation of various machines, such as watermills, textile machines and sawmills. Greeks were able to harness the power in the moving water thousands of years ago when they used water wheels, which picked up water in buckets around a wheel. The water's weight caused the wheel to turn, converting kinetic energy into mechanical energy for grinding grain and pumping water. In the 1800s, the water wheel was often used to power machines such as timber-cutting saws in European and American factories. It was then realized that the force of water falling from a height can turn a turbine which when connected to a generator can produce electricity. Thus, the world's first hydroelectric station (of 12.5 kW capacity) was commissioned on 30 September 1882 on Fox River at the Vulcan Street Plant, Appleton, Wisconsin, USA, used for lighting two paper mills. Hydropower is considered a renewable source of energy as it uses water but does not consume it. It, therefore, offers significant potential for carbon emission reductions. The installed capacity of hydropower by the end of 2013 was 1000 GW which contributed 16.4% of worldwide electricity supply [1]. Hydropower remains the largest source of renewable energy in the electricity generation. Hydropower systems use the energy in flowing water to produce electricity. Although there are several ways to harness the moving water to produce energy, run-of-the-river systems, which do not require large storage reservoirs, are often used for microhydro, and sometimes for small-scale hydro projects.

Hydropower plants (HPPs) today span a very large range of scales, from a few watts to several gigawatts. The largest projects, Itaipu in Brazil with 14,000 MW and Three Gorges in China with 22,400 MW, both produce between 80 and 100 TWh/yr. The great variety in the size of hydropower plants gives the technology the ability to meet both large centralized urban energy needs and decentralized rural needs. The primary role of large hydropower system is in providing electricity generation as part of centralized energy networks. The small hydropower plants can operate in isolation and supply independent systems, often in rural and remote areas of the world. In this chapter, small hydropower for generation of electricity is described.

*Operation and Control of Renewable Energy Systems*, First Edition. Mukhtar Ahmad.

## 12.2  Classification of Hydropower Plants

Classification of HPPs can be based on different factors:
  Depending on Head:

- Low (less than 50 m)
- Medium (between 50 and 250 m)
- High (greater than 250 m)

  Based on exploration and storage:

- With flow regulation (daily or seasonal) reservoir type
- Without flow regulation, run-of-river type

  Depending on the size of HPPs, these are classified as follows:

- Large hydro – having capacity of more than 100 MW and feeding to large grid
- Medium hydro – 15–100 MW usually connected to grid
- Small hydro – 1–15 MW usually connected to grid
- Mini hydro – 100 kW–1 MW and connected to grid
- Micro hydro – less than 100 kW

Small hydro are generally "run-of-river" type as shown in Fig. 12.1 which has very small dam or barrage usually just a weir and little or no water stored. Therefore,

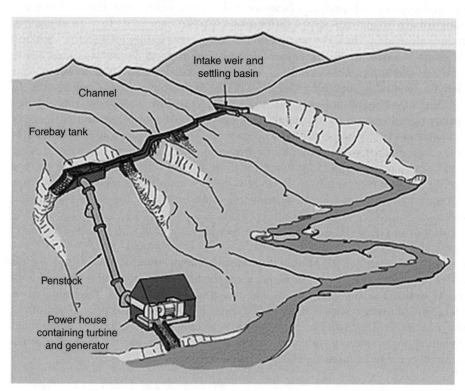

**Figure 12.1**  Run-of-river power plant.

run-of-river installations do not have the same kinds of adverse effect on ecological balance as in large-scale hydro. Small hydro plants have many advantages such as the following:

- High efficiency (70–90%)
- High capacity factor (more than 50%)
- High level of predictability

### 12.2.1  Basics of Hydropower Generation

Hydraulic power can be captured wherever a flow of water is from a higher level to a lower level. The vertical fall of the water, known as the "head", is essential for power generation. When the water flows from higher level to lower level, the potential energy of the water is converted to equivalent amount of kinetic energy. Thus, the height of the water is utilized to calculate its potential energy, and this energy is converted to speed up the water at the intake of the turbine and is calculated by balancing these potential and kinetic energies of water.

$$\text{Potential energy of water } E = mgH$$

$$\text{which is then converted into kinetic energy} = \frac{1}{2}mv^2$$

where $m$ = mass of water in kg, $g$ = acceleration due to gravity
$H$ = effective water head at the turbine, and $v$ = jet velocity of water at the turbine blades

$$\text{Thus } \frac{1}{2}mv^2 = mgH$$

$$\text{Or } v = \sqrt{2gH}$$

Hydro turbines convert water force into mechanical shaft power, which can be used to drive an electricity generator or other machinery. The power available is proportional to the product of head and flow rate. The general formula for any hydro system power output is

$$P = \eta \rho g Q H \tag{12.1}$$

where $Q$ is the volume of water flow passing through the turbine in m³/s, $\rho$ is the density of water and $\eta$ is efficiency of turbine. The best turbines can have hydraulic efficiencies in the range of 80% to over 90%, although this will reduce with size. Micro-hydro systems (<100 kW) tend to be 60–80% efficient. Corresponding energy over a period of time $\Delta t$ will be

$$E = \eta \rho g Q H \Delta t \tag{12.2}$$

## 12.3  Resource Assessment

Adequate head and flow are necessary requirements for hydro generation. It is therefore important to know both the flow of water and the gradient at a particular location to assess the potential of hydroelectric power availability. In many countries, stream flow records are maintained by hydrological institutes. These records are useful in determining the flow at a particular site. However, if the records are not available, the discharge should preferably be directly measured for at least a year. A single measurement of instantaneous flow in a watercourse cannot be used for this purpose. There are numerous methods of stream flow measurement also known as stream gauging. These include direct methods, such as volumetric gauging, and dilution methods, as well as indirect methods involving stage-discharge relations or rating curves [2, 3]. Since the velocity of a stream varies with depth and width across a stream, it is important to know the main quantity to measure when choosing a stream gauging method. If the interest is to measure the stream surface velocity, the float method is well suited. This method involves throwing some buoyant, highly visible object into the stream and measuring the time it takes to float a known distance.

### 12.3.1 Velocity Area Method

The process of measuring stream flow (volume rate of flow), or discharge, is called stream gauging. For obtaining a more accurate stream discharge measurement, the velocity–area method is used. Discharge is the volume of water flowing down a stream or river per unit of time, commonly expressed in cubic metre per second. This is a conventional method for medium-to-large rivers, involving the measurement of the cross-sectional area of the river and the mean velocity of the water through it. Discharge, or the volume of water flowing in a stream over a set interval of time, can be determined with the equation:

The discharge $Q = AV$

It is a useful approach for determining the stream flow with a minimum effort. Stream water velocity is normally measured using a current meter. A current meter consists of a propeller or a horizontal wheel with small, cone-shaped cups attached to it. The cups get filled with water and turn the wheel when placed in flowing water. The number of rotations of the propeller or wheel cup depends on the velocity of the water flowing in the stream. If purely laminar flow is assumed, the stream velocity is expected to vary vertically following a parabolic function because of the zero velocity (no-slip condition) at the bottom of the stream bed. In case of a turbulent flow, it will be a logarithmic function. For this reason, velocity should ideally be measured at several depths for each interval along the river cross section. Alternatively, a single measurement is taken to measure flow at 0.6 times the total depth, which typically represents the average flow velocity in the stream. Velocity also varies within the cross section of a stream, where stream banks are associated with greater friction and hence slower moving water. Thus, it is necessary to take velocity measurements along a cross section of a stream as shown in Fig. 12.2. Since stream channels are rarely straight, it is helpful to measure velocity across an "average" reach of the stream (e.g. average width and depth) with a single channel, a relatively flat stream bed with little vegetation and rocks and few back-eddies that hinder current meter movement.

The stream is divided into sections based on where velocity and stage height measurements were taken in the cross section of the stream. By multiplying the cross-sectional area (width of section × stage height) by the velocity, the discharge for that section of stream can be calculated. The discharge from each section is then added to determine the total discharge of water from the stream.

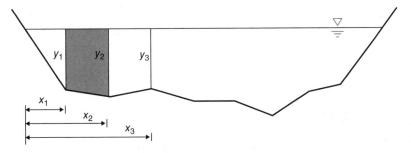

**Figure 12.2** Velocity–area method.

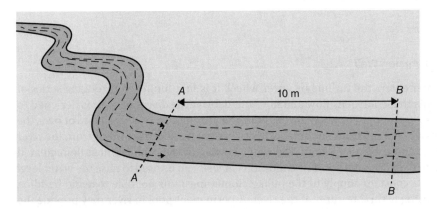

**Figure 12.3** Float method.

### 12.3.2 Float Method

The basic idea of float method is to measure the time that it takes the object to float a specified distance downstream as shown in Fig 12.3. Float method can be used to measure the surface velocity of river flow. Because surface velocities are typically higher than mean or average velocities, the mean velocity is obtained by using a correction factor.

$$V_{mean} = kV_{surface}$$

where $k$ is a coefficient that generally ranges from 0.8 for rough beds to 0.9 for smooth beds (0.85 is a commonly used value).

This method is simple and inexpensive but not as accurate as the velocity–area method. Surface velocity measurements should be carried only on windless days to avoid deflection of the floats due to wind. Even under these conditions, surface floats may be diverted from a direct course between measuring stations because of surface disturbances and crosscurrents. To determine the velocity of the discharge, about 8–30 m long section of the channel that includes the part where the cross section was measured is marked. The length will be dependent upon the speed of the water. In many channels, 8 m length would be too short a distance because the float would travel too fast to get an accurate time estimate. In such streams, a longer distance may be used. The float is gently released into the channel slightly upstream from the beginning of the section. The time taken by the "float" to travel the marked section is measured, and this process is repeated at least three times to calculate the average time. The surface velocity is then computed by dividing the length of the section by the time it took the float to move through the section.

## 12.4   System Components

Small run-of-the-river hydropower systems consist of these basic components [4, 5]:

- Diversion weir, channel or pipeline
- Turbines
- Generators

### 12.4.1   Diversion Weir

Before water enters the turbine or water wheel, it is first funnelled through a series of components that control its flow and filter out debris. The diversion weir is designed to provide assured supply of water throughout the year. It also provides control over the flow and save channel from flooding. A hydro system must extract water from the river in a reliable and controllable way. The water flowing in the channel must be regulated during high river flow and low flow conditions. A weir can be used to raise the water level and ensure a constant supply to the intake. Sometimes, it is possible to avoid building a weir by using natural features of the river. A permanent pool in river may provide the same function as a weir.

Different types of intakes are characterized by the method used to divert the water into the intake. For micro-hydro schemes, only the smaller intakes will be suitable. The following three types of intakes will be discussed here: the side intake with and without a weir and the bottom intake.

#### 12.4.1.1   Side Intake without Weir

It is relatively less expensive, requires no complex machinery for construction, requires little maintenance and repairs. However, during low supply of water, little water will be diverted and therefore not suitable where there is large variation in flow of streams.

#### 12.4.1.2   Side Intake with Weir

The weir used in this configuration can be partly or completely submerged into the water. It has good water level control, requires little maintenance but has the same problem during low flows as with side intake without weir.

#### 12.4.1.3   Bottom Intake

At a bottom intake, the whole weir is submerged into the water. Excess water will pass the intake by flowing over the weir. It is therefore very suitable for fluctuating flows.

## 12.4.2   Water Conductor System or Channels

Water conductor system is a channel that delivers water from weir to de-silting tank and from de-silting tank to forebay. De-silting tank or settling basin is provided to reduce undesirable sediment particles in water from entering the head race tunnel or channel. The main principal is to provide a section wide and long enough so that the resulting reduced flow velocity will allow the sediment to settle out. Such reduction in the velocity also reduces the bed shear stress and the turbulence.

Decrement or reduction in velocity, shear stress and the turbulence, if adequate, stop the bed material from moving part of the suspended material to deposit. The flow into the basin is regulated by gates at intake. The sediment which will be settled is flushed out of the basin through the flushing conduit/tunnel back into the river.

## 12.4.3   Forebay Tank

The forebay tank forms the connection between the channel and the penstock. The main purpose is to allow the last particles to settle down before the water enters the penstock. Depending on its size, it can also serve as a reservoir to store water.

A sluice will make it possible to close the entrance to the penstock. In front of the penstock, a trash rack needs to be installed to prevent large particles to enter the penstock.

## 12.4.4   Penstock

The penstock is the pipe which conveys water under pressure from the forebay tank to the turbine. The major components of the penstock are shown in Fig. 12.8. The penstock often constitutes a major expense in the total micro-hydro budget, as much as 40% is not uncommon in high head installations, and it is therefore worthwhile optimizing the design. The trade-off is between head loss and capital cost. Head loss due to friction in the pipe decreases dramatically with increasing pipe diameter. Conversely, pipe costs increase steeply with diameter. Therefore, a compromise between cost and performance is required.

The design philosophy is first to identify available pipe options, then to select a target head loss, 5% of the gross head being a good starting point. The details of the pipes with losses close to this target are then tabulated and compared for cost-effectiveness. A smaller penstock may save on capital costs, but the extra head loss may account for lost revenue from generated electricity each year.

## 12.4.5   Spillways

Spillways are designed to permit controlled overflow at certain points along the channel. Figure 12.4 depicts a flood spillway in detail, including flow control and channel emptying gates. Flood flows through the intake can be twice the normal channel flow, so the spillway must be large enough for diverting this excess flow.

The spillway is a flow regulator for the channel. In addition, it can be combined with control gates to provide a means of emptying the channel.

The spill flow must be fed back to the river in a controlled way so that it does not damage the foundations of the channel.

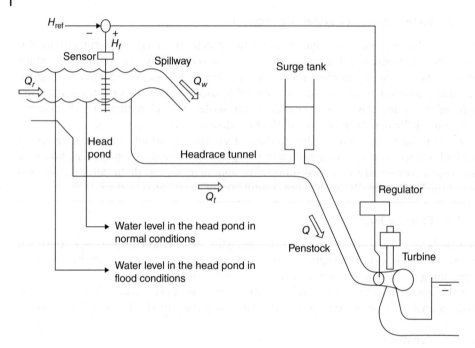

**Figure 12.4** Spillway.

## 12.5 Turbines

Water wheels and turbines can be used to convert the energy of running water into mechanical energy. Turbines are more commonly used nowadays to power small hydropower systems as turbines are more compact in relation to their energy output than water wheels. They also have fewer gears and require less material for construction. The moving water strikes the turbine blades, to rotate a shaft which is connected to a generator. Conventional pumps can also be used as substitutes for hydraulic turbines. When the action of a pump is reversed, it operates as a turbine. Since pumps are mass produced, these are more readily available and less expensive than turbines. However, for adequate pump performance, the micro-hydro site must have fairly constant head and flow. Pumps are also less efficient and more prone to damage.

There are two general classes of turbines: impulse and reaction. The type of turbine selected for a particular project is based on the height of standing water called *head* and volume of water flow [5–7].

## 12.6 Impulse Turbines

Impulse turbines use the velocity of water which moves the runner blades of the turbines. The water stream hits each bucket on the runner. There is no suction on the down side of the turbine, and water remains at atmospheric pressure after and before hitting the runner. An impulse turbine is generally used for high-head and low-flow applications. There are three types of impulse turbines used for small hydro systems; these are (i) Pelton turbine, (ii) cross-flow and (iii) Turgo turbine.

### 12.6.1 Pelton Turbine

A Pelton turbine essentially consists of one or more injectors for generating the high-speed jet and a wheel with a series of split buckets for receiving the jet energy as shown in Fig. 12.5. An injector nozzle converts the pressure energy of the water into the kinetic energy of the high-speed jet. It also regulates the flow rate via a built-in needle which is driven by a servomotor. The jet hits each bucket, and the impact of water on the buckets causes the runner to rotate and produce mechanical energy. Nearly all the energy of water is expended in rotation of the runner, and deflected water is discharged in tail race. The buckets are so shaped that the water enters tangentially in the middle and is split in half. Each half is turned backwards and flows again tangentially in both the directions to avoid thrust on the wheel. Often, two buckets are mounted side by side, thus splitting the water jet in half. This balances the side-load forces on the wheel and helps to ensure smooth, efficient momentum transfer of the fluid jet to the turbine wheel.

The number of jets is not more than two for horizontal shaft turbines and is limited to six for vertical shaft turbines. The flow partly fills the buckets, and the fluid remains in contact with the atmosphere. Therefore, once the jet is produced by the nozzle, the static pressure of the fluid remains atmospheric throughout the machine. Because of the symmetry of the buckets, the side thrusts produced by the fluid in each half balances each other. For maximum power and efficiency, the turbine system is designed such that the water-jet velocity is twice the velocity of the bucket. A very small percentage of the water's original kinetic energy will still remain in the water; however, this allows the

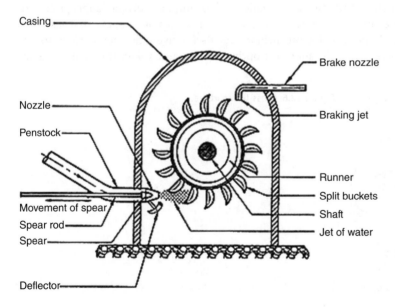

**Figure 12.5** Pelton turbine.

bucket to be emptied at the same rate it is filled, thus allowing the water flow to continue uninterrupted.

For a constant water flow rate from the nozzles, the speed of the turbine changes with changing loads on it. For quality hydroelectricity generation, the turbine should rotate at a constant speed. To keep the speed constant despite the changing loads on the turbine, the water flow rate through the nozzles is changed. To control the gradual changes in load, servo-controlled spear valves are used in the jets to change the flow rate. And for sudden reduction in load, the jets are deflected using deflector plates so that some of the water from the jets does not strike the blades. This prevents over-speeding of the turbine.

Depending on water flow and design, Pelton wheels operate best with heads from 15 to 1800 m, although there is no theoretical limit.

## 12.6.2  Cross-Flow Turbine

A cross-flow turbine as shown in Fig. 12.6 has a drum-shaped runner consisting of two parallel discs connected together near their rims by a series of curved blades. A cross-flow turbine always has horizontal runner shaft 1 (unlike Pelton and Turgo turbines which can have either horizontal or vertical shaft orientation). It uses an elongated rectangular section nozzle directed the against the full length of the runner.

A cross-flow turbine is drum-shaped and uses an elongated, rectangular section nozzle directed against curved vanes on a cylindrically shaped runner. It resembles a "squirrel-cage" blower. The cross-flow turbine allows the water to flow through the blades twice. The first pass is when the water flows from the outside of the blades to the inside; the second pass is from the inside back out. A guide vane at the entrance to the turbine directs the flow to a limited portion of the runner. The cross-flow was developed to accommodate larger water flows and lower heads than the Pelton.

The Banki–Michell turbine is a simple and economic turbine appropriate for micro-hydropower plants. The peak efficiency of this turbine is somewhat less than that of a Kaplan, Francis or Pelton turbine, but its relative efficiency is close to one within a large range, especially above the optimum discharge value. The Banki–Michell

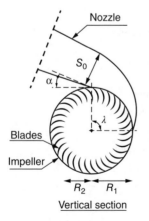

**Figure 12.6**  Cross-flow turbine.

**Figure 12.7** Banki–Michell turbine.

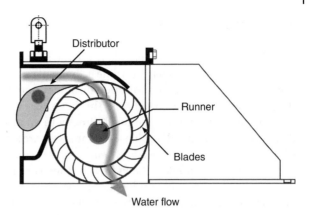

Distributor

Runner

Blades

Water flow

turbine has a drum-shaped runner consisting of two parallel discs connected together near their rims by a series of curved blades as shown in Fig. 12.7. The turbine has a horizontal rotational shaft, unlike Pelton and Turgo turbines, which can have either horizontal or vertical shaft orientation. The water flow enters through the cylinder defined by the two disc circumferences (also called impeller inlet), and it crosses twice the channels confined by each blade couple. After entering the impeller through a channel, the particle leaves it through another one. Going through the impeller twice provides additional efficiency. When the water leaves the runner, it also helps to clean the runner of small debris and pollution. So the cross-flow turbines get cleaned as the water leaves the runner (small sand particles, grass, leaves, etc., get washed away), preventing losses. Other turbine types get clogged easily and consequently face power losses despite higher nominal efficiencies.

### 12.6.3 Turgo Turbine

The Turgo turbine is an impulse machine similar to a Pelton turbine but which was designed to have a higher specific speed. It uses a special nozzle at the end of a pipe to convert the flow of the water into a high-pressure jet. This jet of water is then directed at an angle of about 20° towards the turbine's internal water wheel which uses spoon-shaped blades to capture the jet of water. These blades are specially shaped so that the pressurized water enters the blades on one side and then drops away and exits on the other, converting the kinetic energy of the water jet into rotational shaft power. Therefore, the flow rate is not limited by the discharged fluid interfering with the incoming jet (as is the case with Pelton turbines). As a consequence, a Turgo turbine can have a smaller diameter runner than a Pelton for an equivalent power. With smaller, faster spinning runners, it is more likely to be possible to connect Turgo turbines directly to the generator rather than having to go via a costly speed-increasing transmission.

One of the disadvantages of the Turgo turbine designs is that to generate sufficient nozzle pressure to rotate the turbine runner at high speeds, they require more head height than other designs and have to be connected to a penstock or pipe to channel the water, in order for it to function.

## 12.7    Reaction Turbine

The reaction turbines considered here are the Francis turbine and the propeller turbine. A special case of the propeller turbine is the Kaplan shown in Fig. 12.8. In all these cases, specific speed is high, that is, reaction turbines rotate faster than impulse turbines given the same head and flow conditions. This has a very important consequence that a reaction turbine can often be compiled directly to an alternator without requiring a gear drive system. Combined turbine–generator sets are also available in the market. Significant cost savings are achieved in eliminating the drive, and the maintenance of the hydro unit is very much simpler. The Francis turbine is suitable for medium heads, while the propeller is more suitable for low heads. Micro-hydropower project uses vertical-type Francis turbine system. The turbine mainly consists of water diversion chamber, turbine runner or wheel, water guide vane and draft tubes. With adjustable guide vanes, the water flow is regulated as it enters the runner, and the vanes are usually linked to a governing system which matches the flow to turbine loading. When the flow is reduced, the efficiency of the turbine is reduced. The runner blades are profiled in a complex manner and direct the water so that it exits axially from centre of the runner. In doing so, the water imparts most of its pressure energy to the runner before leaving the turbine via a draft tube. The spiral casing is tapered to distribute water uniformly around the entire perimeter of the runner, and the guide vanes feed the water into the runner at the correct angle.

On the whole, reaction turbines require more sophisticated fabrication than impulse turbines because they involve the use of larger and more intricately profiled blades together with carefully profiled casings. The extra expenses involved are offset by high efficiency and the advantages of high running speeds at low heads from relatively compact machines.

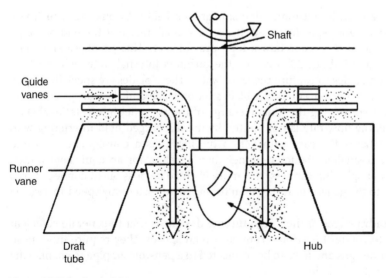

**Figure 12.8**  Kaplan turbine.

Fabrication constraints make these turbines less attractive for use in micro-hydro in developing countries. Nevertheless, because of the importance of low head micro-hydro, work is being undertaken to develop propeller machines which are simpler to construct. Most reaction turbines tend to have poor part-flow efficiency characteristics.

### 12.7.1 The Propeller Turbine

A propeller turbine generally has a runner with three to six blades in which the water contacts all of the blades constantly. It is similar to a boat propeller or aeroplane propeller. The difference between the propeller and Kaplan turbines is that the propeller turbine has fixed runner blades while the Kaplan turbine has adjustable runner blades. The basic propeller turbine consists of a runner, a scroll case, wicket gates and a draft tube. The turbine shaft passes out of the tube at the point where the tube changes direction. The turbine has propeller-like blades but works just reverse. Instead of displacing the water axially using shaft power and creating axial thrust, the axial force of water acts on the blades of the turbine generating shaft power. The propeller has three blades in the case of very low head units, and the water flow is regulated by static blades or swivel gates ("wicket gates") just upstream of the propeller. This kind of propeller turbine is known as a fixed-blade axial-flow turbine because the pitch angle of the rotor blades cannot be changed. The part-flow efficiency of fixed-blade propeller turbines tends to be very poor.

Propeller turbine can be installed in vertical, horizontal or inclined positions.

### 12.7.2 Reverse Pump Turbines

Centrifugal pumps as shown in Fig. 12.9 can be used as turbines by passing water through them in reverse. Reversible machine sets consist of a motor generator and a centrifugal pump turbine that works either as a pump or as a turbine depending on the direction of rotation. The advantages of using these pumps as turbines are low cost due to their mass production and availability of spare parts. Furthermore, a well-designed, compact powerhouse saves equipment and civil costs. With a wide range of specific speeds, pump turbines can be installed at sites with heads from less than 50 m to more than 800 m and with varying unit capacities.

**Figure 12.9** Reverse pump turbine.

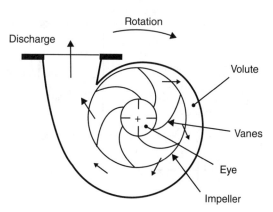

The main problem in using these pumps is that manufacturers do not normally provide characteristic curves of their pumps working as turbines. This makes it difficult to select an appropriate pump to run as a turbine for a specific operating condition.

## 12.8   Generators for Small Hydro Plants

Induction generators or synchronous generators can be used for small hydropower generation [8]. Induction generators are commonly used for small hydro schemes due to advantages such as availability, low cost and robustness. The cost per kilowatt of a single-phase generator is generally higher than a three-phase generator. Hence, a three-phase generator, which produces a single-phase output, is normally used. A three-phase generator can be converted into a single-phase generator, which produces approximately 80% of the machine rating, by connecting two capacitors as shown in Fig. 12.10. In order to further minimize the capital cost, very simple method of controlling voltage and frequency control is used. A capacitor-excited induction generator used with a hydraulic turbine in a stand-alone generating system can provide high-quality voltage and frequency control not matched by other small generating units. This is achieved without a turbine governor by using a controllable additional impedance on the load side. The control is achieved using a static power converter. In these schemes, the generator operates under manual control of the sluice gate, and if the consumer load changes, then the generated voltage and the frequency also vary. If the load is light, the generator speed can increase, leading to runaway condition. The control technique used to maintain the generated voltage and the frequency at its rated value is to maintain the total load connected to near-constant value by connecting a resistive ballast, which maintains the sum of the consumer load and the ballast load at a constant value.

*By using self-excited induction generators rather than synchronous generators, cost savings and reliability improvements can be achieved, due to the simple construction and inherent robustness of cage induction machines. However, until recently, the extra cost and complexity of the voltage and frequency control equipment have more than offset the advantages of using stand-alone induction generators.*

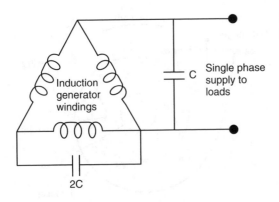

**Figure 12.10** Three-phase generator run as single-phase generator.

## 12.9    Design Considerations of Micro-Hydropower Plants

In the design of a micro-hydropower system, the following parameters are considered. The selection of turbine type, size and speed is based on the net head and maximum power flow rate. Most small hydropower sites are categorized as low- or high-head. The higher head is better because less water is needed to produce a given amount of power and smaller and less expensive equipment can be used. Low head is classified when there is change in elevation of less than 3 m. A vertical drop of less than 0.6 m will not be suitable to make a small-scale hydroelectric system. However, for extremely small power generation amounts, a flowing stream with as little as 35 cm of water can support a submersible turbine, as the type used originally to power scientific instruments towed behind oil exploration ships. Small-scale hydropower systems, as well as **mini-hydro systems** or **micro-hydro systems**, can be designed using either water wheels or impulse-type hydro turbines. Once the turbine power, specific speed and net head are known, the turbine type, the turbine fundamental dimensions and the height or elevation above the tail race water surface that the turbine should be installed to avoid cavitation phenomenon can be calculated. In case of Kaplan or Francis turbine type, the head loss due to cavitation, the net head and the turbine power must be recalculated. In general, the Pelton turbines cover the high-pressure domain down to (50 m) for micro-hydro. The Francis type of turbines cover the largest range of head below the Pelton turbine domain with some overlapping and down to (10 m) head for micro-hydro. The lowest domain of head below (10 m) is covered by Kaplan type of turbine with fixed or movable blades. For low heads and up to (50 m), the cross-flow impulse turbine can also be used. Once the turbine type is known, the fundamental dimensions of the turbine can be easily estimated.

Once the turbine is selected, the mean flow duration curve for the river is obtained to determine power potential of stream. In case of low discharge rivers (less than 4 m³/s), it may be possible to build a weir. It is a low wall or dam across the stream to be gauged with a notch through which all the water may be channelled. A simple linear measurement of the difference in level between the upstream water surface and the bottom of the notch is sufficient to quantify the flow rate (discharge). In order to prevent the trash entering into the entrance flume, bars at certain spacing (called trash rack) are placed in a slanting position (at an angle of 60°–80° with horizontal). The maximum possible spacing between the bars depends on the type of turbine used. Typical values are 20–30 mm for Pelton turbines 40–50 mm for Francis turbines and 80–100 mm for Kaplan turbines. The water carried by the power channel is distributed to various penstocks leading to the turbines through the forebay. Water is temporarily stored in the forebay in the event of a rejection of load by the turbine, and it can be withdrawn when the load is increased. In addition, the forebay acts as a sort of regulating reservoir.

Pipes are used for conveying water from the intake to the power house. They can be installed over or under the ground, depending on factors such as the nature of the ground, the penstock materials, the ambient temperature and the environmental requirements.

When the turbine and generator operate at the same speed and can be placed so that their shafts are in line, direct coupling is the right solution. In this case there, will be no power loss incurred and maintenance is minimal. Turbine manufacturers will recommend either rigid or flexible type of coupling although a flexible coupling that can tolerate certain misalignment is usually preferred. In the lowest power range, turbines can run at less than (400) rpm and speed is increased through gears to (1500) rpm of standard alternator. In the range of powers produced in small and micro-hydro schemes, this solution is always more economical than the use of a direct-coupled alternator. The rotational speed of a turbine is a function of its power and net head. In the small and micro-hydro schemes, turbine selection is made keeping in mind whether it will be coupled directly or through a gearbox to reach the synchronous speed of generator.

Standard generators should be installed when possible, so in each, the runner profile is characterized by a maximum runaway speed.

The runaway speed of a hydraulic turbine is the speed at which the turbine coupled to the generator runs at the maximum possible speed due to loss of load. The runaway speed of a hydraulic turbine is the speed at which the turbine coupled to the generator runs at the maximum possible speed due to loss of load. The runaway speed of the turbine is determined by the turbine designer and is influenced by the maximum discharge of water from the penstock, the combined inertia of the turbine runner and the generator and the flywheel. Depending on the type of turbine, it can attain $2 \rightarrow 3$ times the nominal speed. The cost of generator and gearbox may be increased when the runway speed is higher, since they must be designed to withstand it.

A governor is a combination of devices and mechanisms, which detects the speed deviation and converts it into a change in servomotor position. Several types of governors are available. The purely mechanical governor is used with fairly small turbines. In modern electric–hydraulic governor, a sensor is located on the generator shaft to sense the turbine speed. The turbine speed is compared with reference speed. The error signal is amplified and sent to the servomotor to act in the required sense. To ensure the control of the turbine speed by regulating the water flow, certain inertia of rotating components is required. Additional inertia can be provided by a flywheel, on the turbine, or generator shaft. The flywheel effect of the rotating components is stabilizing, whereas the water column effect is destabilizing. For MHP schemes at remote locations (not connected to the grid), the parameter that needs to be controlled is the turbine speed, which controls the frequency. In an off-grid system, if the generator becomes overloaded, the turbine slows down. Therefore, an increase in the flow of water is needed to ensure that the turbine does not stall. If there is not enough water to do this, then either some of the load must be removed or the turbine will have to be shut down. On the other hand, if the load decreases, then the flow to the turbine is decreased or it can be kept constant, and the extra energy can be diverted into a ballast (dummy) load connected to the generator terminals.

In a run-of-river hydro scheme, the flow of the water is not altered, so its minimum flow rate needs to be the same or higher than that of the proposed turbine output power, ensuring maximum efficiency. The result is that the costs involved for a run-of-river scheme are much lower and have less environmental impact than other small-scale hydro plants. The disadvantage is that the water flow rate is variable throughout the year, and the system is unable to store the water's energy.

The development of a small-scale hydropower electrical scheme which uses a small dam or weir, water storage reservoir (impoundment) or requires a diversion of the rivers, water flow through tunnels or canals requires far more water usage in total as well as more complex civil and ground engineering works to match the site elevation, not to mention the environmental impact that is proportional to the size of the scheme.

### 12.9.1 Example

A small stream drops 20 m down the side of a mountain producing a water flow rate of 500 l/min past a fixed point. How much power could a small-scale HPP generate in kilowatts, if the type of water turbine used has a maximum efficiency ($\eta$) of 85%.

The data given: Head $= 20$ m, flow rate $= 500$ l/min, efficiency $= 0.85$ and gravity $= 9.81$ m/s$^2$. But first we must convert the water flow rate of 500 l/min into m$^3$/s.

1000 l is equal to 1 m$^3$, so 500 l is equal to 0.5 m$^3$. One minute is equal to 60 s, then a flow rate of 0.5 m$^3$/min is equal to 0.00833 m$^3$/s.

$$P = \eta\, g\, Q\, H$$
$$= 0.85 \times 9.81 \times 0.0083 \times 20 = 1.3842\,\text{kW}$$

# References

**1** REN 21, Renewable energy policy network for the 21$^{st}$ century' c/o UNEP 15, Rue de Milan F-75441 Paris CEDEX 09, France

**2** Curtis, D., Langley, W. and Ramsey, R. (1999) *Going with the Flow: Small Scale Water Power Made Easy*, Maya Books, Twickenham, UK, p. 160p.

**3** Fetter, C.W. (2001) *Applied Hydrogeology*, 4th edn, Prentice Hall, New Jersey, p. 598p.

**4** Oliver, P. (2002) Small hydro power: Technology and current status. *Renewable and Sustainable Energy Reviews*, **6** (6), 537–556.

**5** Bokalders, V., Harvey, A., Brown, A. and Edwards, R. (1991) *Micro-hydro power: A guide for development workers*, IT Publications Ltd, London.

**6** Harvey, A. *et al.* (1998) *Micro-hydro design manual guide to small-scale water schemes*, Intermediate Technology, London, p. 374p.

**7** Nasir, B.A. (2014) Design considerations of micro-hydro-electric Power Plant. *Energy Procedia*, **50**, 19–29.

**8** Smith, N.P.A. (1996). Induction generators for stand-alone micro-hydro systems. Proceedings of the 1996 International Conference on Power Electronics, Drives & Energy Systems for Industrial Growth, PEDES'96. New Delhi, India. Part 2. IEEE, Piscataway, NJ.

# 13

# Control of Grid-Connected Photovoltaic and Wind Energy Systems

## 13.1 Introduction

In many countries, the renewable generation from wind and solar has increased substantially during the past few years and forms a significant proportion of the total generation in the grid. At present, this renewable generation is concentrated in few areas, which is quite significant. It therefore requires serious consideration to balance the variability of such generation. With further increase in renewable generation, it is quite clear that some changes may be necessary in the operation of grid for large-scale integration of such variable renewable energy sources (RESs), although the demand for electricity, or load, also varies with time, but because it is slow or small, the power system is designed to handle that uncertainty. The short-term changes in load (over seconds or minutes) are generally small and caused by random events that change the demand in different directions. Over longer periods (several hours), changes in load tend to be more predictable. For example, there is a daily pattern of morning and evening load pattern highly correlated with human endeavours. The main difference is that load variations are more predictable than wind and solar variations. In order to maintain a constant frequency, the generation and load must be matched at every instant. The problem arises when there is mismatch between power supply and demand.

But all the generation from RES is not unpredictable. For example, the electricity production of an individual wind turbine is highly variable, but the aggregate variation of multiple turbines at a single site is less and can be easily monitored. The aggregation of multiple wind generation sites over a large geographic area results in even less variations. Having large number of turbines in many locations results in less variability and enhanced prediction. Variability of generation is also function of time. The variability of large-scale wind power over seconds or minutes is generally small. However, for several hours, it can be large.

Similarly, some aspects of variation in solar energy availability are also predictable (e.g. sunrise and sunset). Other aspects, such as intermittent cloud cover, cannot be predicted. However, the same reduction in variability is observed for the aggregation of solar photovoltaic (PV) plants over a broad geographic area [1, 2]. In this chapter, only the PV system and wind energy sources connected to grid and their control are discussed.

*Operation and Control of Renewable Energy Systems*, First Edition. Mukhtar Ahmad.
© 2018 John Wiley & Sons Ltd. Published 2018 by John Wiley & Sons Ltd.

## 13.2  Operation and Control of Grid-Connected PV System

In general, PV cells can be connected to the grid (grid connection application), or they can be used as isolated power supplies. In isolated applications, PV panels supply local loads which can be residential or commercial. An electrical power generating system that uses a PV array as the primary source of electricity generation and is intended to operate synchronously and in parallel with the electric utility network is known as a grid-connected PV system. About 75% of the total PV systems installed in the world are connected to grid. All PV systems are connected to the utility grid through a voltage-source inverter (VSI) and a boost converter.

Each grid-connected PV system has to perform two functions:

- Extract maximum power output from the PV array(s)
- Supply harmonic-free ac current into the grid

There are numerous ways of injecting synchronized power from PV modules into the utility grid. In each of these approaches, the MPPT and inverters are implemented with different techniques.

Such systems may also include battery storage, other generating sources and may operate on site loads independent of the utility network during outages (isolated mode). Two types of grid-connected PV systems are considered here [3]. These are

1) single-phase grid-connected PV systems and
2) three-phase grid-connected PV systems

### 13.2.1  Control of Single-Phase PV System

For residential applications, single-phase systems are commonly used. A new research trend in the residential generation system is to employ the PV parallel-connected configuration rather than the series-connected configuration to satisfy the safety requirements and to make full use of the PV generated power. For residential applications, typically, the power ratings range from 3 to 10 kW, and thus, a dc/dc converter is necessary to boost the dc voltage level within an allowable range for the PV inverter.

Figure 13.1 shows the block diagram of a PV system, and Fig. 13.2 shows the configuration of the grid-connected PV system for single-phase inverter connected by a dc/dc Boost converter and cascaded with a dc/ac inverter. The dc/dc boost converter performs MPPT for maximizing the output power of PV array to fully utilize the PV power and voltage boost to match that of grid inverter. The boost converter controls the PV array operating point by the switch duty command. The single-phase inverter controls the

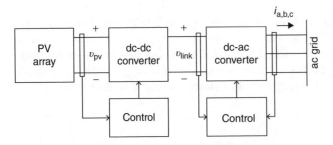

**Figure 13.1** Block diagram of grid-connected PV system.

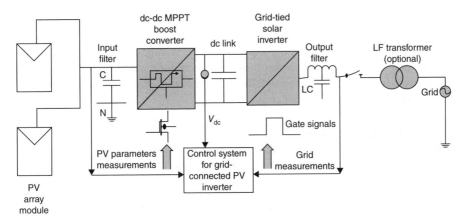

**Figure 13.2** Grid-connected PV system.

dc-link capacitor voltage and controls the output current to be in phase with the grid voltage. Then a single-phase buck dc/ac inverter is used to step down and to modulate the output voltage according to the grid voltage.

Inverter-interfacing PV module(s) with the grid involves two major tasks. One is to ensure that the PV module(s) is operated at the maximum power point (MPP). The other is to inject a sinusoidal current into the grid. Since the inverter is connected to the grid, the standards given by the utility companies must be obeyed. During grid-connected mode, grid controls the amplitude and frequency of the PV inverter output voltage, and the inverter operates in a current-controlled mode. The current controller for grid-connected mode fulfils two requirements – (i) during light load condition, the excess energy generated from the PV inverter is fed to the grid; and (ii) during an overload condition or in case of unfavourable atmospheric conditions, the load demand is met by both PV inverter and the grid. On the other hand, during isolated grid operation, the PV inverter operates in voltage-controlled mode to maintain constant amplitude and frequency of the voltage across the load.

### 13.2.1.1 Control of PV-Side dc/dc Converter

The objective of PV-side controller is to extract the maximum power from the input source considering ambient temperature and solar irradiance. In general, the protection of the dc/dc converter (boost converter) should also be taken into consideration in this controller. The dc-dc boost converter stage could step up the input dc voltage to a required voltage value, which is the next input of the inverter. This stage also provides voltage regulation control on the intermediate dc bus that is not available at the source terminals, which can regulate the dc voltage into the required value. Perturb and Observe method is applied for maximum power tracking. This algorithm is mostly used, due to its ease of implementation. It is based on the following criterion: if the operating voltage of the PV array is perturbed in a given direction and if the power drawn from the PV array increases, this means that the operating point has moved towards the MPP, and therefore, the operating voltage must be further perturbed in the same direction. Otherwise, if the power drawn from the PV array decreases, the operating point has moved away from the MPP, and therefore, the direction of the operating voltage perturbation must be reversed.

The dc/dc controller comprises a three-level, nested-loop structure. The MPP controller forms the outer loop and generates the voltage set point for the voltage controller. The voltage loop is used to control the input capacitor voltage of the boost converter. The voltage loop generates a current set point for the inner current loop. This current regulator is used to regulate the inductor current. The output of the current loop is used to derive the duty cycle for the boost converter.

### 13.2.1.2  Control of Grid-Side Inverter

The output of the boost converter is connected to the dc side of a single-phase VSI via a dc-link capacitor. The grid converter performs the following tasks:

- Control of active power generated to the grid
- Control of reactive power transfer between PVF system and the grid
- Control of dc-link voltage
- Ensure high quality of the injected power
- Grid synchronization

The VSI is regulated by a nested control scheme with an outer voltage loop and an inner current loop. Usually, there is a fast internal current loop, which regulates the grid current, and an external voltage loop, which controls the dc-link voltage and maintains the voltage at the desired level. In synchronous reference frame control, also called $dq$ control, reference frame transformation module, for example, $abc \rightarrow dq$ is used. It transforms grid current and voltage waveforms into a reference frame that rotates synchronously with the grid voltage. The synchronous frame regulator reduces the complexity by transforming the ac system to an equivalent dc system and then implementing proportional integral (PI) controllers for the $d$- and $q$-axis, respectively. In single-phase systems, unlike three-phase systems, not enough information is available to directly convert the ac current into the synchronous frame. In order to create the additional orthogonal-phase information from the single-phase inverter signal, two approaches can be adopted in order to create the additional phase-shifted state variables. These are as follows:

- Differentiating the inverter output voltage and inductor current. This is done by differentiating the measured real component to build another stationary orthogonal component. However, this method is very sensitive to noise. The inverter real-phase output waveform contains harmonics in addition to the fundamental component. Differentiation process thus does not yield a purely orthogonal component.
- Fictive axis emulation: To concurrently generate the required orthogonal current, fictive model of the system, which is the transfer function of the real system in a SRF, hereinafter called the fictive axis emulation can be used. The detailed structural diagram of the FAE is represented in Fig. 13.3. The grid current is assumed to be in line with the α-axis, and the β-axis currents are emulated using the FAE. The measured α and derived β currents are transformed into the $dq$ axis and fed into the synchronous frame regulator. The angle used for the transformation is generated by a phase-locked loop (PLL). The synchronous frame regulator calculates a desired αβ voltage vector set point. This vector is scaled by the peak grid voltage amplitude to generate the modulation index for the modulator. The modulation indices are used to generate the switching signals for the VSI using a unipolar modulator.

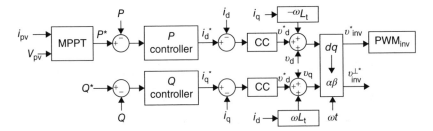

**Figure 13.3** *dq* control.

### 13.2.1.3 Inner Current Loop

The inverter is required to inject a sinusoidal grid current with low THD and in phase with the grid voltage. Thus, the output of the dc link voltage controller, which represents the reference grid current amplitude, is multiplied by a sinusoidal template obtained using a PLL synchronized with the utility voltage. The current controller attempts to match the grid current with this reference sinusoidal current. The most common types of current controllers are the PI with feed-forward controller and the proportional resonant (PR) controller. However, PR controllers have the ability to remove both the current's magnitude and phase steady-state errors without the need of voltage feed-forward unlike the conventional PI controller. Further, applying a Park transformation ($\alpha\beta \rightarrow dq$) leads to the possibility of PI controllers to regulate the injected current, and afterwards, the modulation reference can be obtained by means of the inverse Park transformation ($dq \rightarrow \alpha\beta$). Since the current control loop is responsible for the power quality, it will also be effective and valid in the design of current controllers and the LCL filter. By introducing harmonic compensators (HCs) for the controller and adding passive damping for the filter, an enhancement of the current controller tracking performance can be achieved and the background distortion influence is alleviated.

## 13.3 Grid Synchronization

The main aspect of connection to the grid by power converter is that the magnitude and phase angle of fundamental frequency component of injected current should synchronize with the grid voltage. Different methods to extract the phase angle have been developed based on grid requirements, and comparative studies of some of them have been carried out. A brief description of the main methods is as follows:

- *Zero-Crossing Method*, Simplest synchronization solution is the filtered zero-crossing detection (ZCD) method where the zero-crossing point is detected every one period and the sin-table pointer is reset from zero scratch. However, due to the poor performances mainly when grid voltages register variations such as harmonics, it is not the optimal synchronization method.

PLL technique has been the state-of-the-art method to detect the phase angle of the grid voltage. This algorithm has a better rejection of grid harmonics, notches and any other kind of disturbances. However, for single-phase applications, creation of the orthogonal signal generator system is required. The block diagram of PLL is shown

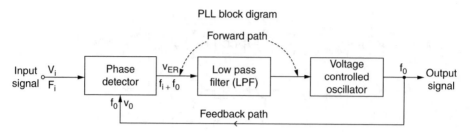

**Figure 13.4** Phase-locked loop.

in Fig. 13.4. In this technique, for the synchronization of time-varying signal, the difference between phase angle of the input and that of the output signal is measured by phase detection. It is then passed through the loop filter, the output of which drives the voltage-controlled oscillator. The VCO generates the output which follows the input signal.

For single-phase system, PLL based on $T/4$ delay can be employed for synchronization. When the fundamental grid frequency time period is $T$, using first-in-first-out (FIFO) buffer, the $T/4$ transport delay technique can be implemented, by setting its size to one-fourth of the number of samples contained in one cycle of the fundamental frequency. In this method, the orthogonal signal output will be perfect only if the grid frequency does not change; otherwise, this will result in errors in synchronization. In additon, it does not have filtering capability to remove the harmonic components in the input signal.

Robust Single-Phase PLL: By using robust PLL technique, the phase, frequency and amplitude information of single-phase signals can be estimated instantly even in frequency variation and harmonic distortion. It consists of two-phase signal generator which produces $v_\alpha$ and $v_\beta$, a vector rotator, a phase synchronizer, a lowpass filter and a multirate sample holder. This method has the following attractive features: (i) the estimation system results in a nonlinear system, but it can be stabilized; (ii) all sub-systems constructing the system can be easily designed; (iii) "two-phase signal generator" can auto-tune a single system parameter in response to varying frequency of the injected signal; (vi) high-order "PLL Controllers" allowing fast tracking can be stably used; and (v) even in hostile envelopments, favourable instant estimates can be obtained.

## 13.4    Control of Three-Phase Grid-Connected PV system

Three-phase grid-connected utility-scale systems can have tens of megawatts of power output under optimum conditions of solar irradiation. These systems are usually ground-mounted and span a large area for power harvesting.

Figure 13.5 shows the block diagram of a three-phase PV system with power feed-in functions. The output of PV array is connected to a boost dc converter that is used to increase the array terminal voltage to a higher value so it can be interfaced to the distribution system grid at 6.6 kV. The dc converter controller also performs the MPPT function. A dc-link capacitor is used after the dc converter and acts as a temporary power storage device to provide the VSI with a steady flow of power. The capacitor's

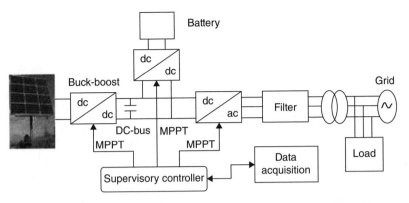

**Figure 13.5** Block diagram of three-phase PV system connected to grid.

voltage is regulated using a dc-link controller that balances the input and output powers of the capacitor. The VSI is controlled in the rotating $dq$ frame to inject a controllable three-phase ac similarly to single-phase system [4]. A PLL is used to lock on the grid frequency and provide a stable reference synchronization signal for the inverter control system, which works to minimize the error between the actual injected current and the reference current obtained from the dc-link controller. An LC low-pass filter is connected at the output of the inverter to attenuate high-frequency harmonics and prevent them from propagating into the power system grid. A second-order LCL filter is obtained if the leakage inductance of the interfacing transformer is referred to the low voltage side. This provides a smooth output current with low harmonics.

## 13.5 Selection of Inverter for PV System

Inverters currently available are typically rated for the following:

- Maximum dc input power, that is, the size of the array in peak watts
- Maximum dc input current
- Maximum specified output power, that is, the ac power output supplied to grid

The inverters must also be able to detect an islanding situation and take appropriate measures in order to protect persons and equipment. Islanding is the continued operation of the inverter when the grid has been removed on purpose, by accident or by a fault. When grid supply is not available, the PV system has to supply the local load.

The inverter may simply fix the voltage at which the array operates or (more commonly) use a MPP tracking function to identify the best operating voltage for the array [4, 5]. The inverter operates in phase with the grid (unity power factor) and generally delivers as much power as it can generate at a particular time to the electric power grid.

### 13.5.1 Central Inverters

The earlier PV systems were based on technology, which used centralized inverters that interfaced a large number of PV modules to the grid as shown in Fig. 13.6. The PV modules were divided into series connections (called a string), so that sufficient voltage can

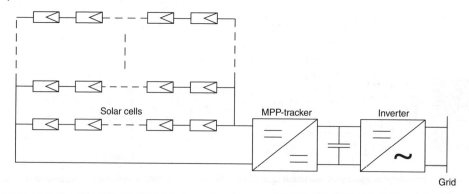

**Figure 13.6** Centralized inverter.

be generated to avoid further amplification. These series connections were then connected in parallel, through string diodes, in order to reach high power levels. Typically, centralized inverter handles all tasks by itself such as MPPT (usually by the microcontroller), grid current control and voltage amplification also if required. The inverter must be designed to handle a peak power of twice the nominal power.

These inverters are characterized by high efficiency and lower cost. However, they have some severe limitations, such as high-voltage dc cables between the PV modules and the inverter, power losses due to a centralized MPPT, power losses due to mismatch in the PV modules, losses in the string diodes. In addition, the reliability of the plant is limited due to the dependence of power generation on a single inverter. Centralized configuration is mainly used in PV plants that have nominal power higher than 10 kW. Thus, the central inverters are mainly three-phase inverters.

### 13.5.2 String Inverter

For small domestic applications in the range of 0.5–1 kW, string inverters are most suitable. The string inverter, shown in Fig. 13.7, is a reduced version of the centralized inverter, where a single string of PV modules is connected to the inverter called *string inverter*. They are based on a modular concept, where PV strings, made up of series-connected solar panels, are connected to separate inverters. The string inverters are paralleled and connected to the grid. The input voltage may be high enough to avoid voltage amplification. This requires roughly 15–16 PV modules in series with

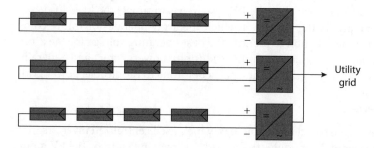

**Figure 13.7** String inverter.

one inverter per string. The total open-circuit voltage for 16 PV modules may reach as high as 720 V. Fewer PV panels can also be used, but then a dc-dc converter or a line-frequency transformer is needed for a boosting stage. The advantages compared to the central inverter are as follows:

- There are no losses associated with the string diodes as no string diode is necessary.
- Separate MPPT can be used for each string.

### 13.5.3  ac Module Inverter

In a module inverter, one inverter is used for each module. An ac module is made up of a single solar panel connected to the grid through its own inverter, as shown in Fig. 13.8. The advantage of this configuration is that there are no mismatch losses, due to the fact that every single solar panel has its own inverter and MPPT, thus maximizing the power production. The power extraction is much better optimized than in the case of string inverters. One other advantage is the modular structure, which simplifies the modification of the whole system because of its "plug and play" characteristic. One disadvantage is the low overall efficiency due to the high-voltage amplification, and the price per watt is still higher than in the previous cases. But this can in the future maybe be overcome by mass production.

**Figure 13.8** Module inverter.

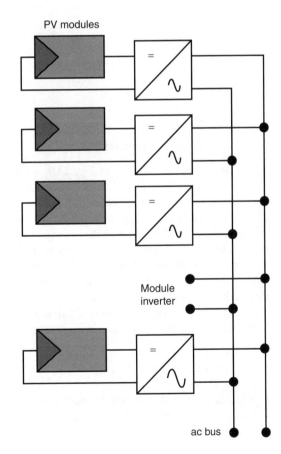

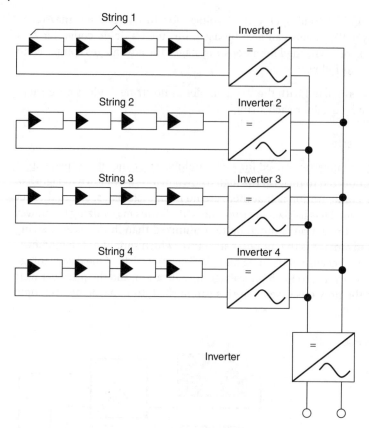

**Figure 13.9** Multi-string inverter.

### 13.5.4 Multi-String Inverters

Multi-string inverter shown in Fig. 13.9 is an intermediate solution between string inverters and module inverters. A multi-string inverter combines the advantages of both string and module inverters, by having many dc-dc converters with individual MPPTs, which feed energy into a common dc-ac inverter. Here each PV module or string is connected to a dedicated dc-dc converter that is connected to a common dc-ac inverter. In this case, the task for each dc/dc converter is MPPT and, normally, the increase of the dc voltage. The dc/dc converters are connected to the dc link of a common dc/ac inverter, which takes care of the grid-current control. This is beneficial since a better control of each PV module/string is achieved, and that common dc/ac inverter may be based on a standard variable-speed drive technology.

The main features of the multi-string technology are as follows:

- Optimum energy yield
- Optimum monitoring of strings
- Low specific costs of PV converter
- Minimum costs of PV system installation
- Nominal power of converter unit not limited
- Modular extendibility

## 13.6  Power Decoupling

The power fluctuations between dc and ac grid for single-phase inverters has to be decoupled by energy storage. Power decoupling is normally achieved by means of an electrolytic capacitor forming a dc link. Because to minimize the ac side of the inverter switching frequency, dc-link capacitor is essential. A very simple solution to mitigate its negative impact is to use bulky electrolytic capacitors in the dc link so that they can act as buffers to the ac-side ripple power. However, those electrolytic capacitors are known to have high equivalent series resistance (ESR) and low ripple current capability, and their lifetime is also relatively short (several thousand hours) when stressed with the nominal voltage and the ripple current. Thus, it should be kept as small as possible and preferably substituted with film capacitors. The capacitor is placed either in parallel with the PV modules or in the dc link between the inverter stages in multi-link inverter.

Recently, some active power decoupling methods have been proposed to cope with this problem, and the fundamental principle behind them is to introduce an extra active circuit in the system, so that the ripple power can be shifted away from the dc link and stored by other components with expanded lifetime, for example, inductors and film capacitors, in a more efficient and effective way.

## 13.7  Isolation Between Input and Output

Conventionally, a classification of PV topologies is divided into two major categories: PV inverters with dc/dc converter (with or without isolation) and PV inverters without dc/dc converter (with or without isolation). Galvanic isolation can be either on the dc side in the form of a high-frequency dc-dc transformer or on the grid side in the form of a big bulky ac transformer. Both of these solutions offer the safety and advantage of galvanic isolation, but the efficiency of the whole system is decreased due to power losses in these extra components. The most important advantages of transformerless PV systems can be observed in higher efficiency and smaller size and weight compared to a system with transformers. However, the resulting galvanic connection between the grid and PV array introduces ground leakage current path due to the effect of solar panel parasitic capacitance, for example, 10–100 nFd to the PV systems that have galvanic isolation (on either the dc side or the ac side).

One disadvantage of transformerless systems is that the missing line-frequency transformer can lead to dc currents in the injected ac current by the inverter, which can saturate the core of the magnetic components in the distribution transformer, which may lead to overheating.

## 13.8  Transformers and Interconnections

Grid-connected PV systems, particularly low-power single-phase systems (up to 5 kW), are becoming more popular worldwide. Issues such as reliability, high efficiency, small size and weight and low price are of great importance to the conversion stage of the PV system. Quite often, these grid-connected PV systems include a line transformer in

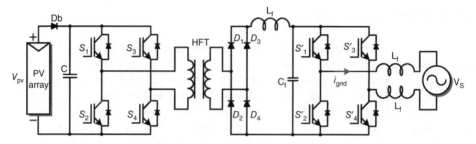

**Figure 13.10** PV system with high-frequency transformer.

the power-conversion stage, which guarantees galvanic isolation between the grid and the PV system, thus providing personal protection. Furthermore, it strongly reduces the leakage currents between the PV system and the ground. However, now it is possible to implement both ground-fault detection systems and solutions to avoid injecting dc current into the grid within the inverters.

The transformer can then be eliminated without impacting system characteristics related to personal safety and grid integration [1, 5–8]. In addition, the use of a string of PV modules allows MPP voltages large enough to avoid boosting voltages in the conversion stage. This conversion stage can then consist of a simple buck inverter, with no need of a transformer or boost dc-dc converter, and it is simpler and more efficient. But if no boost dc-dc converter is used, the power fluctuation causes a voltage ripple in the PV side at double the line frequency. This in turn causes a small reduction in the average power generated by the PV arrays due to the variations around the MPP.

The efficiency drop is 2% more if line-frequency transformer is used instead of high-frequency transformer. Hence, when grid isolation is mandatory, the incorporation of a high-frequency transformer is preferred. This implies the need for a dc-dc converter in the structure of the PV power system. A grid-connected inverter with high-frequency link of a PV system is shown in Fig. 13.10. The high-frequency inverter converts dc to 20 kHz square-wave ac. The high-frequency transformer has appropriate turn ratio to change the voltage to the desired level. This voltage is then rectified by diode bridge converter. The rectified dc is then converted to power frequency by an inverter, and the output voltage is connected to the grid. The isolated full-bridge dc-dc converter is usually used at power levels above 750 W, to perform both the MPPT and the galvanic isolation. Commonly, its efficiency ranges from 92% to 93% under a 45% to 100% load condition. However, it is not suitable for large-power industrial application.

### 13.8.1 Transformerless PV Inverter Topologies

Since a transformer imposes an efficiency drop and as *high efficiency* is one of the most important characteristics of a PV inverter; thus, whenever possible, these inverters are non-isolated electronic circuits. Hence, many trasformerless systems have been designed as discussed later [7]. Transformerless PV inverters use different solutions to minimize the leakage ground current and improve the efficiency of the whole system.

When no transformer is used in a grid-connected PV system, a galvanic connection between the grid and the PV array exists. As a consequence, a common-mode resonant

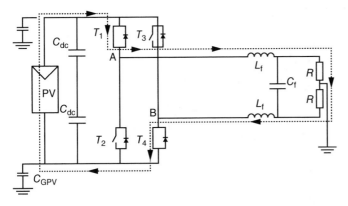

**Figure 13.11** H-bridge inverter.

circuit appears, consisting of the stray capacitance between the PV modules and the ground, the dc and ac filter elements and the grid impedance. A varying common-mode voltage can excite this resonant circuit and generate a common-mode current. In order to avoid these leakage currents, different inverter topologies that generate no varying common-mode voltages, such as the half-bridge and the bipolar pulse-width modulation (PWM) full-bridge topologies, have been proposed. One of them is to connect the midpoint of the dc-link capacitors to the neutral of the grid, such as the half-bridge, neutral-point clamped (NPC) or three-phase full bridge with a split capacitor topology, thereby continuously clamping the PV array to the neutral connector of the utility grid. Half-bridge and NPC type of converters have very high efficiency, above 97%, as shown in [7]. The H-bridge is a well-known topology, and it is made up of two half-bridges as shown in Fig. 13.11. Most single-phase HB inverters use unipolar switching in order to improve the injected current quality of the inverter, which is done by modulating the output voltage to have three levels with twice the switching frequency. Moreover, this type of modulation reduces the stress on the output filter and decreases the losses in the inverter. A disadvantage of half-bridge and NPC type of converters is that, for single-phase grid connection, they need a 700-V dc link.

The bipolar PWM full-bridge requires a lower input voltage but exhibits a low efficiency. In the bipolar PWM, the diagonal pairs of switches $S_1$ and $S_4$ and $S_2$ and $S_3$ (Fig. 13.12) are controlled in a complementary manner. Therefore, two output voltage levels $-V_{dc}$ and $+V_{dc}$ can be obtained. The common-mode voltage is nearly constant because zero-voltage state does not exist. The common-mode voltage varies only with low-frequency grid voltage. The RMS value of the leakage current is very low. Unfortunately, because of bipolar switching, this scheme imposes high voltage stress and high inductor current ripple, which is associated with additional losses. The high power losses are due to two factors. The first is the internal reactive power flow inside the inverter. The second is the double switching frequency required to obtain the same inductor current ripple frequency. It is therefore desired to develop new topologies to combine the advantages of the bipolar PWM (low leakage current level) and those of the unipolar PWM (high efficiency, low current ripple and three-level inverter output voltages). This can be done by adding extra switches to the full-bridge topology. These extra switches would disconnect the PV array from the grid during freewheeling period.

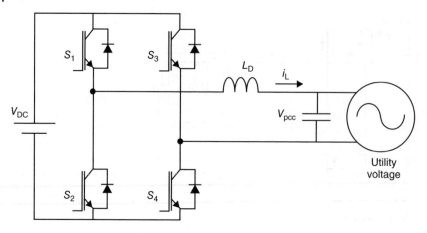

**Figure 13.12** Bipolar inverter.

The most widely and simplest topology used in three-phase systems is the full-bridge inverter, which consists of three legs, each leg with two switches.

## 13.9 Filters for Grid-Connected PV Inverters

In a single-phase current-source inverter (CSI), the pulsating instantaneous power of twice the system frequency generates even harmonics in the dc-link current. These harmonics reflect onto the ac side as low-order harmonics in the current and voltage. Undesirably, these even harmonics affect MPPT in PV system applications and reduce the PV lifetime. In order to mitigate the impact of these dc-side harmonics on the ac side and on the PV, the dc-link inductance must be large enough to suppress the dc-link current ripple produced by these harmonics.. Practically, large dc-link inductance is not acceptable, because of its cost, size, weight and the fact that it slows MPPT transient response. To reduce the necessary dc-link inductance, a parallel resonant circuit tuned to the second-order harmonic is employed in series with the dc-link inductor. The filter is capable of smoothing the dc-link current by using relatively small inductances. Even though the impact of the second-order harmonic is significant in the dc-link current, the fourth-order harmonic can also affect the dc-link current, especially when the CSI operates at high modulation indices. Therefore to improve the parallel resonant circuit, a double tuned resonant circuit tuned to double and fourth order frequency can be used.

## 13.10 Islanding Detection Methods

Islanding condition of grid-connected PV inverters refers to the condition when "a portion of the utility system that contains both load and distributed resources" remains energized while it is isolated from the remainder of the utility system. In case of low-power PV systems, it must be stopped within 2 s of the formation of the unintentional island in order to avoid possible damages to local electrical loads or the

PV inverter during the grid reconnection. Moreover, this condition can be mandatory in order to guarantee the safety of workers during maintenance. Therefore, it is desirable to incorporate detection functions into PV inverters for protection.

The islanding detection methods can be broadly classified into active, passive and communication-based techniques. Passive methods were proposed in the initial phases when the interconnection of PV system was made to the grid. Then as the technology advanced, there has been increase in the number of active methods. Active methods have been proposed to overcome the shortcomings of passive methods. Communication-based methods are reliable but not economical as compared to passive and/or active methods. A reliability measure of a good islanding detection scheme is non-detection zone (NDZ). Good islanding detection method should have negligible NDZ.

In passive methods, under/over-voltage and under/over-frequency relays are placed on the distribution feeders for the detection of various types of abnormal conditions. Now for islanding condition, these relays must cease the operation of the PV system when it is isolated from utility. The behaviour of the system at the time of utility disconnection depends on the change in active and reactive power at the instant before the island is formed. It relies on the detection of an abnormality in the voltage at the point of common coupling (PCC) between the PV inverter and the utility. The parameters vary greatly at the point of the PCC when the system is islanded. The difference between a normal grid-connected condition and an islanding condition is based on the threshold setting of the system parameter.

Active methods actively attempt to create an abnormal PCC voltage when the utility is disconnected, but the action is taken on the utility side of the PCC. In this case, a controlled disturbance is introduced at the PCC, and when the islanding condition occurs, the disturbance results into the detection threshold, minimizing the NDZ.

The concept of this method is that small-disturbance signal will become significant upon entering the islanding mode of operation in order to help the inverter to cease power conversion. Hence, the values of system parameter will be varying during the cessation of power conversion, and by measuring the corresponding system parameters, islanding condition can be detected.

Diverse methods, such as impedance measurement or detection slip mode frequency shift, active and reactive power variations and active frequency drift, have been proposed but they have drawbacks, as they introduce a disturbance at the PCC and interaction between PVs must be considered.

Communication-based methods involve a transmission of data between the inverter or system and utility systems, and the data is used by the PV system to determine when to cease or continue operation.

## 13.11 Operation and Control of Grid-Connected Wind Energy System

Conventional power stations are usually connected to the high-voltage or extra-high-voltage system, while wind turbines may be connected to ac system at various voltage levels, including the low-voltage, medium-voltage, high-voltage as well as to the extra-high-voltage system. The suitable voltage level depends on the amount of power generated. In the earlier stage, the technology used in wind turbines was based on a squirrel-cage induction generator connected directly to the grid. But in this case, power pulsations in the wind are almost directly transferred to the electrical grid. Furthermore, there was no control of the active and reactive power except from some capacitor banks, which are important control parameters to regulate the frequency and the voltage in the grid system. As the power range of the turbines increased, these control parameters became very important, and power electronics is introduced as an interface between the wind turbine and the grid.

Nowadays, wind turbines are connected to variable-speed generators with power electronics interface. The variable-speed operation of a wind turbine can be achieved depending on the large or a narrow wind speed range [8]. The main difference between wide and narrow wind turbine speed range is the energy production and the capability of noise reduction. A broad speed range gives larger power production and causes reduction of the noise in comparison to a narrow speed range system. Variable-speed operation is typically achieved by using one of two different configurations. The first employs a synchronous generator (SG) that spins at variable speeds and uses a full power converter to ensure that the produced power matches in frequency and phase to that of the utility grid. The second, and most common way of achieving variable-speed operation, is to use a doubly fed induction generator (DFIG). Variable-speed operation can only be achieved by decoupling the electrical grid frequency and mechanical rotor frequency of wind turbine. To achieve this, power electronic converters such as an ac-dc-ac converter combined with advanced power control techniques are used. Presently, DFIG equipped with partial-scale power converter is the most commonly employed configuration. However, the configuration with SG with full-scale power converter is becoming the preferred technology choices in the best-selling power ranges of the wind turbines. Both these technologies are already discussed in the earlier chapters.

### 13.11.1  Grid Integration of Wind Turbine System

The fluctuation and unpredictable nature of wind energy are undesirable for grid operation. Integration of large-scale wind power may have severe impacts on the power system operation. Traditionally, wind turbines are not required to participate in frequency and voltage control. However, in recent years, attention has been increased on wind farm performance in power systems. Consequently, some grid codes have been defined to specify the steady and dynamic requirements that wind turbines must meet in order to be connected to the grid. Most of the countries have strict requirements for the behaviour of wind turbines, known as grid codes, which are updated regularly. Basically, the grid codes are always trying to make the WTS to act as a conventional power plant from the electrical utility point of view. That means the WTS should not only be a passive power source simply injecting available power from the wind, but also act as an active generation unit. The WTS must be able to manage the delivered active/reactive power according to the demands and provide frequency/voltage support for the power grid.

According to most of the grid codes, the individual wind turbines must be able to control the active power at the PCC. Normally, the active power has to be regulated based on the grid frequency, so that the grid frequency can be somehow maintained.

Similarly, the reactive power delivered by the WTS also has to be regulated in a certain range depending on the active power. This will lead to larger MVA capacity when designing the full power converter system. In addition, the reactive power range of the WTS is also specified according to the grid voltage levels. Most modern turbine generators are decoupled from grid frequency through power electronics. Therefore, the inertia of the generator and the turbine rotor do not automatically participate in the grid inertial response as would traditional SGs. Further, because of this decoupling, changes in grid frequency do not elicit an automatic governor response that is common with conventional generation sources. In countries and regions with relatively isolated grids and relatively large levels of wind penetration, participation in grid frequency regulation by wind turbines and wind plants is crucial.

The various active power control requirements laid out by grid operators have resulted in technological development in wind turbine control. The advancements have been made at both individual turbine and wind plant scales. Technologies to provide response in all of the inertial, primary and secondary timescales have been developed at the individual turbine. Similarly, methods for providing active power control of entire wind plants have also been developed. Performing active power control collectively across a wind plant is intended to provide faster response and recovery to grid frequency deviations than can be achieved by performing active power control on individual turbines separately.

To reduce the impact of the wind gust and ensure the security and stability of grid operation, the wind turbines or wind farms are preferred to be configured as distributed generation networks (with many small-size power stations connected to the medium voltage distribution grid), whose power flows and electrical behaviours are different compared with the traditional centralized generation networks (with only a few large-sized power stations connected to the high-voltage transmission grid). Therefore, the protection schemes of the future grid utilities with more wind penetration should be also changed – resulting in a more distributed protection structure and may allow the islanding operation of some wind turbine units and local loads, which are composed as microgrids.

### 13.11.2 Power Electronics in Wind Energy System

Variable-speed wind turbines are able to operate at an optimal rotation speed as a function of the wind speed. The power electronic converter may control the turbine rotation speed to get the maximum possible power by means of a maximum power point tracking (MPPT) algorithm. In this way, it is also possible to avoid exceeding the nominal power if the wind speed increases. At the same time, the dc-link capacitor voltage is kept as constant as possible, achieving a decoupling between the turbine-side converter and the grid-side converter. A simple grid-connected wind energy system is shown in Fig. 13.13. Cage induction generators and SGs may be integrated into power systems with full rated power electronic converters. The wind turbines with a full-scale power converter between the generator and the grid give the added technical performance. Usually, a back-to-back voltage-source converter (VSC) is used in order to achieve full control of the active and reactive power. The grid-connected converter will work as an inverter, generating a PWM voltage whose fundamental component has the grid frequency and also being able to supply the active nominal power to the grid.

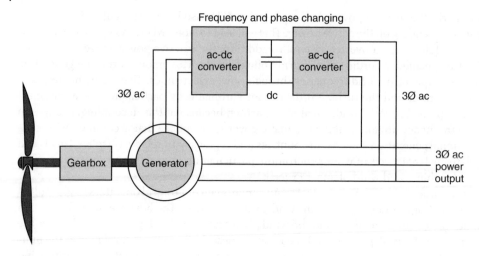

**Figure 13.13** Wind energy system.

### 13.11.3 Control of Doubly Fed Induction Generator–Based Wind Turbine Systems

At present, the configuration of DFIG equipped with partial-scale power converter is dominating the market, The DFIG-based WTS configuration with partial-scale back-to-back power converters is shown in Fig. 13.14 The stator side of the DFIG is directly connected to the grid, while a partial-scale power converter is responsible for the control of the rotor frequency as well as the rotor speed [9]. The power rating of the partial-scale converter settles the speed range to about ±30% of synchronous speed. Moreover, the converter performs as a reactive power compensator and a smooth grid interconnection. Smaller power converters are economical compared to full-scale converter. However, the main drawbacks are the use of slip rings, some protection schemes and the controllability under grid faults. The rotor-side converter (RSC) controls the DFIG's rotor speed and power, while the grid-side converter (GSC) controls the dc-bus voltage and performs reactive power compensation as well.

The objective of the grid-side converter or the GSC is to permit the active power flow, regulating the dc-link voltage to a constant level. Close-to-unity power factor operation

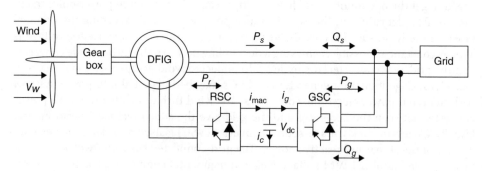

**Figure 13.14** DFIG-based wind energy system.

is usual, but it is also possible to control the reactive power flow between the converter and the stator/grid. Vector control is one of the most widely used control schemes for the rotor-side converter in DFIG-based WTS. The *dq* reference frames in a vector control can be aligned either to the grid voltage or to the stator flux. Upon aligning the direct axis of the reference frame with the stator voltage position, $v_{qs}$, becomes zero, and $v_{ds}$ is then equal to the amplitude of the terminal voltage. Then the active power depends on $i_{ds}$ and reactive power on $i_{qs}$, thus independent control of the active and reactive power flowing between the grid and the GSC can be obtained. Since the stator currents are related to the rotor currents by adjusting the rotor voltage appropriately, the desired rotor currents and, hence, the desired stator currents corresponding to the optimal electromagnetic torque and the desired Var flow/power factor can be achieved.

The grid-side PWM converter should have the ability to control the dc-link voltage, in order to ensure that the active power from the GSC can be delivered into the grid. At the same time, the ability of controlling the reactive power is required by the grid codes as well. The control of GSC in DFIG is similar to the control of other grid-connected converter, the *d*-axis is aligned with the grid voltage vector and the dc-link voltage is controlled with the *d*-axis output current $i_{ds}$, while the reactive power $Q$ is controlled by the *q*-axis current $i_{qs}$.

In order to work effectively, the power converter must be controlled in collaboration with the wind turbine pitch control. In other words, given a particular wind speed, there is a unique rotational speed required to achieve the goal of maximum power tracking (MPT). At below the rated wind speed, the wind turbine operates in the variable-speed mode, and the rotational speed is adjusted such that the maximum value of $C_p$ is obtained. With increasing wind speed, the rotational speed of wind turbine increases. Once the rotor speed exceeds its upper limit, the pitch controller will begin to increase the pitch angle to shed some of the aerodynamic power.

### 13.11.3.1 Control of a DFIG under Unbalanced Grid

Under unbalanced network condition, there are four rotor current components, two positive sequence and two negative sequence. Under unbalanced conditions, the positive- and negative-sequence currents of both RSC and GSC should be controlled independently, so as to meet the reactive support requirement of the grid code. It is clear from that the unbalanced grid voltage can bring double-grid-frequency oscillations on the GSC power and dc-link voltage. As the dc-link voltage fluctuations may cause harmonics in the converter output, as well as the over- or under-voltage protection, resulting in tripping of the converters, the GSC must be made to eliminate them. In this case, the aim of the control system is no longer to regulate the grid voltage.

Producing appropriate rotor voltage and current by the control of the rotor-side voltage and current is the key to enhancing DFIG unbalanced operation capacity, thus the control goal can be succeeded. One of the recent techniques is described as follows.

The double-synchronous rotating frame (SRF) control strategy is based on the symmetrical component method, where the rotor's positive- and negative-sequence currents in the positive and negative SRF are controlled respectively based on the DFIG's mathematical model in positive and negative SRF. The rotor's positive-sequence current reference set is based on the controlling function of the DFIG's average active power (average torque) and average reactive power, and the rotor's negative-sequence current set is based on the unbalance control target. The positive- and negative-sequence vector orientation of stator voltage is used respectively in the positive-synchronous rotating coordinate system and the negative-synchronous rotating coordinate system.

### 13.11.4 PMSG-Based Wind Energy Conversion System

In a variable-speed WECS, the generator can be an asynchronous generator, an electrically excited synchronous generator (WRSG) or a permanent magnet synchronous generator (PMSG). The variable-speed WECS offers advantages over the constant speed approach, such as MPPT capability and reduced acoustic noise at lower wind speeds.

PMSGs have the advantages of being robust in construction, very compact in size, not requiring an additional power supply for magnetic field excitation and requiring less maintenance [9]. These systems require full rated PE converters and can work on full speed.

A direct power conversion PM generator connected to grid is shown in Fig. 13.15. It can be used with variable-pitch variable speed with or without gearbox wind turbine system. The three-phase variable voltage, variable-frequency output from the wind turbine is rectified using a converter. The presence of dc link performs control of decoupling between the turbine and the grid. The dc link also gives an option of connecting batteries as energy storage and their charging. The dc signal is then inverted by grid-side converter into a three-phase, 50/60 Hz waveform. This waveform can then be scaled using a transformer to voltage levels required by the utility's ac system. The generator is decoupled from the grid by a voltage-sourced dc-link; therefore, this PE interface provides excellent controllable characteristics for the wind energy system. The power electronic devices are used to control the speed of wind turbine, to decouple the PMSG from the power grid, and the WECS does not need to synchronize its rotational velocity with the electrical network frequency.

The generator-side converter is used to regulate the velocity of WTG, which enables optimal speed tracking for the optimal power capture from any particular wind speed. The grid-side converter is used to stabilize the dc-link system voltage, to deliver the energy from the PMSG sides to the electric network system and to set a UPF of WFS during wind variation. For this reason, direct power control (DPC) is adopted to control

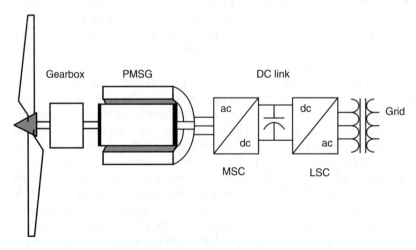

**Figure 13.15** PM-generator-based wind energy system.

instantaneous values of reactive power and active power of network connection, respectively. Moreover, the input reactive power and active power are controlled in $d$-$q$ synchronous reference frame. Accordingly, the PI control loops are used. Normally, a dc chopper is introduced to prevent overvoltages of dc link in case of grid faults when extra turbine power needs to be dissipated as the sudden drop of grid voltage.

The MPPT controller generates the reference speed of the generator, which when applied to the velocity control loop of the PMSG-side converter control system, maximum power will be produced by the VS-WECS. For this reason, vector control scheme is used as the control strategy for the generator-side converter with double closed-loop regulation. For the GSC, the current flowing in the generator stator should be controlled to adjust the generator torque and consequently the rotating speed. This will contribute to the active power balance in normal operation when the maximum power is extracted [10].

For the line-side (grid-side) converter (LSC), it should have the ability to control the dc-link voltage, $v_{dc}$, in order to ensure that the active power from the GSC can be delivered into the grid. At the same time, the ability of controlling the reactive power is required by the grid codes as well. Similarly to the control of other grid-connected converter, the $d$-axis is aligned with the grid voltage vector and the dc-link voltage $v_{dc}$ is controlled with the $d$-axis output.

So, in the inside loop, the current controllers are used to regulate $q$-axis and $d$-axis stator current to follow the command, whereas a velocity controller is used in the outside loop to regulate the WTG speed in order to follow the command value and produces corresponding $q$-axis current command.

It is important to note that it is fundamental to confine the converted mechanical power during high wind velocities and if the turbine extracted power reaches the nominal power. Then, power limitation can be realized by pitch control, stall control or active stall system.

### 13.11.4.1 Current-Source-Based PMSG

In most of the wind energy converter systems using PMSG, voltage control inverters are used. However, CSCs have some distinct advantages and are being considered for applications in WEC systems [11]. A CSC requires a dc-link inductor to provide a smooth dc current for operation. The grid-side converter in a CSC is a CSI which converts the dc-link current to three-phase ac currents that can be accepted by the grid. Contrary to VSIs which are voltage-buck converters, CSIs are essentially voltage-boost converters. Several CSC topologies can be used in the WECS. The grid-side converter can be chosen between a PWM CSI and a phase-controlled thyristor converter; whereas for the generator side, diode rectifier, thyristor rectifier and PWM current-source rectifier (CSR) are possible choices.

Here a back-to-back PWM current-source converter (CSC) topology is described for high-power wind energy applications. Compared with VSC configurations, PWM CSCs provide a simple topology solution and excellent grid integration performance, such as sinusoidal current and fully controlled power factor. The dc-link reactor provides natural protection against short-circuit fault, and therefore, the fault ride through strategy required by the grid code can be integrated easily into the system. As shown in Fig. 13.16, the proposed configuration consists of a PMSG, a full-power back-to-back CSC and a grid-connected transformer. The back-to-back converter consists of a generator-side

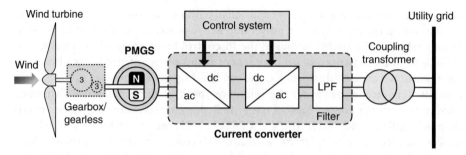

**Figure 13.16** Current source converter–based PMSG.

converter, a grid-side converter and filter capacitors at both sides. The generator-side and grid-side converters are connected via a dc-link choke. Filter capacitors are connected in parallel at both sides to assist current commutation as well as filter out switching harmonics. A step-up transformer is employed to connect the converter to the grid, providing isolation and grid integration.

The operation of the CSCs requires a constant current source, which could be maintained by either a generator-side or a grid-side converter. Generally, the grid-side converter controls the dc-link current based on the assumption of a stiff grid. But the actual dc-link current is determined by the power difference of both sides. The power disturbances of the generator output, mainly due to the disturbances of wind speed, are not simultaneously reflected by the grid-side converter control.

## 13.12 Summary

In this chapter, wind turbine systems with different generators and power electronic converters are described. Different types of wind turbine systems have quite different performances and controllability. The direct-drive PMSG system has its average efficiency of 2.3% and 1.6% higher than the fixed-speed SCIG system at the 500 kW and 3 MW rated power, respectively. The multiple-stage geared-drive DFIG concept is still dominant in the current market. Additionally, the interest in the direct-drive or geared-drive concepts with a full-scale power electronic converter is emerging. Current developments of wind turbine concepts are mostly related to offshore wind energy; variable-speed concepts with power electronics will continue to dominate and be very promising technologies for large wind farm. The performance of PMs is improving and the cost of PMs is decreasing in recent years, which make variable-speed direct-drive PM machines with a full-scale power converter more attractive for offshore wind power generations.

## References

1 Blaabjerg, F. *et al.* (2006) Overview of control and grid synchronization for distributed power generation systems. *IEEE Transactions on Industrial Electronics*, **53** (5), 1398–1409.

2 Carrasco, J.M. *et al.* (2006) Power-electronic systems for the grid integration of renewable energy sources: A survey. *IEEE Transactions on Industrial Electronics*, **53** (4), 1002–1016.

3 Kowalaska, T.O. *et al.* (eds) (2014) *Advanced and Intelligent Control in Power Electronics and Drives*, Springer.

4 Khalifa, A. S. and El-Saadany, E. F. (2009) Control of 3-phase grid connected photovoltaic power systems. Proceedings International Conf. I PEMC.

5 Kjare, S.B. *et al.* (2005) A review of single-phase grid-connected inverters for photovoltaic modules. *IEEE Transaction on Industry Applications*, **41** (5), 1292–1306.

6 Jain, S. and Agarwal, V. (2007) A single-stage grid connected inverter topology for solar PV systems eith maximum power point tracking. *IEEE Transactions on Power Electronics*, **22** (5), 1928–1940.

7 Gonzalez, R. *et al.* (2007) Transformerless inverter for single-phase photovoltaic systems. *IEEE Transaction on Power Electronics*, **22** (2), 693–697.

8 Siegfried, H. (1998) *Grid Integration of Wind Energy Conversion Systems*, John Wiley & Sons.

9 Cardenas, R. *et al.* (2013) Overview of control systems for the operation of DFIGs in wind energy applications. *IEEE Transactions Industrial Electronics*, **60** (7), 2776–2798.

10 Errami, Y (2011) Modelling and control strategy of PMSG based variable speed wind energy conversion system. International Conference on Multimedia Computing and Systems (ICMCS).

11 Dai, J. (2009) A novel control scheme for current-source-converter-based PMSG wind energy conversion systems. *IEEE Transaction on Power Electronics*, **24**, 963–972.

# 14

# Renewable Energy Sources Integration in Microgrid

## 14.1   Microgrid

As discussed in the earlier chapters, the concern for climate change has resulted in the use of various renewable energy sources (RESs) to reduce greenhouse gases. These renewable sources are generally connected at the distribution level and are known as distributed generators or distributed energy resources (DERs). Using DER in distribution system reduces the physical and electrical distance between generators and loads. The presence of energy sources near the load contributes to enhancement of voltage profile, reduction in losses and postponing the investment in new transmission systems and large-scale generation. However, the effect of penetration of a large number of DERs is to change the pattern of power flow in the electric distribution systems. Currently, the relatively low penetration levels of renewable systems cause few problems. As penetration becomes greater, the available wind and solar energy become a greater problem requiring central generation to provide the power backup. Since these sources are intermittent sources and can cause stability problems found with intermittent loads such as roiling mills and arc furnaces, central generation or distributed generation (DG) or storage is required to provide this backup energy. Without storage and/or local generation, there is a technical limit to the amount of wind generation that can be added to the grid system, perhaps as much as 20% of the peak demand is possible.

To realize the potential of integrating the distribution system with DER, a system approach which views generation, storage, protection and loads as an integral part of the distribution system is required. Such integration must not depend on fast, complex command and control systems. Each active component of the new distribution system must react to local information such as voltage, current and frequency to correctly change its operating point or faults. Distribution systems can efficiently handle disturbances using extended microgrid concepts. These concepts require that some of the generation, storage and corresponding loads need to separate from the distribution system to isolate sensitive loads from the disturbance (and thereby maintaining service) without harming the integrity of the remaining T&D system.

In order to improve the reliability of power supply, with better quality of power and increased stability margins, concept of *smart grid* and *microgrid* are being considered and implemented. The microgrid is defined as follows [1]:

*Operation and Control of Renewable Energy Systems*, First Edition. Mukhtar Ahmad.
© 2018 John Wiley & Sons Ltd. Published 2018 by John Wiley & Sons Ltd.

- The term *"Microgrid"* has become popular nowadays and is used to describe the concepts of managing energy supply and demand using an isolated grid that can work in islanded mode or as connected to the utility's distribution grid.
- Another definition of microgrid presently being used: A microgrid is group of interconnected loads and DERs within clearly defined electrical boundaries that acts as a single controllable entity with respect to the grid and that connects and disconnects from such grid to enable it to operate in both grid-connected or island mode.

A microgrid is basically a cluster of interconnected distributed generators, loads and intermediate energy storage units that can be collectively treated by the grid as a controllable load or generator. This approach allows for local control of DG, thereby reducing or eliminating the need for central dispatch. During disturbances, the generation and corresponding loads along with storage can separate from the distribution system to isolate the critical loads from the disturbance. Therefore, dynamic islanding is a key feature of a microgrid. There are a number of benefits that can be obtained from this ability to island for events such as faults and voltage sags. Thus, with the availability of a large number of DER sources, a microgrid is an important part of the distribution system. However, the control of microgrid with the presence of distributed energy sources (DERs) requires a major change in the control strategy of grid control.

If the present energy management system (EMS) is considered as reference, the following changes may be required in monitoring and control of microgrids.

- Forecasting: It is very difficult to forecast the power generating capacity especially when large number of DERs such as solar and wind power are connected.
- The generation needs to respond to various events such as voltage drops, faults, blackouts and so on and switch to island operation using local information only. This will require an immediate change in the output power control of the microgenerators as they change from a dispatched power mode to one controlling frequency of the islanded section of network along with load control. Controlled islanding operation and resynchronization are special requirements for microgrids.

From the point of view of the control structure of microgrid EMS, microgrid can be divided into centralized control and decentralized control.

A centralized control system achieves intelligence from a particular central location, which depends on the network type and could be a switch, a server or a controller. It is easy to operate a centrally controlled network. However, the centralized control system requires a single control device that processes all measured data. This unique controller point could cause several communication problems.

In decentralized control, all devices are able to control themselves independently as opposed to a single controller. This kind of control strategy believes in providing full autonomy to the local controllers of the DERs as it trusts that they are intelligent and smart enough and can communicate with each other to form a larger intelligent, smart and efficient unit. In the decentralized control scheme, the main focus is on the improvement of the overall performance of the system. Environmental conditions, weather conditions and so on are the main deciding factors for the decentralized scheme, and hence, the multi-agent system is used in this control strategy.

## 14.2   Types of Microgrids

A microgrid can be built using ac or dc network. Despite the fact that the ac microgrid system has a benefit to utilize existing ac grid technologies, protections and standard, its application involves drawback of low efficiency due to the number of power conversions required within the crucial current path from the main grid to the loads. Since the RESs such as PV cells and fuel cells produce dc and wind energy system produces ac of variable frequency, dc-to-ac and ac-dc-ac conversion is required in ac microgrid. One solution of this problem is the application of a dc microgrid as an efficient method to combine high reliability and the possibility to reduce the losses [2]. It can eliminate dc/ac or ac/dc/ac power conversion stage and thus has advantages in terms of efficiency, cost and system size. dc microgrid also has an added advantage that it does not require reactive power.

In most of the countries, energy infrastructure is old and vulnerable to extreme weather events. During the last few years, power failures have increased. Because a microgrid is localized, it can mitigate power disruptions by continuing to operate – providing electricity to its local customers – when the main grid supply is not available to consumers.

Microgrids can meet the needs of a wide range of applications in commercial, industrial and institutional settings. Larger microgrid applications include communities ranging from large neighbourhood to small towns to military bases. Another application of microgrid is in isolated mode in areas where people live without regular access to electricity. These "off-grid" areas are currently served (if at all) by diesel generators or similar small-scale electricity generating equipment.

A microgrid operates in either grid-connected or autonomous mode depending on the main grid condition/existence. During disturbances, the generation and corresponding loads can separate from the distribution system to isolate the load serviced by microgrid from the disturbance and thereby maintaining service without harming the transmission grid integrity. Operation of the microgrid assumes that the power electronic controls of current DERs are modified to provide a set of key functions, which currently do not exist. These control functions include the ability to: regulate power flow on feeders; regulate the voltage at the interface of each DER; ensure that each DER rapidly picks up its share of the load when the system islands. In addition to these control functions, the ability of the system to island smoothly and automatically reconnect to the bulk power system is another important operational function. In addition, fast and flexible action of under-voltage and under-frequency relays is required in islanded mode for keeping the MG stable. The main challenge in operating an ac microgrid with different types of DER system is the coordination of the numerous generators for sharing the real and reactive power output and the control of power system frequency and voltage. Some of these problems are not faced in dc grid.

## 14.3   dc Microgrid

dc microgrids eliminate the waste of energy associated with the conversion of dc to ac and ac to dc to ac and then again to dc which is required for so many of the electrical loads found these days. LED lights, variable-speed motors, computers, televisions and countless other forms of consumer electronics are loads that account for a steadily increasing fraction of the electricity consumed requiring dc supply for their operation. Moreover, many distributed RESs (PV cells and fuel cells) produce dc power which is now converted to ac to connect it the ac system. This ac power is again converted to dc for many end applications described earlier. One possible solution is to use the dc power directly in a dc microgrid. At present, local dc grids are in use within data and telecommunications centres or, on a much smaller scale, within automobiles, ships and aeroplanes. But it is expected that more dc microgrids will be coming up soon, at least in certain settings. A broad class of traditional dc distribution applications, such as traction, telecom, vehicular and distributed power systems, can be classified under dc MG framework. dc microgrids may also prove beneficial in many energy-intensive manufacturing operations. These include paper and pulp production and the smelting of aluminium, which now wastes more than 6% of the total energy consumed in the conversion of ac to dc.

The rapid development of power electronics technology which made dc voltage regulation a simple task, in addition to the increasing penetration of dc loads and sources, encouraged researchers to reconsider dc distribution for at least portions of today's power system to increase its overall efficiency. One of the great benefits of dc microgrids is their capability of static storage integration. Most of storage elements such as batteries and ultra-capacitors use dc supply. Moreover, flywheels, even though they are mechanical energy storage systems (ESS), are mostly coupled to a permanent magnet synchronous machine (PMSM) that is integrated to the distribution system through a dc link [3]. Adding dc storage to a dc microgrid is a comparatively simple compared to the complications of integrating dc storage in the ac domain where additional hardware is required.

A simple dc microgrid is schematically shown in Fig. 14.1. As for a typical dc microgrid, the main components of a microgrid are as follows: (i) DG sources such as photovoltaic panels, small wind turbines, fuel cells, diesel and gas microturbines; (ii)

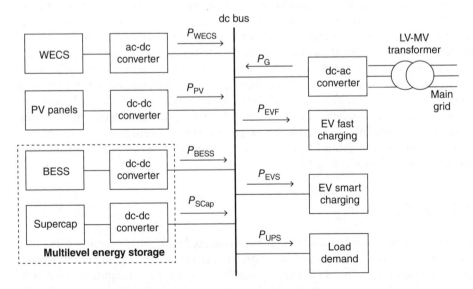

**Figure 14.1** Layout of a dc microgrid.

distributed energy storage devices such as batteries, supercapacitors, flywheels; and (iii) critical and non-critical loads. Energy storage devices are employed to compensate for the power shortage or surplus within the microgrid. They also prevent transient instability of the microgrid by providing power in transient. The transient power shortage in a microgrid can be compensated for by fast energy storage devices in the microgrid or by the utility grid through a bidirectional power converter when operating in grid-connected mode. A microgrid is connected into the utility grid through a bidirectional power converter that continuously monitors both sides and manages power flow between them. There is a single point of connection to the main distribution utility called point of common coupling (PCC). If there is a fault in the utility grid, the power converter will disconnect the microgrid from the grid, creating an islanded energy system. The microgrid can continue to operate in the islanded mode that is primarily intended to enhance system reliability and service continuity.

### 14.3.1  Control Methods for dc Grid System

In the microgrid shown in Fig. 14.1, the wind energy system is connected to the dc bus via an ac-dc converter. PV panels are connected to the dc bus via a dc-dc converter. The energy storage can be realized through flow battery technology and super capacitor connected to the dc bus via a dc-dc converter. The supercapacitor has much less energy capacity than the battery storage. It is aimed mainly at compensating for fast fluctuations of power. The multilevel energy storage helps in managing the intermittent and volatile renewable power outputs. It also mitigates the impacts on the main grid due to EV fast charging. The dc microgrid is connected to power distribution network through a centralized voltage-source bidirectional ac/dc converter G-VSC and a switch. The key point of power management in dc microgrid is to maintain the active power balance between power sources, energy storage battery, loads and distribution grid under any condition, which is represented by the stable dc bus voltage. The common dc voltage must be maintained with a limited variation band [4]. An abnormal dc-link voltage can disrupt normal operation or even cause the whole system to collapse. Furthermore, a constant dc voltage indicates balanced active power flow among the multi-sources and consumers (loads). Thus, the active power flow must be balanced within the grid under all conditions.

The main control objective for the individual devices can be summarized as follows:

The operation of the PMSG-based wind turbine is very much straightforward. It usually operates at the maximum power point tracking (MPPT) mode to extract maximum power from the wind although wind curtailment might be required under certain conditions (e.g., strong wind coupled with light load). Wind curtailment can be achieved by using the turbine's power and pitch control systems as described in Chapter 7.

The PV generation unit also works on MPPT mode. PV is always working in the state of maximum power output. When the sunlight is sufficient, the output power of PV can supply the load and export power to ac grid with batteries. When the system load decreases and dc bus voltage decreases, it operates in voltage control mode.

When the dc microgrid is grid-connected, the aim of the converter is to maintain a constant dc voltage by controlling its power exchange between the ac and dc systems to ensure that active power is balanced within the dc microgrid.

### 14.3.2 Energy Storage System

The supercapacitor is only responsible for fast and frequent access to stored energy at high roundtrip efficiencies of the order of 90%. The supercapacitor absorbs and levels the pulse power in the dc grid which is caused by the load current change, especially by the charging of EV. Response times are of the order of milliseconds, and both ESSs have a very high cycle life of charge and discharge operations.

One of the most flexible methods for the superior performance of BESS in dc grid is to connect the battery by a proper dc/dc converter. Under different microgrid conditions, the BESS operates at charging, discharging or floating modes, and the modes are managed according to the dc bus voltage condition at the point of BESS coupling. Consequently, the BESS is required to provide necessary dc voltage control under different operating modes of the microgrid. However, during abnormal conditions (e.g. ac grid fault or islanding), the ability of the grid-side converter for dc voltage control is likely to be severely affected or completely lost. Consequently, the battery ES system is required to provide necessary dc voltage.

Under normal condition, the loads operate based on their own requirements. However, appropriate load management, which involves load shedding based on predefined load priority levels, may be necessary during abnormal or island conditions.

### 14.3.3 Operational Modes of dc Microgrid

dc microgrid should operate in both grid-connected and islanding states. In each of these two states, there are different operating modes, and the control and management system should regulate the dc bus voltage. The satisfactory operation of the dc microgrid during the variations of wind and solar generation, load and grid connection conditions results in different operating modes that need to be considered in order to ensure a secure and reliable power supply. The following four operation modes are considered [5].

#### 14.3.3.1 Mode 1: Islanding Mode (Battery Discharge)

During utility grid disturbances, microgrid is transferred from the grid-connected to the islanded mode, and a reliable and uninterrupted supply of consumer loads is offered by local DERs. Mode 1 corresponds to islanding and subsequent island operation. Due to the disconnection to the external ac network, the dc microgrid becomes an island system, and the grid-tied power converter releases control of the dc-link voltage level, and one of the converters in the microgrid must take over that control. During the islanded mode, the battery plays the main role in regulating the dc-link voltage level, and the supercapacitor plays a secondary role in responding of the sudden power requirement as an auxiliary source/sag, that is, for peak shaving during transients. If the generated PV/wind power is less than local the load demand, dc bus voltage is regulated by battery discharging.

$$\text{Or } P_B = P_L - P_G \tag{14.1}$$

where $P_B$ is the power supplied by batteries, $P_L$ is local load and $P_G$ is local generation.

In case the power generated is much less than load demand, appropriate load shedding may be required. Similarly, if prolonged operation in island mode is required, it may result in low battery energy storage, and in order to guarantee power supply to the most critical loads, appropriate load shedding also becomes necessary. If the required power from the battery ES system is smaller than its maximum power rating, the dc voltage can be fully controlled, and no load shedding is required. However, there are various conditions where load shedding becomes necessary.

If the required power from the battery ES system exceeds its maximum power rating, the ES system operates at current/power limit, and consequently, the dc voltage cannot be fully controlled and will continue to decrease from the desired value. In order to prevent the total collapse of the dc microgrid, load shedding is implemented based on the dc voltage measurement without the need for communication. Two different voltage levels can be used for each load, and the loads are prioritized to ensure those with lower priority being tripped first. The load with the highest priority will have the disconnected only as last resort.

### 14.3.3.2    Mode 2: Islanding Mode (Excess Power Available)

In this mode, the dc microgrid operates in islanding mode and the dc bus voltage is regulated by the battery ES system through charging. In this mode, the dc bus voltage is regulated by dc/dc converters for PV and WECS. Since the maximum power generated by PV and wind in this mode is greater than the local load demands, the dc/dc converters for PV modules and WECS may not work with MPPT. Depending on the power generated by PV and wind, the power control mechanism will be applied. If the battery has been fully charged, the dc/dc converter for battery is disabled; otherwise, the dc/dc converter is enabled. In islanded mode, the grid-tied power converter releases control of the dc-link voltage level, and one of the converters in the microgrid must take over that control. Since each converter of DG sources is used for optimal control of its belonging source, only the converters of the energy storage elements are free to regulate the dc-link voltage level. During the islanded mode, the battery plays the main role in regulating the dc-link voltage level,

The operation of the grid in different modes is based on the dc grid voltage. Microgrid works in modes 1 and 4 when the reference voltage is set to 0.9 and 0.95 times of normal grid voltage, respectively. The microgrid enters in modes 3 and 2 when the reference voltage is 1.0 and 1.05 times the normal grid voltage, respectively. The reference voltage is decided on the following basis:

1) The voltage difference between different modes must not be too small; otherwise, the malfunction of mode switching will occur due to the sampling inaccuracies and external disturbance.
2) The voltage difference between different modes should not be too large; otherwise, the converters will be required to work at low voltage (LV) and high current.

### 14.3.3.3    Mode 3: Grid-Connected Mode (Power Taken from Grid)

In this mode, the dc microgrid operates with connection to ac grid through bidirectional dc/ac converter. In the grid-connected operation mode, the grid-tied power converter has control over the dc-link voltage level.

If the output sum of the power of the DG systems is sufficient to charge the storage devices, any excessive power is supplied to the utility grid. If the sum of the power of the DG and storage systems is less than the total load demand, the required power is supplied from the utility grid. In the grid-connected mode, power management is performed in a complementary manner between storage devices, and as a result, a dc microgrid can operate safely and efficiently. If the total load is more than the power available from wind and solar system and the grid is also not able to supplement it, the battery storage supplies the power. For the battery ES system, its dc voltage and charging level are monitored by its battery energy-management system whose aim is to maintain a certain amount of energy storage by providing appropriate levels of battery charge/discharge current. The control is achieved by regulating the duty ratio of the bidirectional dc-dc converter and various approaches, such as the predictive method which can be used.

### 14.3.3.4 Mode 4: Grid-Connected Mode (Power Supplied to Grid)

In this mode, the dc microgrid operates with connection to ac grid through grid-side converter. The dc bus voltage is regulated by the grid-side converter through inversion, which means the generated PV power and wind is greater than the local load. The grid converter works in constant power mode. The dc bus voltage is regulated by PV and wind interface converter based on droop control. The battery ESS can be charged with the constant power, and the surplus power is injected to the utility power grid. The converters for PV modules and WECS work with MPPT. If the battery has been fully charged, the dc/dc converter for battery is disabled. The wind turbine operates at the MPPT mode where its output power is set according to wind speed. The converter of WECS is directly connected to the wind generator which is PMSG. The converter controls the generator current to ensure that the output power is equal to reference value. The PV interface dc-to-dc converter also works on MPPT. The G-VSC controls the common dc voltage. A closed-loop PI-regulator-based dc voltage control system whose input is the difference between the desired and actual dc voltages is usually used. The output from the dc voltage controller is the active power order.

### 14.3.4 Application of dc Microgrids

In most applications, the dc microgrid has been limited to just dc lighting and the distribution for the lighting. This is a direct result of the dc industry focusing on this development area due to the demand for lighting in all sectors while trying to increase market awareness. The other area where application of dc grid is being considered is data centres due to both their high energy demand and their almost entirely dc load profile. Data centres require electric power with high availability and, with a possibility of reduced electric losses, the need for cooling. High reliability can be achieved by using local energy sources, and by using a dc power system, the number of conversion steps, and therefore also the losses, can be reduced. The dc microgrid can also supply closely located sensitive ac loads during outages in the ac grid. Residential buildings particularly apartment complexes can have dc grid for lighting, LCD TV, refrigerators and so on.

## 14.4 ac Microgrid

As discussed earlier in this chapter, the interconnection of small modular generation systems (PV, fuel cells, micro turbines, small wind generators) and storage devices to LV distribution grids leads to a new energy system [1, 6]. The main microgrid components include loads, DERs, master controller, smart switches, protective devices, as well as communication, control and automation systems. Microgrid loads are commonly categorized into two types: fixed and flexible (also known as adjustable or responsive). Fixed loads cannot be controlled and must be satisfied under normal operating conditions while flexible loads are controllable. DERs consist of DG units and distributed ESSs which could be installed at electric utility facilities and/or at consumers' premises. Microgrid DGs are either dispatchable or non-dispatchable. Dispatchable units can be controlled by the microgrid master controller and are subject to technical constraints depending on the type of unit, such as capacity limits. Non-dispatchable units, on the contrary, cannot be controlled by the microgrid master controller since the input source such as solar and wind is uncontrollable.

The connection of small generation units with power ratings less than a few tens of kilowatts to LV networks results in the following benefits.

- Ensures energy supply with power quality and reliability for critical loads.
- Promotes demand-side management and load levelling.
- Promotes community energy independence and allows for community involvement in electricity supply.

It also brings additional benefits for global system operation and planning, particularly regarding investment reduction for future grid expansion. The ac microgrid such as dc microgrid can operate in the following two different operating conditions.

### 14.4.1 Interconnected or Grid-Connected Mode

Figure 14.2 presents schematic diagram of a generic multiple-DER microgrid. In grid-connected mode, no direct voltage and frequency control is required. The utility grid provides a reference voltage and frequency source. The PCC voltage is dominantly determined by the main grid, and the main role of the microgrid is to accommodate the real and/or reactive power generated by the DG units and the load demand.

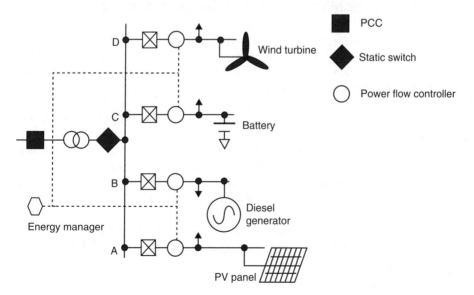

**Figure 14.2** ac microgrid with multiple DER.

In this mode, the main objective is to export a controlled amount of active and reactive power into an established voltage which is done through the control of active and reactive components of current. Interface converters are Voltage-Source Converters (VSCs) which are operated in current-controlled mode (CC-VSC) in synchronization with the grid voltage, according to P-Q control strategy. Thus, the inverter is operated in constant current control mode using *d-q* axis–based current control.

Under normal operation, each DG system in the microgrid usually works in a constant current control mode in order to provide a preset power to the main grid. All the power generating sources follow the value of active power reference generated by an internal control system such as a *MPPT2* system. The overall control structure of interface converters in the grid-connected mode is the current control mode in stiff synchronization with the grid [7, 8].

### 14.4.2 Islanded Mode

A microgrid can be islanded from the utility grid by upstream switches at the PCC [9]. Islanding could be introduced for economic as well as reliability purposes. During utility grid disturbances, microgrid is transferred from the grid-connected to the islanded mode, and a reliable and uninterrupted supply of consumer loads is offered by local DERs. In the islanded mode of operation, the DER systems are mainly controlled to regulate the microgrid voltage magnitude and frequency. This process must be fast and reliable and is the function of the local control. In the absence of a connection to the utility grid, a sustained islanded mode operation also implies that the sum of DER system power outputs equals the aggregate load power. This is strictly and rapidly ensured by the local control.

## 14.5 Control of ac Microgrid in Grid-Connected Mode

As shown in Fig. 14.2, each generation unit is interfaced to the microgrid at its respective point of connection and the microgrid is connected to the main grid at the PCC. Integration of DER units, in general, introduces a number of operational challenges that need to be addressed in the design of control and protection of distribution systems. The most relevant challenges in microgrid protection and control are as follows: the interconnection system must be able to measure the relevant indicated voltages and frequencies at the PCC or the point of connection of DR and disconnect within a given allowed time all local power generating units in the microgrid when these measured voltages or frequencies fall within a specified range. In addition, because of connection of DERs, which are normally power electronics interfaced, the fault current can be bidirectional. It may also result in reduction in fault current capacity, disruption in fault detection and protection sensitivity.

The main variables used to control the operation of a microgrid are voltage, frequency and active and reactive power. Reactive power injection by a DG unit can be used for

1) power factor correction,
2) reactive power supply, or
3) voltage control at the corresponding point of coupling.

In grid-connected mode, the main converter of MG operates in the PQ mode. The power is balanced by the utility grid. The battery is fully charged. ac bus voltage is maintained by the utility grid, and dc bus voltage is maintained by the main converter. Each DG system is usually operated to provide or inject preset power to the grid, which is the current control mode in synchronization with the grid. In this mode, the function of the overall control is thus to issue the real- and reactive-power commands for the DER systems.

The commands by overall control are based on a variety of criteria, such as economic operation of the microgrid; optimal operation and stability of the microgrid; and microgrid internal conditions and requirements.

An important aspect to consider in grid-connected operation is synchronization with the grid voltage. For unity power factor operation, it is essential that the grid current reference signal is in phase with the grid voltage. This grid synchronization can be carried out by using a PLL. There are two approaches for the overall control applied to power system: (i) centralized control or (ii) decentralized.

A fully centralized control is shown in Fig. 14.3. In a centrally controlled microgrid, the communication network is necessary to communicate control signals to the microgrid components. The central controller with the help of extensive communication between the central controller and controlled units performs the required calculations and determines the control actions for all the units at a single point. On the other hand, in a fully decentralized control, each unit is controlled by its local controller, which only receives local information and does not consider system-wide variables or other controller actions. In a microgrid with decentralized control, the communication network enables each component to talk with other components in the microgrid, decides on its operation and further reaches predeterminned objectives. Since interconnected power

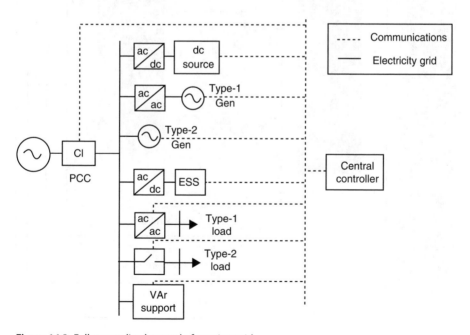

**Figure 14.3** Fully centralized control of ac microgrid.

systems are spread over a large area, a fully centralized control is complicated as it requires extensive communication.

At the same time, a fully decentralized approach is also not possible due to the strong coupling between the operations of various units in the system, requiring a minimum level of coordination that cannot be achieved by using only local variables. It is therefore desired that the controllers of various distributed resources are also distributed to avoid the communication delay and the single point of failure. A compromise between fully centralized and fully decentralized control schemes can be achieved by means of a hierarchical control scheme consisting of three control levels: primary, secondary and tertiary as shown in Fig. 14.4. Although microgrids may not be spread over large area as conventional power systems, this type of control is now suggested for the microgrid, because of the large number of controllable resources and stringent performance requirements.

This three-level hierarchical control is organized as follows. The primary control deals with the inner control of the DG units by adding virtual inertias and controlling their output impedances. The secondary control is conceived to restore the frequency and amplitude deviations produced by the virtual inertias and output virtual impedances. The tertiary control regulates the power flows between the grid and the microgrid at the PCC. These control levels differ in their (i) speed of response and the time frame in which they operate and (ii) infrastructure requirements (e.g., communication requirements).

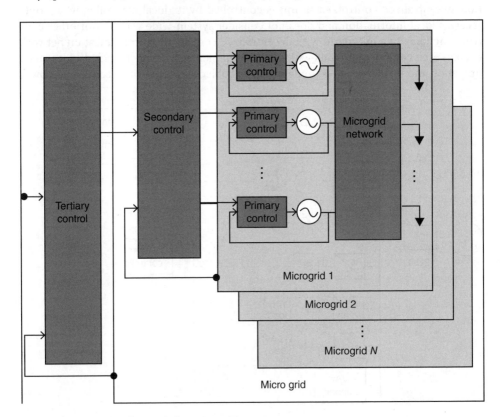

**Figure 14.4** Hierarchical control of ac microgrid.

### 14.5.1 Primary Control

Primary control, also known as local control or internal control, is the first level in the control hierarchy, featuring the fastest response. This control is based exclusively on local measurements and requires no communication. Given their speed requirements and reliance on local measurements, islanding detection, output control and power sharing (and balance) control are included in this category. Primary control ensures the fastest response to load changes and changes in the power supply to the grid. The operating range of this regulation is within milliseconds and is essential for ensuring the operational stability of the grid. The use of primary control is also provided for controlling the performance of ESSs such as batteries. For this purpose, it is necessary that the contribution of the active power has to be in accordance with the availability of power and the state of battery charge.

This control level adjusts the voltage reference provided to the inner current and voltage control loops. Primary control is the droop control method used to share load between converters. As a main control loop, inverters are programmed to act as generators by including virtual inertias by means of the droop method. It specifically adjusts the frequency or amplitude output voltage as a function of the desired active and reactive power. Voltage-Source Inverters (VSIs) used as interface for dc sources, or as part of back-to-back converters, require a specially designed control to simulate the inertia characteristic of synchronous generators and provide appropriate frequency regulation. For this purpose, VSI controllers are composed of two stages: DG power sharing controller and inverter output controller. Power sharing controllers are responsible for the adequate share of active and reactive power mismatches in the microgrid, whereas inverter output controllers should control and regulate the output voltages and currents. Inverter output control typically consists of an outer loop for voltage control and an inner loop for current regulation. Power sharing is performed without need for communication by using active power–frequency and reactive power–voltage droop controllers that emulate the droop characteristics of synchronous generators.

### 14.5.2 Secondary Control

The main role of secondary regulation is the control of voltage and frequency. In microgrid, the secondary control is involved in maintaining the stability within a tolerance of grid operational parameters. The secondary control ensures that the frequency and voltage deviations are regulated towards zero after every change of load or generation inside the microgrid. In addition, this control can be used for microgrid synchronization to the main grid before performing the interconnection, transiting from islanded to grid-connected mode.

The initial time of secondary control is higher than in primary control due to the availability of primary resources and also the capacity of the energy storage. It takes from seconds to minutes as compared to the primary control which is fast in order to (i) decouple secondary control from primary control, (ii) reduce the communication bandwidth by using sampled measurements of the microgrid variables and (iii) allow enough time to perform complex calculations.

Originally, frequency deviation from the nominal measured frequency grid brings to an integrator implementation. The frequency and amplitude levels in the microgrid are sensed and compared with the reference frequency and amplitude; the errors are then processed through compensators and sent to all the units to restore the output voltage.

A central controller is required to ensure that the power system operation is as seamless as possible during major disturbances such as transition from grid-connected mode to islanded mode. The optimal operation is sought through the implementation of a market environment using a multi-agent system (MAS), where the individual DER units are controlled by local agent heat exchange information with a central controller to determine most economic conditions.

### 14.5.3 Tertiary Control

Tertiary control ensures economic optimization, based on energy cost and electricity market. Tertiary controller enhances exchange of information or data with the distribution system operator in order to optimize the microgrid operation within the utility grid. When the microgrid is operating in grid-connected mode, the power flow can be controlled by adjusting the frequency (changing the phase in steady state) and amplitude of the voltage at the PCC. When the grid is present, the synchronization process can start from references of the microgrid with the frequency and amplitude of the mains grid. After the synchronization, these signals can be given by the tertiary control. Tertiary control can be considered part of the host grid and not the microgrid itself.

This tertiary control is also responsible for coordinating the operation of multiple microgrids interacting with one another in the system and communicating needs or requirements from the host grid (voltage support, frequency regulation, etc.). For example, the overall reactive power management of a grid that contains several microgrids could be accomplished by properly coordinating, through a tertiary control approach, the reactive power injection of generators and microgrids at the PCC, based on a centralized loss minimization approach for the entire grid. This control level typically operates in the order of several of minutes, providing signals to secondary level controls at microgrids and other sub-systems that form the full grid.

## 14.6 Autonomous Operation of Microgrid

During normal grid operation, the DG provides a constant power to the grid. The main grid is supplying or absorbing the power difference between the DGs and the local load. The islanding phenomenon that results in the formation of a microgrid can be due to either pre-planned or unplanned switching incidents. Islanding of the MG can take place by unplanned events such as faults in the MV network or by planned actions such as maintenance requirements called intentional islanding [10]. When the main power grid is out due to faults or switching, the DG delivering preset or available maximum power to the microgrid can create voltage and frequency transients which are dependent on the degree of the power difference. In the case of a pre-planned microgrid formation, appropriate sharing of the microgrid load amongst the DG units and the main grid may be scheduled prior to islanding. Thus, the islanding process results in minimal transients and the microgrid continues operation, as an autonomous system.

Prior to islanding, the operating conditions of microgrid could be widely varied, for example, the DG units can share load in various manners and the entire microgrid portion of the network may be delivering or importing power from the main grid. Furthermore, the disturbance can be initiated by any type of fault and line tripping may be followed up with single or even multiple reclosure actions. Thus, the severity of the transients experienced by the microgrid, subsequent to an unplanned islanding process, is highly dependent on (i) the pre-islanding operating conditions, (ii) the type and location of the fault that initiates the islanding process, (iii) the islanding detection time interval, (iv) the post-fault switching actions that are envisioned for the system and (v) the type of DG units within the microgrid. However, the switching transient will have great impact on MG dynamics. During islanding, the power balance between supply and demand does not match at the moment. As a result, the frequency and the voltage of the microgrid will fluctuate, and the system can experience a blackout unless there is an adequate power-balance matching process. A microgrid composed only of RESs and conventional CHP units, such as the diesel generator, gas engine and microturbine, is hard to make sure the good dynamic performance of microgrid during islanded operation due to the intrinsic characteristics of these generation systems. The RES has an intermittent nature since their power outputs depend on the availability of the primary source, wind, sun and so on, and therefore, they cannot guarantee by themselves the power supply required by loads. On the other hand, CHP systems are limited by their insufficient dynamic performance for load tracking especially, the frequency of the microgrid may change rapidly due to the low inertia present in the microgrid. Therefore, local frequency control is one of the main issues in islanded operation.

If there are no synchronous machines to balance demand and supply, through its frequency control scheme, the inverters should also be responsible for frequency control during islanded operation. In addition, a voltage regulation strategy is required; otherwise, the MG might experience voltage and/or reactive power oscillations. In grid-connected mode, the inverters may be operated in PQ control mode. But with large number of microsources connected to form a microgrid, PQ control cannot be applied because voltage regulation is necessary for local stability and reliability. However, by using a VSI to provide a reference for voltage and frequency, it is thus possible to operate the MG in islanded mode, and a smooth movement to islanded operation can be performed without changing the control mode of any inverter. The advantage of using VSI is that it can react to network disturbances based only on information available at its terminals. Thus, a VSI can provide a primary voltage and frequency regulation in the islanded MG. After identifying the key solution for MG islanded operation, two main control strategies are possible: (i) single-master operation (SMO) or (ii) multi-master operation (MMO). In both cases, a convenient secondary load-frequency control during islanded operation must be considered to be installed in controllable MS.

### 14.6.1 Islanding Detection

The main islanding detection techniques used may be broadly classified as remote and local techniques. Local techniques can be divided into passive and active detection methods. Remote islanding detection techniques are based on the communication between utilities and DGs. Supervisory control and data acquisition (SCADA) has been used to determine whether the distribution system is islanded or not. These methods are reliable but not economic to implement for small systems. As for the local techniques, the core of passive method is that some of the system parameters (voltage, frequency, etc.) change greatly with islanding but not much in normal running when connected with grid. This character can be helpful in islanding detection by continuously monitoring the parameters of the system without bringing disturbance. During the grid-connected operation, the DGs are operated to provide the optimum power to the grid according to many factors such as the availability of energy, energy cost. The main grid is supplying or absorbing the power difference between the DG and the local load demand. When the main power grid is out (power outage), the DG that continues to inject predetermined optimum power can cause voltage and frequency transients, depending on the degree of power difference. The power difference makes the voltage and frequency drift away from the nominal values. When the voltage and frequency drifts have reached certain levels, it is deemed that an islanding is occurring. This is the method that has been used to detect islanding. This methodology is enough for islanding detection. However, this method would cause large non-detection zones because islanding cannot be detected under a perfect match of generations and loads in the island system.

With active methods, islanding can be detected even under the perfect match of generation and load. Active methods directly interact with the power system operation by introducing perturbations. The idea of an active detection method is that this small perturbation will result in significant change in system parameters when the microgrid is in islanded mode, whereas the change will be negligible in grid-connected mode. Some of the active detection techniques are as follows [11].

### 14.6.1.1 Impedance Measurement Method

Impedance measurement method is similar to passive method which measures the change in impedance caused by islanding. In an active direct method, however, a shut inductor is momentarily connected across the supply voltage from time to time, and the short-circuit current and supply voltage reduction are used to calculate the power system source impedance. A large number of impedance detection methods have recently been proposed because of the belief that this method has no NDZ in the single-inverter case.

### 14.6.1.2 Slip-Mode Frequency Shift (SMS) Method

The method applies positive feedback to the phase of the voltage as a method in shifting the phase and, subsequently, the short-term frequency. The SMS issued to detect the islanding condition because of the easy implementation of the method is caused by the involvement of only a slight modification of a required component.

### 14.6.1.3 Active Frequency Drift Method

The principle of the active frequency drift (AFD) or frequency bias method is forcing variations in the inverter output using positive feedback to accelerate the frequency of the inverter current. The AFD uses the waveform of the inverter current, shown in Fig. 14.5, along with an undistorted sinewave for comparison. When this current waveform is applied to a resistive load in an island situation, its voltage response will follow the distorted current waveform and go to zero in a shorter time than it would have under purely sinusoidal excitation. This causes the rising zero-crossing of voltage at

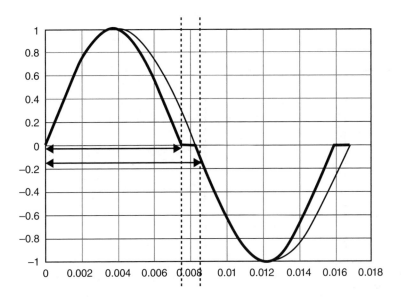

**Figure 14.5** Inverter current waveform.

node (PCC) to occur sooner than expected, giving rise to a phase error between this voltage and PV inverter current. The PV inverter then increases the frequency of the current to attempt to eliminate the phase error. The voltage response of the resistive load again has its zero-crossing advanced in time with respect to where it was expected to be, and the PV inverter still detects a phase error and increases its frequency again. This process continues until the frequency has drifted far enough from 0 to be detected by the over/under-frequency protection. The waveform drift of the grid is not present as a stabilizing influence.

#### 14.6.1.4 Sandia Frequency Shift (SFS)

Sandia frequency shift (SFS) is an extension of the frequency bias method and is another method that utilizes positive feedback to prevent islanding. In this method, it is the frequency of voltage at node (PCC) to which the positive feedback is applied.

There are two cases that must be considered. They are as follows: (i) the inverter is bidirectional and (ii) the inverter is unidirectional. To implement the positive feedback, the "chopping fraction" defined in Fig. 14.6 is made to be a function of the error.

The chopping function is given by

$$C_f = \frac{2t_z}{\text{Time period of utility voltage}} \tag{14.1}$$

This is the third method that uses positive feedback to prevent islanding. Sandia voltage shift (SVS) applies positive feedback to the amplitude of voltage of PCC node. If there is a decrease in the amplitude of (usually it is the RMS value that is measured in practice), the PV inverter reduces its current output and thus its power output. If the utility is connected, there is little or no effect when the power is reduced. In the islanded mode, the microgrid operates as an independent entity and must provide voltage and frequency control, as well as real- and reactive-power balance.

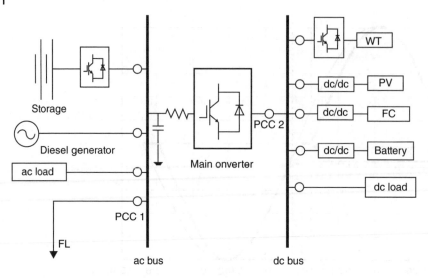

**Figure 14.6** ac/dc microgrid.

For example, if the net load demand is less than the total generation, the microgrid central controller should decrease the net generated power. This is done by assigning new set points to the DER units. On the other hand, if the power generated within the microgrid cannot meet the load demand, either non-critical load shedding or activation of battery charging may be initiated.

### 14.6.2 Stability Issues

Local oscillations may emerge from the interaction of the control systems of DG units, requiring a thorough small-disturbance stability analysis. Moreover, transient stability analyses are required to ensure seamless transition between the grid-connected and stand-alone modes of operation in a microgrid. In addition, microgrids might show a low-inertia characteristic, especially if there is a significant share of power-electronic-interfaced DG units. Although such an interface can enhance the system dynamic performance, the low inertia in the system can lead to severe frequency deviations in stand-alone operation if a proper control mechanism is not implanted. While the different sorts of loads have wide and frequent dynamic range, the bus voltage will fluctuate in wide range. In order to keep the stability, the energy storage equipments have to provide a quick response.

## 14.7 Load Frequency Control in Microgrid

The goal of LFC in a microgrid system is to maintain the system's frequency within acceptable limits around nominal value under various conditions, such as fluctuating power demand, and/or contingency situation such as unexpected loss of one or more of the system's generating units, in order to ensure system's stable operation. In case of small and isolated microgrid systems, however, the stability of the microgrid system is an issue of much greater significance as there are no means of connecting to primary grid power. The transition to islanded operation mode and the operation of the network in islanded mode require micro generation sources to particulate in active power frequency control, so that the generation can match the load. During this transient period, the participation of the storage devices in system operation is indeed very vital, since the system has very low inertia, and some micro sources (micro turbine and fuel cell) have very slow response to the power generation increase. The power necessary to provide appropriate load increase is obtained from storage devices. Knowing the network characteristics, it is possible to define the maximum frequency droop. To maintain the frequency between acceptable limits, the V/f inverter connected to storage device will adjust the active power in the network. It will inject active power when frequency falls from the nominal value and will absorb active power if the frequency rises above its nominal value.

### 14.7.1   Secondary Load-Frequency Control

There are two methods of performing secondary load-frequency control of microgrid:

i) Locally using PI controller at each microsource
ii) Centralized supervised by centralized microgrid controller

   In both cases, target values for active power outputs of the primary energy sources are defined based on the frequency deviation error. If the MG frequency stabilizes in a value different from the nominal one, the storage devices would keep on injecting or absorbing active power whenever the frequency deviation differs from zero. This should be only admissible during transient situations, where storage devices are responsible for the primary load-frequency control. Storage devices (batteries or flywheels with high capabilities for injecting power during small time intervals) have a finite storage capacity and can be loaded mainly by absorbing power from the LV grid. Therefore, correcting permanent frequency deviations during any islanded operating conditions should then be considered as one of the key objectives for any control strategy.

## 14.8   Combined ac/dc Microgrid

There were historical reasons that led ac to dominate over dc power systems. However, these are not valid any longer. In power generation, many DERs, such as photovoltaic systems, and fuel cells have emerged that generate power in dc form. Many of the modern electrical loads, as well as ESSs, are either internally dc or work with dc power and connect to the ac systems through converters. Thus, application of dc power would allow the elimination of many ac-dc and dc-ac conversion stages, which would in turn result in considerable decrease in component costs and power losses and increase in reliability. Therefore, dc microgrid is more suitable for the integration of distributed RESs. However, for a comprehensive microgrid, where the various sources as complementary should be integrated to overcome the environmental influence and reduce the interruption maintenance time, a pure dc grid would be deemed inappropriate. In addition, because of widespread use of ac appliances as well as ac distribution systems that prevail now, it is worthwhile to ingrate dc into existing ac system in microgrid.

Therefore, a hybrid ac/dc microgrid should be assumed to fully demonstrate the advantages of ac and dc distribution networks in view of easier renewable energy integration.

In hybrid ac/dc microgrid, ac sources and loads are connected to the ac network whereas dc sources and loads are tied to the dc network. ESSs can be connected to dc or ac links as shown in Fig. 14.6. The hybrid grid can operate in a grid-tied or autonomous mode.

### 14.8.1 Operation and Control of Hybrid ac/dc Grid

Generally, a hybrid ac/dc microgrid consists of three main parts: (i) ac sub-microgrid, (ii) dc sub-microgrid and (iii) power electronics interfaces between ac and dc buses Fig. 14.6 shows a general architecture of hybrid ac/dc microgrid connected to a utility ac grid.

In the system, ac and dc portions are connected through interlink bidirectional ac/dc converters (IC) with a proper control system and power management. Since the ac microgrid can be directly connected to utility grid through a simple circuit breaker, the ac sub-microgrid is generally dominant in the hybrid ac/dc microgrid to provide a stable voltage. The ac power generators, such as wind turbine and small diesel generators, and the ac loads, such as ac motors and traditional lighting, can be connected to the ac sub-microgrid. On the other hand, dc power sources such as photovoltaic panels, fuel cells and batteries can be connected to dc sub-microgrid through simple dc/dc converters. Besides, the ac loads with variable-frequency operation requirements, such as adjustable speed motors could be connected to dc sub-microgrid.

The hybrid grid can operate in two modes [12, 13]. In grid-tied mode, the bidirectional main converter is to provide stable dc bus voltage and required reactive power and to exchange power between the ac and dc buses. The dc-dc converters of PV system and the wind energy converter are controlled to provide maximum power. When the output power of the dc sources is greater than the dc loads, the converter acts as an inverter and injects power from dc to ac side. When the total power generation is less than the total load at the dc side, the converter injects power from the ac to dc side. When the total power generation is greater than the total load in the hybrid grid, it will inject power to the utility grid. Otherwise, the hybrid grid will receive power from the utility grid. In the grid-connected mode, the battery converter is not very important in system operation because power is balanced by the utility grid.

In autonomous mode, the droop control method used for ac grid can be used separately. For dc subgrid also the droop control can be used. But sharing the load demands in both ac and dc sub-microgrids simultaneously cannot be simply realized by means of droop-controlled distributed sources. The power sharing in both sub-microgrids would heavily depend on the control strategy of interlinking converter.

Once it is known how the power is shared between the various sources within each grid, the next requirement is to coordinate the power sharing of whole microgrid. If energy storage is available, it plays a very important role for voltage control and stability. dc bus voltage is maintained stable by a battery converter or boost converter according to different operating conditions. The main converter is controlled to provide a stable and high-quality ac bus voltage. Both PV and WTG can operate on MPPT or off-MPPT mode based on system operating requirements. When dc voltage or ac frequency (or both) is low, the renewable sources should operate in the MPPT mode and the controllers of the battery should operate in the discharging mode or diesel generator will produce more power. The main converter is controlled to transfer power between ac and dc links based on load and the resource conditions on the two sides.

When dc voltage and ac frequency are high, which indicates that there is energy surplus from both sides, the diesel generator will produce less power, the controllers of battery should operate in the charging mode to store energy. The main converter is controlled to transfer power between ac and dc links based on load and the resource conditions on the two sides, if the batteries are fully charged and voltage and frequency are still high, the PV and wind energy sources may be operated in off-MPPT mode.

### 14.8.2 Modelling

The three-phase transmission lines are modelled assuming balanced conditions and constant-power loads. However, these do not necessarily hold valid for microgrids as the microgrid is formed on distribution system.

In order to counter these challenges, the control system of microgrid must include the following:

*Output Control*: Output voltages and currents of the various DG units must track their reference values and ensure properly damped oscillations.

*Power Balance*: DG units in the microgrid must be able to accommodate sudden active power imbalances (excess or shortage), keeping frequency and voltage deviations within acceptable ranges. Demand-side management may be required to attain this objective.

## 14.9 Summary

The renewable sources are generally connected at the distribution level and are known as distributed generators or DERs. Using DER in distribution system reduces the physical and electrical distance between generators and loads. The effect of penetration of large number of DERs is to change the pattern of power flow in the electric proliferation of DER units in the form of DG, distributed storage (DS) or a hybrid of DG and DS units that has brought about the concept of microgrid. A microgrid is defined as a cluster of DER units and loads, serviced by a distribution system, and can operate in (i) the grid-connected mode, (ii) the islanded (autonomous) mode and (iii) ride-through between the two modes. In addition, a microgrid can be a dc microgrid or ac microgrid or hybrid ac/dc microgrid.

In this chapter, the operation and control of dc microgrid are first considered in grid-connected as well as isolated mode. It is followed by operation and control of ac microgrid in these modes, and finally, hybrid ac/dc microgrid is described.

## References

1 Hatziargyriou, N. *et al.* (2007) Microgrids-an overview of ongoing research, development, and demonstration projects. *IEEE Power and Energy Magazine*, 78–94.
2 Kai, S. *et al.* (2014) dc microgrid for wind and solar power integration. *IEEE Journal of Emerging and Selected Topics in Power Electronics*, **2** (1), 115–126.

**3** Lee, J. *et al.* (2010) dc micro-grid operational analysis with detailed simulation model for distributed generation, Energy Conversion Congress and Exhibition ECCE.

**4** Panbao, W., *et al.* (2013) An autonomous control scheme for dc micro-grid system, 39th annual Conference of IEEE Industrial electronics society IECON.

**5** Xu, L. and Chen, D. (2011) Control and operation of a dc micro grid with variable generation and energy storage. *IEEE Transaction on Power Delivery*, **26** (4), 2513–2522.

**6** Ahmed, M. *et al.* (2015) *Integration of Renewable Energy Sources in Microgrid, Energy and Power Engineering*, Scientific Research Publishing.

**7** Lasseter, R.H. and Paolo, P. (2004) Microgrids: A conceptual solution, 35th IEEE Power Electronics Specialists Conference PESC.

**8** Green, T.C. and Prodanovic, M. (2007) *Control of Inverter Based Microgrids, Electric Power System Research*, Elsevier.

**9** IEEE PES Taskforce on Microgrid control (2014) Trends in microgrid control. *IEEE Transaction on Smart Grid*, **5**, 1905–1919.

**10** Guerrero, J.M. (2010) Hierarchical control of droop-controlled ac and dc microgrids—a general approach toward standardization. *IEEE Transactions on Industrial Electronics*, **58**, 158–172.

**11** Lopes, J.A.P. *et al.* (2006) Defining control strategies for microgrids islanded operation. *IEEE Transactions Power Systems*, **21**, 916–924.

**12** Xiang, L. *et al.* (2011) A hybrid ac/dc microgrid and its coordination control. *IEEE Transactions on Smart Grid*, **2**, 278–286.

**13** Guangqian, D. *et al.* (2014) Control of hybrid ac/dc microgrid under islanding operational conditions. *Journal of Modern Power Systems and Clean Energy*, **2**, 223–232.

# Index

*Operation and Control of Renewable Energy Systems,* First Edition. Mukhtar Ahmad.
© 2018 John Wiley & Sons Ltd. Published 2018 by John Wiley & Sons Ltd.